World Soils Book Series

Series editor

Prof. Alfred E. Hartemink
Department of Soil Science, FD Hole Soils Laboratory
University of Wisconsin–Madison
Madison
USA

The *World Soils Book Series* publishes books containing details on soils of a particular country. They include sections on soil research history, climate, geology, geomorphology, major soil types, soil maps, soil properties, soil classification, soil fertility, land use and vegetation, soil management, soils and humans, soils and industry, future soil issues. The books summarize what is known about the soils in a particular country in a concise and highly reader-friendly way. The series contains both single and multi-authored books as well as edited volumes. There is additional scope for regional studies within the series, particularly when covering large land masses (for example, The Soils of Texas, The Soils of California), however, these will be assessed on an individual basis.

More information about this series at http://www.springer.com/series/8915

Mohammad Hassan Roozitalab
Hamid Siadat · Abbas Farshad
Editors

The Soils of Iran

Springer

Editors
Mohammad Hassan Roozitalab
Soil and Water Research Institute
Tehran
Iran

Abbas Farshad
ITC
University of Twente
Enschede
The Netherlands

Hamid Siadat
Soil and Water Research Institute
Tehran
Iran

ISSN 2211-1255 ISSN 2211-1263 (electronic)
World Soils Book Series
ISBN 978-3-030-09863-6 ISBN 978-3-319-69048-3 (eBook)
https://doi.org/10.1007/978-3-319-69048-3

Printed on acid-free paper

This Springer imprint is published by Springer Nature
The registered company is Springer International Publishing AG
The registered company address is: Gewerbestrasse 11, 6330 Cham, Switzerland

Preface

Iran has a long history of ancient civilizations and a rich traditional knowledge in domesticating wild plants and using soil and water resources for production of food, feed, and fiber. The development of early soil knowledge in the Iranian history goes back to the Achaemenid dynasty in 550 BC. However, the modern and formal scientific attempt to conduct soil research and studies in the country was in the early 1950s in conjunction with modernization of agriculture sector and construction of big dams.

In 1964, the first English book on soils of Iran and the accompanied maps at a scale of 1:2,500,000 was published after several years of field studies and soil characterization through the joint collaboration between Soil Institute of Iran and FAO. Since then, soil survey projects have been conducted in many regions, and soil maps at reconnaissance and semi-detailed scale of 1:50,000 are available for most of the agricultural plains. Moreover, the land resource and potentiality maps at the scale of 1:250,000 have been prepared for all the provinces. In addition, universities and research institutes have carried out many studies and research projects on different aspects and properties of soils in different regions of the country. In addition, Soil and Water Research Institute made an initiative to prepare a new soil map of the country at a scale of 1:1000,000 using remote sensing and geographic information system. The project was completed in 2000 after several years of field works under the leadership of a prominent soil scientist, Mohammad Hassan Banaei (1938–2005).

In this initiative to prepare a new book on soils of Iran for publishing worldwide by Springer International Publishing AG, we have benefited from the valuable contribution of many distinguished scientists from various universities and institutions in and outside of Iran who accepted our invitation to prepare the various chapters of the book. Without their deep dedication and valuable support, we certainly could not fulfill our mission to prepare this comprehensive book on soils of Iran. We are highly indebted and grateful to these distinguished colleagues and friends who kindly accepted our invitation to join this noble endeavor. Unfortunately, during the preparation of the book, one of our distinguished colleagues and friends, Prof. Najafali Karimian (1943–2015), who is the lead author of the chapter on soil fertility (Chap. 10), passed away in March 2015. He was a respected professor at the Shiraz University and a prominent soil scientist. Prof. Karimian made a great contribution to the advancement of soil fertility and plant nutrition in the country.

This book provides a wide range of information on the soils of Iran. Preparation of the book also initiated organization of a few scientific meeting in the form of workshops at the Soil and Water Research Institute (SWRI), or as discussion/consultation meeting at the Iranian Soil Science Society (ISSS), where soil-related topics and some application of new techniques were presented. The main target groups and beneficiaries of this book are soil, water, geography, and environmental scientists, planners, and policy makers, international, regional, and national institutions as well as graduate students of soil, earth sciences, geography, and environmental sciences working and interested in the soil resources of Iran.

We highly hope that publication of this book shall contribute to the dissemination of information and advancement of knowledge on soil resources of Iran and lead to strengthening collaboration among various stakeholders and beneficiaries. It is also our intent to encourage and promote regional and international collaboration with the local scientists and organizations for generating suitable technologies to use and manage soil resources of the country in a sustainable manner and to meet the challenges ahead such as climate change, land degradation, and water scarcity.

Tehran, Iran — Mohammad Hassan Roozitalab
Tehran, Iran — Hamid Siadat
Enschede, The Netherlands — Abbas Farshad

Contents

About the Editors

Dr. Mohammad Hassan Roozitalab, Ph.D. is a distinguished soil scientist with a wide range of scientific and working experiences in Iran and abroad. He received his M.Sc. (1974) and Ph.D. (1978) from Oklahoma State University, Stillwater, Oklahoma, USA. His major research and teaching activities at Tehran University have focused mainly on soil genesis and classification and clay mineralogy. He has contributed to the advancement of soil science in Iran as President of Soil Science Society for 17 years from 1992 to 2009 and Director General of Soil and Water Research Institute from 1983 to 1995. Dr. Roozitalab also served as Deputy Head of Agricultural Research, Education and Extension Organization (AREEO) in Iran from 1995 to 2006. He has contributed to strengthening the regional and global agricultural research collaboration as President of Association of Agricultural Research Institutions in the Near East and North Africa, AARINENA (1998–2000), Coordinator of Regional Highland Research Network of ICARDA (2008–2011), and Chairman of the Global Forum on Agricultural Research (GFAR) from 2002 to 2006.

Dr. Hamid Siadat, Ph.D. is a distinguished soil scientist and editor-in-chief of two scientific periodicals in Iran: Journal of Soil Research and Journal of Water Research in Agriculture. He is a retired research professor and has been an invited professor at The Academy of Sciences, Islamic Republic of Iran for more than 20 years, and a member of the recently formed Council of Luminaries in Agricultural Research, Education and Extension Organization. He served as Deputy Director General of Soil and Water Research Institute from 1983 to 1995. His main field of expertise is soil–water–plant relationship.

Dr. Abbas Farshad, Ph.D. is a senior and retired scientific member of the ESA Department of ITC, Twente University, in the Netherlands. Backed with over 10 years of field experience in Iran, he joined ITC in 1980, where he taught soil-related topics, doing research and supervising students for 30 years. His job also made him cooperating with various institutions in different countries, namely Australia, Bangladesh, Belgium, Botswana, Cape Verde, Egypt, Ghana, Iran, Kenya, Mexico, Morocco, Nigeria, Portugal, Spain, Sri Lanka, Thailand, Tunisia, Turkey, UAE, and USA.

Introduction

1

Mohammad Hassan Roozitalab, Hamid Siadat, and Abbas Farshad

Abstract
Iran covers an area of about 165 million hectares, a population of about 80 million, and, for a great part, is located in the arid and semiarid regions of the world. This chapter provides introductory information on the contents of the various parts of the book. The main topics include the history of soil research and the prevailing conditions and environment (climate, geology, vegetation, and other soil-forming factors) where the soil resources of the country are developed. The book also provides detailed information on the kinds of soils and their classification and how the soil resources of the country are used and managed. The main physical and chemical properties as well as fertility and biology of the soils are also described. Furthermore, the book contains information on the main agroecological zones, paleosols, and past climate change. In the end, the main challenges facing the soil resources of the country are discussed.

Keywords
Soils of Iran • Soil Information • Soil Resources • Soil Research

Iran, with a total land area of about 165 million hectares, a population of about 80 million (in 2015), lies between 25° 00′ and 39° 47′N and 44° 02′ and 63° 20′E. The country is for a great part located in the arid and semiarid regions of the world, and its deserts are among the driest areas on the earth. The mean annual precipitation of the country is less than 250 mm ranging from less than 50 mm in the central desert region to more than 1800 mm in the eastern part of the Caspian Sea (see Chaps. 3 and 9).

The climatic conditions are very diverse, which make production of various crops and fruits possible during the year. The land is mostly mountainous (56%), and the remaining are undulating piedmonts or flat plains (see Chap. 4). The average elevation is about 1500 m above sea level, and 73% of the land is defined as highland with elevation of more than 800 m above sea level (Roozitalab et al. 2013). The main agricultural plains in the arid region are situated in central part with the altitude of more than 1000 m above sea level. The highest plain in the central plateau is located in the south Zagros mountain with elevation of 1710 m asl, and the lowest plain is the Lut Desert (Dasht-e Lut), which was inscribed to the UNESCO World Heritage Site list in 2016, with lowest elevation of 185 m asl.

In Farsi (the main language), *khak* (soil) is more than a habitat for growing crops, as it also connotes home country that is something heartfelt and respectful. This rich tradition is rooted in the two belief systems of Zoroastrianism and Islam, which have supported legally and culturally the socio-technical agricultural systems where soil and water have played a prominent role (Farshad and Zinck 1998). Classically, soil is known to function as *provider* (of food, feed, and fiber), regulator (biological activity and molecular exchanges), filter (of water and air), and an *archive* (history book of the landscape) (Farshad et al. 2015). Over the past fourteen hundred years, since the Iranian became Muslim, these two belief systems have coexisted and, to a large

M.H. Roozitalab (✉) · H. Siadat
Soil and Water Research Institute, Tehran, Iran
e-mail: mroozitalab@gmail.com

H. Siadat
e-mail: hamidsiadat@yahoo.com

A. Farshad
Twente University, Enschede, The Netherlands
e-mail: farshad@itc.nl

M.H. Roozitalab et al. (eds.), *The Soils of Iran*, World Soils Book Series,
https://doi.org/10.1007/978-3-319-69048-3_1

degree, have profoundly interacted to shape human–nature relationship (Balali et al. 2011).

This book deals with the various subjects and important topics related to the soil resources of Iran as presented below.

Chapter 2 comprises an overview of history of soil research in Iran. The authors briefly discuss the development of early soil knowledge in the rich Persian history since the Achaemenid dynasty dated back to 700 BC. They also review the early development of traditional soil knowledge when a Persian king was motivated to use broad soil knowledge to collect taxes from the owners of arable land based on soil fertility and the yield produced. Modern soil research period began in mid-twentieth century in conjunction with the modernization of agriculture and construction of the first generation of big dams with the aim of expanding irrigated agriculture in the arid and semiarid regions of the country. During this period, several research institutes and many universities were established to conduct training and research on different areas of soil science.

Chapter 3 is allocated to climate of Iran. The country is characterized by an extreme diversified climate, ranging from extra-arid to per-humid type. In this chapter, the effect of main factors on Iran's climate is explained and the main climatic parameters including precipitation, temperature, humidity, solar radiation, wind regime, and prevailing winds are classified. According to the modified De Martonne's classification system, 64% of the whole country is arid and 20% semiarid. Hence, Mediterranean to per-humid climate type is only limited to 16%. The status of climatic condition during 2050s and 2080s is forecasted to be drier and warmer.

In Chap. 4, the authors discuss in detail the geological zones and their related geomorphic features in Iran. The Iranian Plateau is divided into eight structural zones. The Zagros belt contains the most important sedimentary basins comprised mainly of calcareous and detrital sediments. Karst landforms are the major expression of Zagros landscape. Alborz zone is composed of sedimentary rocks including limestone, dolomites, and clastic rocks. Large areas in northern Iran are covered by loess deposits with a thickness of 30–70 m. Central Iran is composed of a series of inland basins, and more than 60 playas have been identified in the region. Makran zone, in east and southeast of Iran, is composed of uplands with varying width and altitude and irregular in trend and physiography. Alluvial fans and fluvial terraces in this region are subdivided into bedded piedmont gravels of Pliocene/Quaternary age, high-level piedmont fans and valley terraces, low-level piedmont fans and valley terraces, silty outwash deposits, and river alluvium. Case studies from different regions of Iran were reviewed in this chapter to show the close relationship between soils and geomorphology.

The vegetation and land use of Iran is the topic of Chap. 5. Because of the diversified climate, topography, and edaphic conditions, the country is covered with various types of vegetation, and in a very different and heterogeneous condition. A total of 167 families of vascular plants, 1215 genera, some of them only by one species and some by about 800 species are reported. The authors have attempted to present a generalized picture of the correlation between vegetation cover and soil distribution for the whole country.

In Chap. 6, the authors describe soil-forming factors and processes in the various regions of the country. Soil-forming processes and the formation of diagnostic horizons are mainly influenced by the overall arid to semiarid climatic zones and the calcareous/gypsiferous parent materials prevailing in these regions. Calcification, salinization, and alkalization are the major processes, and salt-affected soils are widespread. Illuviation and formation of argillic horizon, vertic features, and gleization are among the other processes in different climatic zones.

In Chap. 7, the authors discuss geographical distribution, classification, and properties of major soils developed in major agro-climatic zones of the country including (i) arid zone, (ii) semiarid region, (iii) Caspian Sea region, and (iv) warm semiarid tropic of the South. More than 97% of the soils are developed in the arid to semiarid regions where formation of calcic, gypsic, cambic, salic, petrocalcic, and/or petrogypsic horizon is a common process. Soils classified in three orders of Aridisols, Entisols, and Inceptisols constitute about 97% of the total soil resources. Mollisols, Alfisols, and Ultisols are mainly developed in the Caspian Sea Region and constitute less than 3% of the total soil resources. Vertisols have only developed in very small and scattered parts in arid to semiarid region. The authors also define the extent of the soil orders, suborders, and dominant great groups in each region and broadly describe their potentials and constraints for agricultural development.

Chapter 8 is allocated to paleosols, which have received less attention than modern soils of the country, although Quaternary deposits and exposures are widespread. The authors provide a short introduction into the general nature of buried and relict paleosols and on methodological aspects how to recognize, characterize, and date paleosols. The focus is on paleosols formed in loess-soil sequences in Northern Iran. These paleosols most probably correlate with interglacial and interstadial periods of the last glacial cycles suggesting that climate change in Northern Iran was in phase with well-known cycles recorded in the temperate zone of the northern hemisphere.

The authors of Chap. 9 describe the agroecological zones of the country. The chapter provides a generalized approach for defining agroecological zones using GIS procedures. It is based on the combination of climatic, land use/land cover, terrain, and soil patterns, which constitute defining elements of each AEZ. Each zone can be further characterized through additional datasets, using the capabilities of GIS. The generalized AEZ map shows the diversity of agricultural environments in Iran. The detailed AEZ map is able to convey the full extent of agroecological diversity, within an overall dryland context, created by the adaptation of land use systems to site-specific climatic, terrain, and soil conditions.

Chapter 10 is allocated to soil fertility and fertilizer recommendation. The authors review the research carried out on soil fertility and balanced soil fertilization in various parts of Iran. They report that soil fertility evaluation on more than 20,000 soil samples has revealed that more than 50% of cultivated soils suffer from the deficiency of one or more nutrients. Furthermore, low soil organic matter content is the main limitation factor of sustainable soil fertility and production. With the introduction of chemical fertilizers some 50 years ago, their use gained popularity in such a way that the amount of chemical fertilizers needed by Iranian farmers exceeded three million tons in 2015. Fertilizer recommendation in Iran is mainly based on the sufficiency range approach. In recent years, fertilizer recommendation based on a general and regional recommendation of fertilizers has gradually replaced by soil test recommendation, which is based on the fertility status of each farm and expected crop production.

In Chap. 11, the authors attempt to define the importance of soil biology and the biological properties of the soils. After a brief history of soil biology in Iran, biological characteristics associated with soil characteristics, climate, and vegetation are described. Furthermore, biological properties of the dominant soils of the country are discussed. The authors provide information on recent research findings and soil biological potentials, especially on production and commercialization of bio-fertilizer technology in Iran. The characteristics of useful soil microorganisms that exist in the country are reported. Finally, the potentials and capacities of production and consumption of bio-fertilizers as well as the priorities and challenges in the field of soil biology in the country are presented.

Chapter 12 is on human-induced land degradation. After a short description of the environmental conditions, the authors discuss the use and management of soil and water resources in the framework of features of the agricultural systems, including land ownership, farm size, and land preparation. While a number of human-induced degradation, including erosion, salinization, compaction, and desertification are discussed, the GLSOD maps (scales of 1:000000 and 1:5000000) and those of landslide and erosion features and intensity (PSIAC-based) of Iran are presented. The chapter is concluded by some strategies, such as land reform, dam construction, choosing the appropriate irrigation method that are followed to improve the situation.

Chapter 13 begins with a short description of the general concepts of land management, followed by a historical overview of soil and water management in Iran from the beginning of agriculture to present time. Making use of geopedologic-derived indicators, carbon dating, mineralogy, archaeological studies, and history has led to propose interesting theories about the genesis of the various soilscapes, which have controlled the land use and soil and water management. The authors also report the land setup systems devoted to soil and water conservation, many of them originally invented in Persia. Furthermore, the ongoing efforts toward the adoption of conservation agriculture technologies are shortly described.

Finally, in the Chap. 14, the authors discuss the soil resources of Iran, challenges, and recommendations. Considering the long history of agriculture, it can easily be understood why soil resources have been degraded to different degrees, particularly in recent decades when intensified utilization of soils started. The agricultural land of the country is fragile and rather limited as only about 10% of the total area of the country is now under agricultural production and about 8% of the total land area is under annual crops and orchards. The authors outline the major challenges facing the soil resources. They propose recommendations for strengthening the soil research and technology development system and creating more incentive for various stakeholders to participate in the national dialogue in promoting sustainable use of soil resources.

This book is the outcome of not only the field studies and research projects carried out during the recent decades in the country, but it has also benefited from wide knowledge and deep experiences of the authors of the various chapters. The editors sincerely wish that the contents of various chapters of the book will satisfy and help the readers to the most and hope that further research and information in the future will enhance the value of the scientific material provided here.

References

Balali MR, Keulartz J, Korthalz M (2011) Reflexive land and water management in Iran, linking technology, governance and culture. Part 1: Land and water management paradigms. Iran J Water Res Agric 24(2). Soil and Water Research Institute, Karaj, Iran

Farshad A, Zinck JA (1998) Traditional irrigation water harvesting and management in semiarid western Iran, a case study of the Hamadan Region. Water Inter 23:146–154

Farshad A, Mohammadi M, Masihabadi MH, Farzaneh A (2015) Geopedology; application of geomorphology, remote sensing (RS) and geographic information system (GIS) in soil studies (in Persian). Soil and Water Research Institute of Iran, Karaj, Iran

Roozitalab MH, Serghini H, Keshavarz A, Eser V, De Pauw E (2013) Elements of research strategy and priorities for sustainable agricultural development of highlands in Central, West Asia and North Africa. International Center for Agricultural Research in the Dry Areas (ICARDA)

Author Biographies

Dr. M.H. Roozitalab is a distinguished soil scientist with wide range of scientific and working experiences in Iran and abroad. He received his M. Sc. (1974) and Ph.D. (1978) from Oklahoma State University, Stillwater, Oklahoma, USA. His major research and teaching activities at Tehran University have focused mainly on soil genesis and classification and clay mineralogy. He has contributed to the advancement of soil science in Iran as President of Soil Science Society for 17 years from 1992 to 2009 and Director General of Soil and Water Research Institute from 1983 to 1995. Dr. Roozitalab also served as Deputy Head of Agricultural Research, Education and Extension Organization (AREEO) in Iran from 1995 to 2006. He has contributed to strengthening the regional and global agricultural research collaboration as President of Association of Agricultural Research Institutions in the Near East and North Africa, AARINENA (1998–2000), Coordinator of Regional Highland Research Network of ICARDA (2008–2011), and Chairman of the Global Forum on Agricultural Research (GFAR) from 2002 to 2006.

Dr. H. Siadat is a distinguished soil scientist and editor in chief of two scientific periodicals in Iran: Journal of Soil Research and Journal of Water Research in Agriculture. He is a retired research professor and has been an invited professor in The Academy of Sciences, Islamic Republic of Iran for more than 20 years and a member of the recently formed Council of Luminaries in Agricultural Research, Education and Extension Organization. He served as Deputy Director General of Soil and Water Research Institute from 1983 to 1995. His main field of expertise is soil–water–plant relationship

Dr. A. Farshad is a senior and retired scientific member of the ESA Department of ITC, Twente University, in the Netherlands. Backed with over 10 years of field experience in Iran, he joined ITC in 1980, where he taught soil-related topics, doing research, and supervising students for 30 years. His job also made him cooperating with various institutions in different countries, namely Australia, Bangladesh, Belgium, Botswana, Cape Verde, Egypt, Ghana, Iran, Kenya, Mexico, Morocco, Nigeria, Portugal, Spain, Sri Lanka, Thailand, Tunisia, Turkey, UAE, and USA.

History of Soil Research

2

Mohammad Reza Balali, Hamid Siadat,
and Mohammad Hassan Roozitalab

Abstract

This chapter outlines development of early soil knowledge in the Persian history since the Achaemenid dynasty dated back 625 BC. Historical development of soil knowledge in Iran could be divided into two periods of indigenous soil knowledge, which continued until early 1950, and the modern soil science period began after that long period. The traditional soil knowledge was evolved in such a way that in the fifteenth century, farmers could use the information to characterize and categorize the agricultural land for production. Modern soil research period began in mid-twentieth century in conjunction with the modernization of agriculture and construction of the first generation of big dams. During this period, research institutes and universities were established. The authors briefly reviews the main achievements in various fields of soil science and the attention of soil science community to the emerging complex issues, which the country is confronting, and the need for more integrating approach in conducting, multi- and trans-disciplinary research to cope with the future challenges.

Keywords

Persian civilization • Traditional knowledge • Agricultural knowledge
Modern soil research

2.1 Ancient History of Soil Knowledge

As far back as the fourth century AD, agricultural practices such as irrigation, improvement of soil fertility, and construction of terraces to control soil erosion reflected development of various levels of soil knowledge and management in different parts of the world (Brevik and Hartemink 2010). This was the case in Iran too. Many innovations were generated to develop agriculture and manage the natural resources (Fig. 2.1) during the sixth century AD to produce the food and feeds needed for the increasing human population and livestock.

M.R. Balali (✉) · H. Siadat · M.H. Roozitalab
Soil and Water Research Institute, Tehran, Iran
e-mail: mrbalali68@gmail.com

H. Siadat
e-mail: hamidsiadat@yahoo.com

M.H. Roozitalab
e-mail: mroozitalab@gmail.com

M.H. Roozitalab et al. (eds.), *The Soils of Iran*, World Soils Book Series,
https://doi.org/10.1007/978-3-319-69048-3_2

The long history of Persian civilization is full of scientific and technological innovations and achievements including those related to agriculture as well as soil and water resources such as qanats (well-known underground tunnels), small dams, water distribution networks, and cultivation of crop varieties and trees (Figs. 2.1 and 2.2). Farshad and Zinck (1998) illustrated the political periods and trends of agricultural development including irrigation systems, indigenous crop cultivation, and agricultural diversity from the Achaemenid Empire dated back to 625 BC to twentieth century. A rich indigenous knowledge recognized both the ecological realities of the arid and semi-arid climates and the social aspects of conserving and distributing water in a way that ensures its benefits to all the stakeholders, including peasants and urban residents.

To collect tax of arable land, Iranian government in the time of Anoushirvan, the 23rd ruler of the Sassanid Dynasty (531–579 AD), introduced Dehghan (farmer) system to the Iranian villages and motivated them to use soil knowledge and to invest in this noble issue

Fig. 2.1 Bahman weir dam 2000 years old (left) and Amir weir dam 1000 years old (right) (Farhangi 1993)

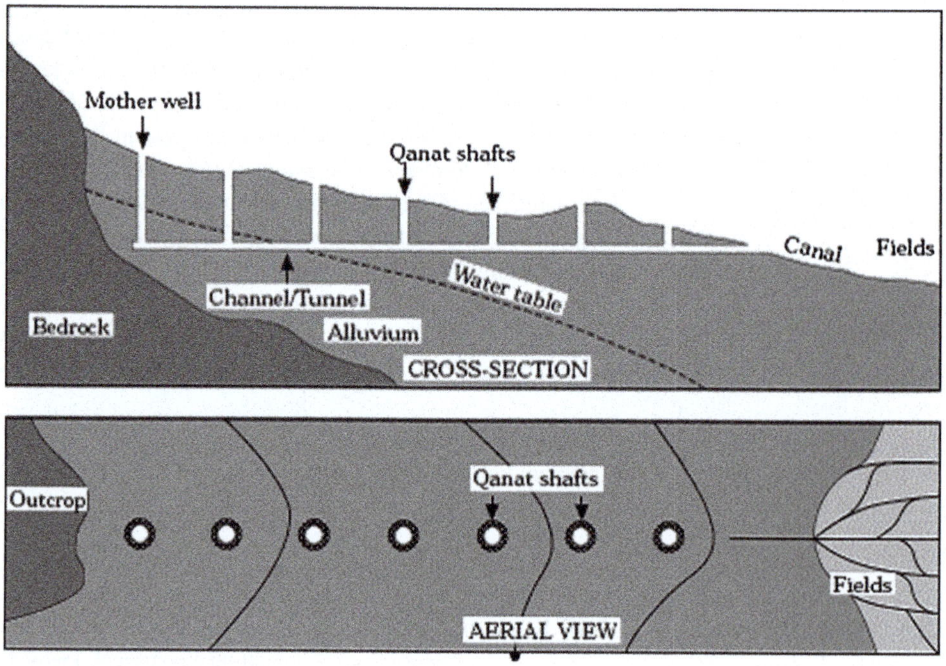

Fig. 2.2 Qanat irrigation system (Lightfoot 1996)

(Farshad 1997). To collect taxes based on the yield,[1] the land were distinguished as arable and non-arable. They carefully determined the size of the arable land according to the soil fertility. Then, one-sixth to one-third, and sometimes one-tenth to half of the production were taken as a tax by the government (Adl 1964).

This accumulated traditional soil knowledge can be traced in the book written in Persian by Abunasr Heravi entitled *Ershado-al-zeraeha* "Farming Guidance" in the fifteenth century (Abunasri Heravi 1500).

The author of the book collected agricultural knowledge of the time and explained the principles and methods of cultivation of plants and trees. Notably, the first chapter of the book emphasizes the importance of pedology and shortly describes the proper land preparation and agricultural suitability of 12 categories of land ranging from those without any kind of limiting factor to the land with maximum limiting factors for the plant growth. These categories include: Rigbum (gravelly loam), Shakhrig (coarse textured), Siahrig (black gravel), Sangchal (stony), Zehsaz (poorly drained), Sefidkhareh (white soil), Shakh (light soil), Zardkhak (yellow soil), Siahkhak (black soil), Sorkhehkhak (red soil), Shoorkhak (saline soil), Shoorrig (saline gravel). The rest of the chapters describe the way of cultivation, protection, and harvesting time of crops regarding plant characteristics and climate condition in a holistic approach (Fig. 2.3).

Before land reform of 1960s, when use of synthetic fertilizers started to expand in the country to increase soil

[1] http://parseed.ir/index.php?linkid=3211&qlang=en (last accessed 10 Aug 2016).

Fig. 2.3 Historic book of Ershado-al-zeraeh written in 1500 AD

fertility and agricultural production, organic fertilizers (derived from animal manure, human excreta, or crop residues) were the main source of soil fertility improvement. Among them, the farmers used pigeons dung in gardens and for producing cash crops. In the abovementioned book, Abunasri Heravi (1500) reported that "…all bird dung is good for agriculture and tree, except duck dung which is not good, best of all is pigeon dung…". Thus, dovecote or Doo cot (*Kabutarkhaneh*) was built to collect pigeons dung from 1200 years ago until recent decades around the country, especially in the central Iranian plateau (Fallah Rad 2012). Dovecotes were made near the agricultural fields or in the four corners of the gardens (Fig. 2.4).

2.2 Modern History of Soil Research

In Iran, along with the modernization of agriculture in 1950s, construction of the first generation of dams was a milestone toward development of soil science. Indeed, the earlier dams did not generate the expected economic outcome, mainly because the soils of the farmlands irrigated by these dams were not suitable for agriculture. Besides, uncontrolled irrigation of the lands resulted in grave problems such as rising groundwater and soil salinity. Therefore, the Iranian government, Food and Agriculture Organization (FAO) signed a joint project to investigate the cause of the problems emerged due to irrigation. M. L. Dewan from FAO and his Iranian counterparts J. Famouri and A. F. Mahdavi (Fig. 2.5) led the project.

In 1942, a schematic soil map of Iran at a scale of 1:6,000,000 (Fig. 2.6),[2] was prepared by Russian scientists V. A. Kovda and Y. P. Lebedev (Dewan and Famouri 1964). In 1952, Bongah Mostaghel Abiyari "Independent Irrigation Company" was established for development of water resources and irrigation projects throughout the country. Its Soil Group was mandated to study the soils and prepare the irrigation capability maps of the land. The first project to prepare the soil map of a region using aerial photographs at an approx. scale of 1:50,000 was started in the Karkheh River Basin in Khuzestan province, southwest of Iran, by a group of scientists who organized and participated in a special training course on soils in the winter of 1952. During 1952–1961, several field studies were conducted on major agricultural plains and major soils were characterized. Subsequently, a decision was made to prepare the first soil map of the country based on the field survey. Finally, a soil map of Iran at the scale of 1:2,500,000 was prepared through a joint collaboration between Soil Institute of Iran and FAO. Consequently, in 1964, a book entitled "The Soils of Iran" was published in English and the Farsi translation was published in 1970 (Fig. 2.7) (Balali et al 2002).

In 1961, the Soil Group was restructured and affiliated to the Ministry of Agriculture. The new institution was named the "Soil Institute" and A. Zaringhalam was appointed as the first director general of the organization (Fig. 2.5).

In parallel with the above activities, to promote the use of chemical fertilizers in agriculture, in 1957, Khuzestan Development Authorities started a soil fertility program in southwest of Iran. Considering the results of this project and the experiences of Iranian experts, the "Soil Fertility Office" was set up in 1960 at the Ministry of Agriculture with collaboration of FAO.

In 1966, the Ministry of Agriculture founded the "Soil Institute of Iran" by integration of the Soil Fertility Office with the "Soil Group," which was affiliated to the "Independent Irrigation Company," the forerunner of the Ministry of Power. The main objective of the new institute was to pursue the task of identifying and mapping the country's agricultural soils and promote the efficient use of soil and water resources. With the growth of this institute and the expansion of the mandates and tasks, younger staffs were employed and provided with opportunity to continue their education abroad. The generated relation with FAO, and several European, and American universities led to hosting and receiving attention of internationally known soil

[2]At the web site of ISRIC, http://isric.org/content/search-library-and-map-collection (last accessed 10 Aug, 2016).

Fig. 2.4 Dovecote built in the four corners of a garden (left) and a renewed dovecote in Esfahan (right) (Fallah Rad 2012)

Fig. 2.5 First directors of Soil Institute of Iran during 1950–1970s; from left to right: J. Famouri, A. Zaringhalam, and A. F. Mahdavi

scientists like Professors R. Dudal and R. Tavernier. Since then, soil research at the Institute has been extended to include other fields such as irrigation, soil chemistry, fertility and plant nutrition, biology, and soil conservation. The Soil Institute was restructured and renamed Soil and Water Research Institute in 1983.

Parallel to this, other institutions affiliated to Agricultural Research, Education, and Extension Organization (AREEO) of the Ministry of Agriculture[3] were established and mandated to conduct thematic soil and water researches. Table 2.1 shows the names and dates of establishment of the institutes under the umbrella of AREEO whose research agenda includes some aspects of soil resources.

2.3 Education in Soil Science

In 1963, the first department of soil science was established in the Agricultural Faculty of Tehran University and Mr. Ibrahim Naji (Fig. 2.8) was appointed as the first head of the new department. This group started to offer soil courses at the BSc level. In 1966, graduate studies at the MSc level was

[3]Presently called Ministry of Agriculture Jihad.

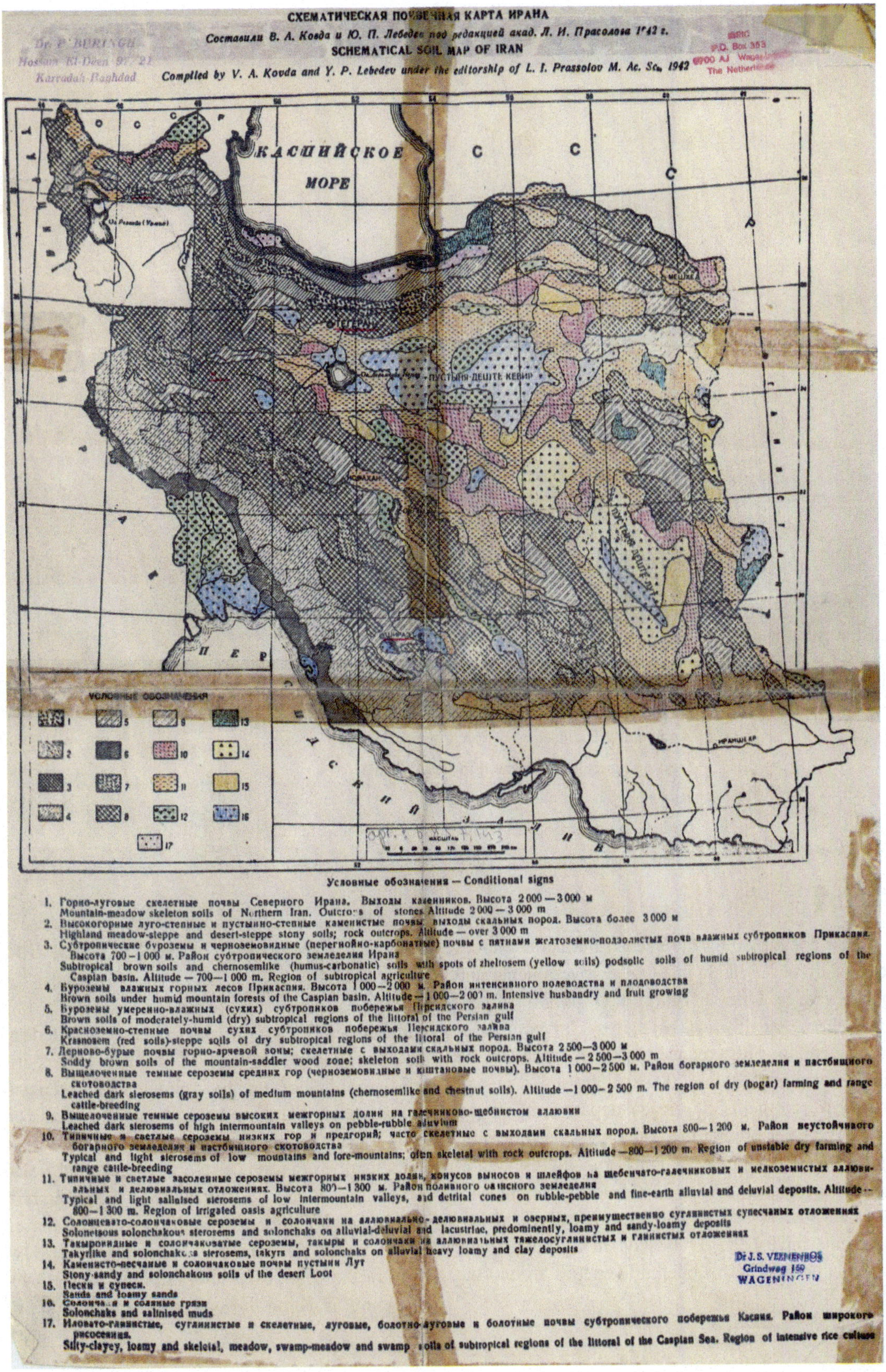

Fig. 2.6 First soil map of Iran by V. A. Kovda and Y. P. Lebedev published in 1942, at scale of 1:6 million. *Source* (ISRIC Web site - http://isric.org/content/search-library-and-map-collection, last accessed on 10 Aug. 2016)

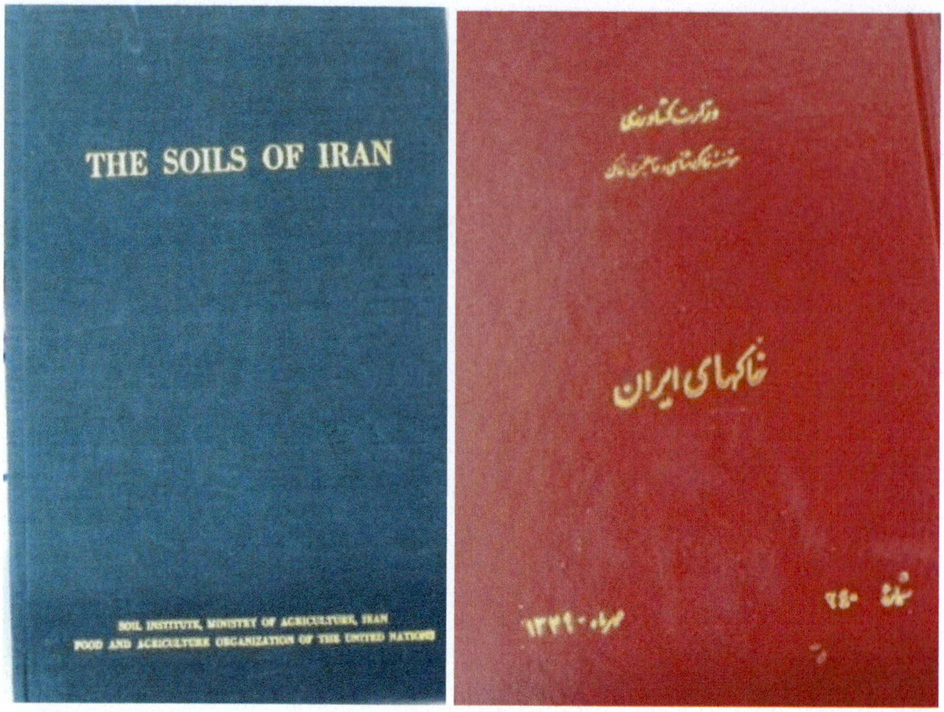

Fig. 2.7 First book on the soils of Iran in English and Farsi

Table 2.1 Research institutes affiliated to AREEO conducting research on soil and water

Institute	Year
Soil and water research Institute(SWRI)	1960
Research institute of forest and rangeland (RIFR)	1968
Agricultural engineering research institute(AERI)	1990
Dryland agricultural research institute (DARI)	1991
Rice research institute of Iran (RRII)	1993
Soil conservation and watershed management research institute (SCWMRI)	1993
National salinity research center (NSRC)	2000

started as a base for providing the needed human resources for scientific staff of research institutes, government, and development agencies. Later in 1990, Ph.D. programs were initiated at Tehran and Tarbiat Modarres Universities as another movement toward expansion of soil research and education in the country (Balali 2004).

Establishment of other soil science departments in the agricultural universities around the country in the following years led to further development of education and research in soil science. According to the statistical information in 2012 by Soil Science Society of Iran (SSSI), presently, soil science departments have been established in 27 public universities and many branches of open universities around the country (SSSI 2012).

2.4 Establishment of Soil Science Society (SSSI)

Soil Science Society of Iran (SSSI) was founded in 1973 in the Agricultural College of Tehran University with the principal goal of greater solidarity between students and educational scientific members to communicate on the issues such as educational programs, student field trips, and holding scientific conferences (Mohammadi and Balali 2002). In 1973, the first Iranian Soil Science Congress (Fig. 2.9) was held in Tehran University and the second Congress was held in the Shiraz University. The third Congress was held in 1994 in the Campus of the Agricultural College of Tehran University in Karaj through joint collaboration of SSSI and

Fig. 2.8 Ibrahim Naji, the first head of soil science department at Tehran University

Soil and Water Research Institute. Since 2015, SSSI has organized 14 biennial soil science congresses in collaboration with various universities and research institutes to share the research findings and technologies developed in different fields of soil science. The total number of papers presented in the first congress in 1973 was 24, which has increased to about 1200 papers in the fourteenth congress in 2015.

In addition to the biennial congresses, since 2002, SSSI has organized 10 annual meetings (Fig. 2.10) jointly with universities and research institutes on various issues related to soil research, education, and development with the aim of promoting dialogue and interaction among research institutes, universities, development agencies, and other stakeholders.

Publication of the Soil and Water Journal[4] by Soil and Water Research Institute (SWRI) in collaboration with SSSI in 1985 and a biannual Soil Bulletin (Fig. 2.11) by SSSI in 1994 was a right step toward disseminating research outputs in the academic and public sphere and showing the importance of soil in agricultural development and protection of the environment. In addition, the number of peer-reviewed articles on soils-related issues published every year has grown. In 2014, about 100 journals in relation to the sphere of agriculture and natural resources were published nationwide of which about 20% were directly related to soil science. The assessment of soil science in the period of 2006–2010 by SSSI shows that the academic staff in soil science departments of public universities has increased from 135 to 170 or about 26%. The number of students attending BSc, MSc, and PhD programs was increased significantly during this period. During these years, the increase in female undergraduate students in the soil departments of the public universities was about 70%. The number of published soil papers in journals and conferences was increased from 529 to 982 in the same period (SSSI, 2014).

Publication of Iranian soil policies and its technical and organizational principles (Roozitalab 1987) provided further steps toward expansion of soil research and encouraged the government to request the Ministry of Jihad-e-Agriculture (MOJA) to prepare the draft for the comprehensive soil law during the third five-year socio-economic development program of the country. The draft document needed by the government was prepared by MOJA in partnership with SSSI in 2005 and has been submitted to the parliament.

Since the adoption by FAO in 2012 of 5 December as World Soil Day, SSSI in collaboration with relevant stakeholders have celebrated the day every year (Fig. 2.12). This is an important opportunity to increase public awareness and draw attention of politicians and decision makers to more investment and sustainable use of the soil resources.

2.5 Research Projects and Achievements

As indicated above, to identify and classify soil resources of the country for optimum land-use and to specify scientific and economical ways for utilizing soil and water resources in agriculture, research activities in soil and water sciences has expanded since 1952. The focus in the early stage was on land classification for irrigation and cartography. Later, in 1967, multiple land evaluation to assess the land potential for general land uses such as irrigated system, dryland farming system, rangeland, and forest started. The main achievements and outputs until 2014 are as follows:

2.5.1 Soil and Land Classification Maps

Preparation of soil and land classification maps at scales of 1:100,000–1:50,000 for about 24 million hectares of the country has been one of the main tasks of Soil and Water

[4]Presently being published as two separate journals entitled "Journal of Soil Research" published in cooperation with SSSI, and "Journal of Water Research in Agriculture".

Fig. 2.9 First Iranian soil congress at Karaj campus, University of Tehran, in 1973

Fig. 2.10 First Iranian annual soil meeting held in 2002 in Karaj, University of Tehran

Research Institute since its inception. The soil survey projects were initiated in 1952 with the main aims of classifing the soils and prepare the land classification maps for irrigation (Fig. 2.13). The land classification for irrigation in Iran has been based on the third revision of the "Manual of Land Classification for Irrigation" prepared by P.J. Mahler, FAO expert, and local scientists (Mahler 1979).

Preparing maps of land resources and potentialities of the country, mostly at the scale of 1:250,000, were another achievement. The project started by Soil and Water Research Institute in 1967 and completed in 1995. The Manual of Multipurpose Land Classification was developed through a joint project (Soil Institute, 1970) and was used as the base to prepare the reports and maps of the various regions of the country (Fig. 2.14).

In recent years, attempts have been made to carry out soil surveys in several pilot sites in a more cost-effective way using modern remote sensing, geopedology, and modeling within GIS environment. A general soil map of Iran based on the US Soil Taxonomy and using geographical information system (see Chap. 7) was prepared by SWRI in 2000 at a scale of 1:1,000,000 (Banaei et al. 2005).

2.5.2 Soil Fertility and Plant Nutrition

These studies that aimed to develop and promote use of chemical fertilizers started in 1960 with cooperation of FAO. The project mainly focused on assessing nutritional needs of major crops based on soil and plant analysis and time, rate, and method of fertilizer application. In 1961, these studies were carried out in 14 regions of the country in the farmers' fields. They were performed in great numbers and gradually covered the entire country with emphasis on N and P. Based on the results of more than 10 thousand experiments in different regions of the country (5343 irrigated, 5175 dry farming fields), the first estimation of the country's fertilizer requirement was made in 1968 (Gomshadzehi and Abraham 1968).

In the 2nd period, soil fertility and plant nutrition studies extended simultaneously with the development of soil

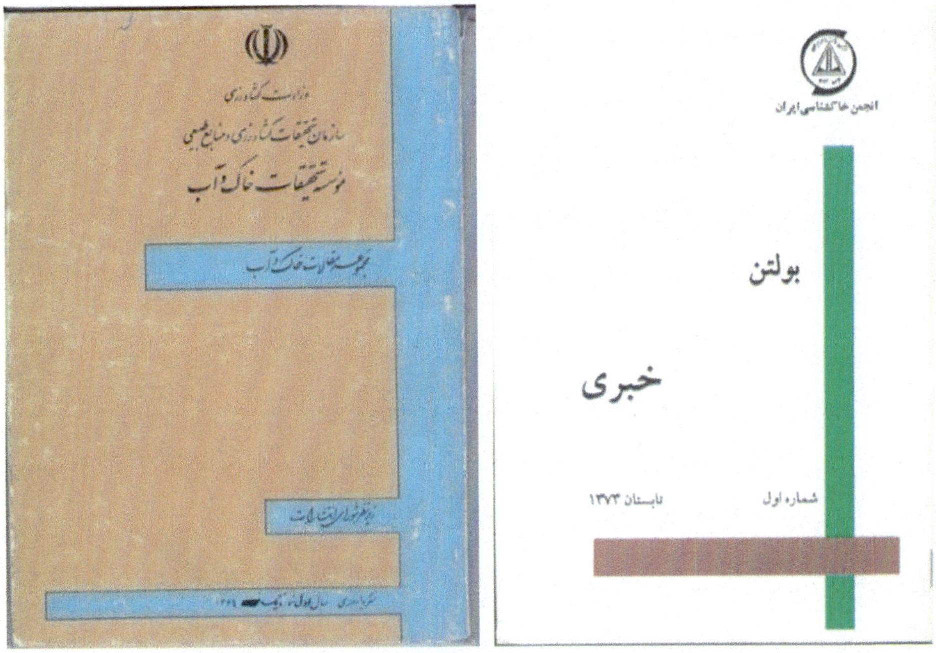

Fig. 2.11 Cover page of the first Soil and water journal published by SWRI in 1985(left) and the first soil bulletin published by SSSI in 1994 (right)

Fig. 2.12 Poster of the first world soil day ceremony (left) and the published stamp (right) in 2012

survey and land classification for irrigation. The main aim was to promote the use of chemical fertilizers for increasing yield in the newly irrigated land brought under cultivation by construction of dams and ground water resources. Therefore, effects of physicochemical properties of soils on crops yields and nutrients concentrations were determined to recommend chemical fertilizers use. The soil fertility research in this period developed widely in different aspects including geographical area and research issues (see Chap. 10).

In the 1990's, with the launching of the national wheat program in the Ministry of Agriculture and the request for more accurate fertilizer recommendations and suitable fertilizers distributions in the country, fertilizer recommendation based on soil testing approach was put in the research

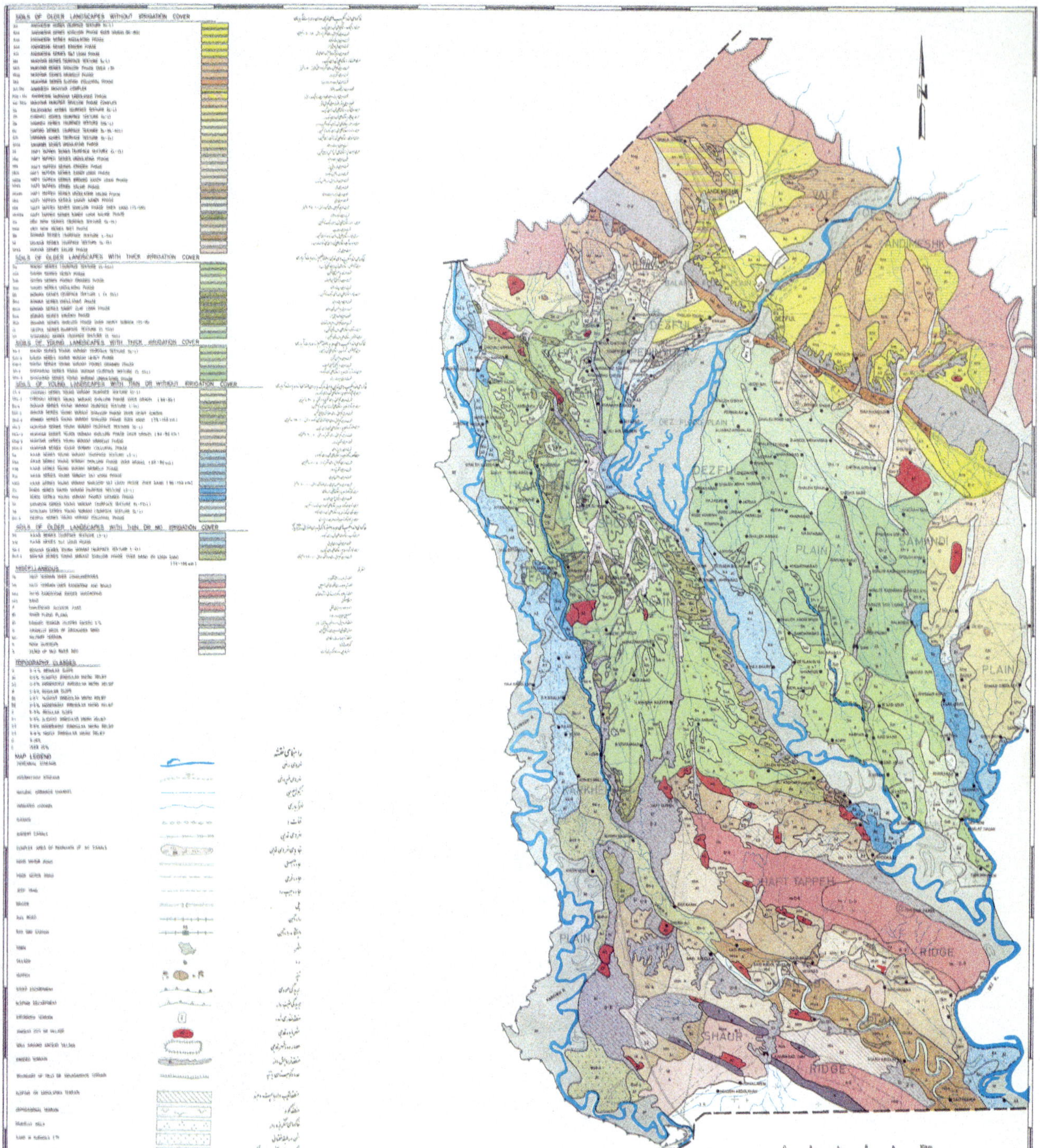

Fig. 2.13 Semi-detailed soil map of Dezful area prepared in 1958 (Courtesy of SWRI)

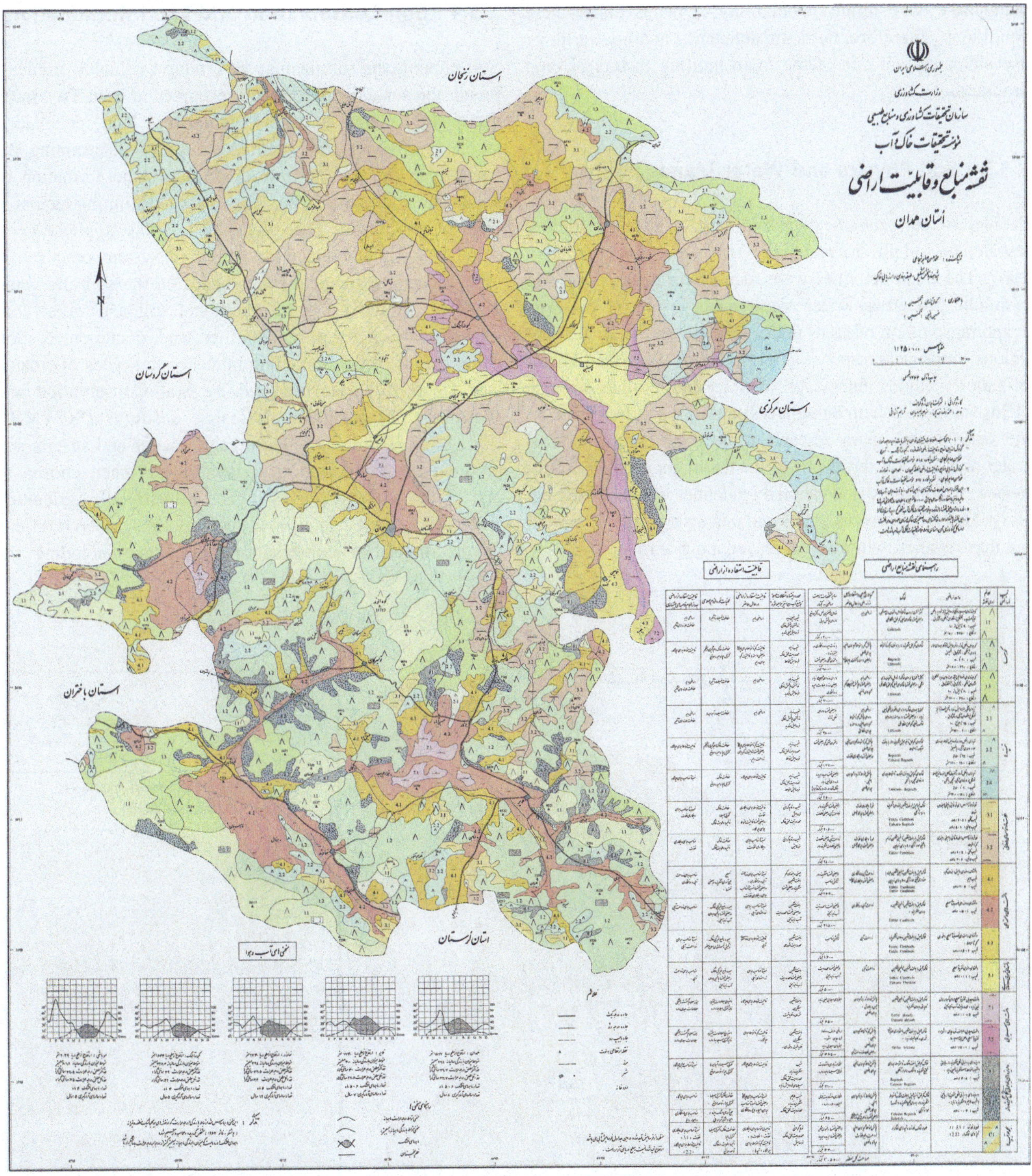

Fig. 2.14 Land resources and potentialities map of Hamadan province at the scale of 1:250,000 (Courtesy of SWRI)

agenda for a shift from general and regional scales to site-specific recommendation. In addition, private soil test laboratories were established around the country to facilitate the program. After 40 years of experience in the study of soil fertility, plant nutrition, and suitable fertilizer recommendation in the farmers' fields throughout the country, the first user-friendly model of fertilizer recommendation was prepared in 2000 (Khademi et al 2006). The majority of Iranian soil resources is calcareous, and essential plant nutrients such as phosphorus, iron, zinc, copper, and manganese have limited availability to plants under these conditions. Therefore, nutrients deficiency and low fertilizer use efficiency are one of the main limiting factors in crop production.

2.5.3 Soil Physics and Water Management

Parallel to soil classification and land evaluation studies, research on soil physics and field water management started in 1967. The main activities were focused on determining soil hydraulic properties, water requirements of crops, potential evapotranspiration of crops under different agro-climatic conditions, crops yield response to water, the efficiency of different irrigation methods, interaction of water and fertilizer on crop yields, and research on drought-resistant crops and varieties. In the same period, many research projects on optimization of water and mineral fertilizer use for the major crops and fruit trees were conduced and national guidelines were published in early 2000s Research institutes and universities mainly carried out these studies, while SWRI played a major role.

2.5.4 Soil Conservation and Land Reclamation

Soil erosion and salinization are among the major problems facing the sustainability of soil resources in Iran. To control these phenomena, soil conservation and land reclamation studies started in 1967 with the aims of determining the cause and the main factors contributing the formation of saline and sodic soils, which are important limiting factors in Iranian agriculture. The studies led to the application of suitable soil amendments and farming system practices in different agro-ecological zones based on the soil and water qualities. Moreover, some traditional soil and water conservation practices were identified and documented. Soil erosion atlas of Iran including intensity and types of erosion (see Chap. 12) was prepared by Soil Conservation and Watershed Management Research Institute (SCWMRI) (Fig. 2.15). To improve soil organic matter and sustain soil productivity, conservation agriculture has been chosen as one of the main strategies by the Ministry of Agriculture since 2007 (see Chap. 13). In recent years, research on different aspects of conservation agriculture including the

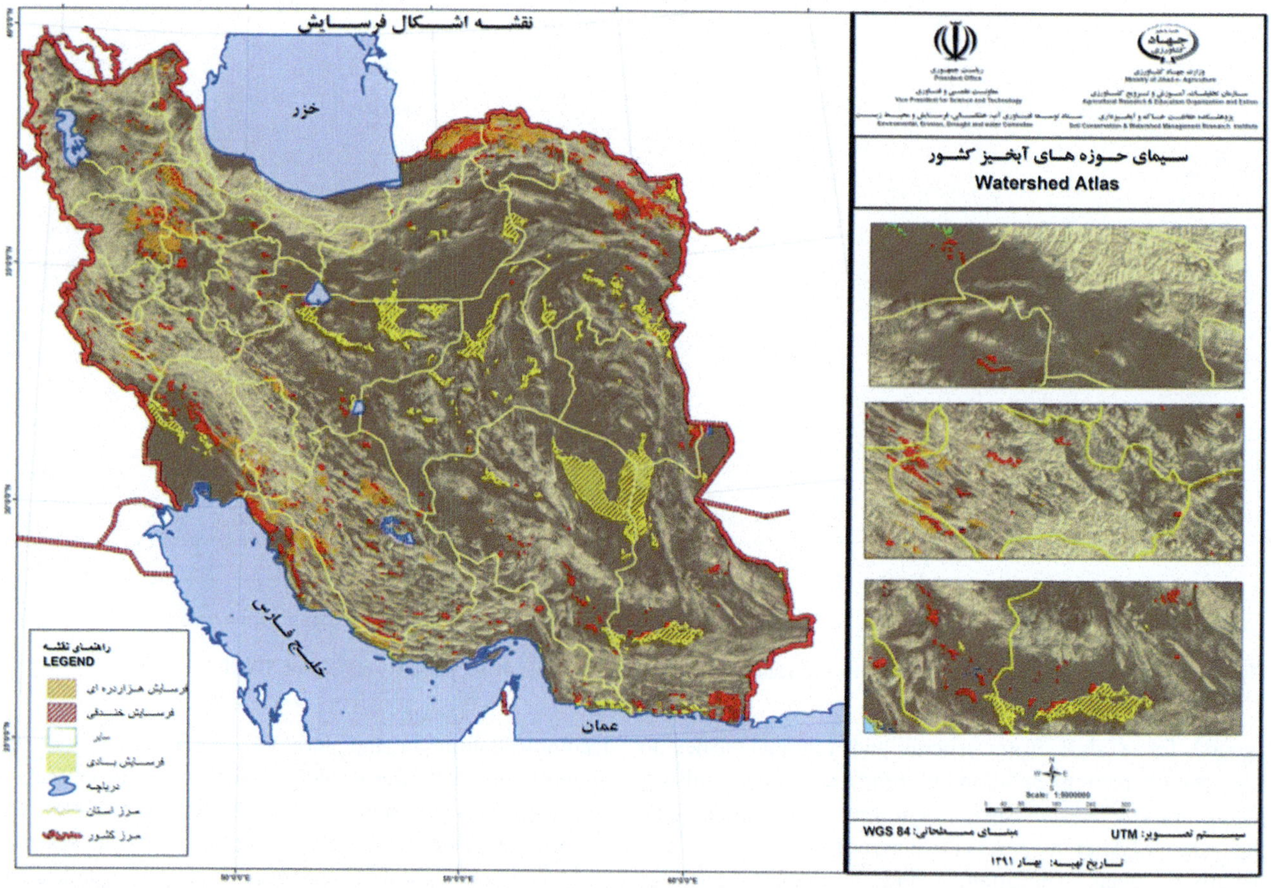

Fig. 2.15 Map of soil erosion types prepared by soil conservation and watershed management research institute (SCWMRI)

suitable machinery and technology for the conditions of Iranian agriculture has started by Agricultural Engineering Research Institute (AERI) (Sharif Rohani et al 2015).

2.5.5 Soil Biology

By 1995, soil biological studies were mainly focus to maximize the biological nitrogen fixation potential in soils and to optimize chemical fertilizer use (see Chap. 11). The main interests were on determining the nitrogen fixation capacity of indigenous strains of various food and forage legumes and preparing liquid inocula for use under leguminous crops. These studies include collecting the strains of different indigenous rhizobia to improve the nitrogen nutrition of plants with the use of free-living bacteria associated with Graminae, Actinorhiza, and Azolla. Isolating phosphorus-solubilizing microorganisms with emphasis on various mycorhiza strains, producing liquid P-solubilizing inoculants to reduce application of phosphate fertilizers are other also important topics of interest. Evaluating the beneficial effects of earthworms, rhizobacteria, sulfur-oxidizing bacteria, and other microorganisms responsible for nitrogen mineralization have also received due attention (see Chap. 11).

2.5.6 Dryland Research Activities

As large areas of agricultural land in the country are under rainfed production system, crop improvement methods are the main opportunity to sustain soil productivity. Thus, Dryland Agricultural Research Institute (DARI), besides plant breeding activities, focuses on identifying soils suitable for rainfed agriculture, crop management improvement methods including site-specific nutrient management, nitrogen management, biofertilizer application, crop rotation, and minimum and zero tillage on these soils.

2.5.7 Soil Quality Monitoring

The research area that is gaining increasing interest and attention of many scientists working at various universities and research institutes is soil health and food safety. The application of chemical fertilizers, use of polluted and wastewater for irrigation of crops, and indiscriminate dumping of untreated urban and industrial waste materials, and by-products in certain areas are becoming a major concern of public and private organization. While some research has been conducted by SWRI for monitoring soil resources, there is an increasing demand for continuous monitoring of soil resources for assessing soil erosion, pollution, salinity, and the changes occurring in soil quality due to improper soil management and human induced land degradation. In addition, research on other areas such as carbon sequestration, conservation agriculture, and impact of climate change on soil qualities and application of modern technologies to improve soil productivity is receiving more attention.

There is also an increasing need to link soil science to other disciplines, especially environmental and socio-economic sciences. To this end, there is an increasing trend and awareness to integrate the rich indigenous knowledge with modern soil sciences to promote sustainability of the environment and production system while enhancing soil and ecological services.

Acknowledgements We would like to thank and express our appreciation to K. Eftekhari, Assistant Professor of SWRI, M. Arabkhedri Associate Professor of SCWMRI, and M. Gorji, Professor of Soil Science of Tehran University for providing useful information, maps, and pictures.

References

Abunasri Heravi G (1500) Ershado-al-zeraeha. Tehran University, 1134

Adl SAH (1964) Agriculture. In: Iran-Shahr, a survey of Iran's land, people, culture, government and economy. Vol II, Tehran University Press, published with the assistance of UNESCO

Balali MR (2004) Research institutions in relation of soil science in Iran. Soil Science Society of Iran (SSSI), bulletin No 19

Balali MR, Malakouti MJ, Banaie MH, Khademi Z, Masihabadi MH (2002) Soil and water research institute of Iran: past, present and future (50th Anniversary Report). Karaj, Iran

Banaei MH, Momeni A, Bybordi M, Malakouti MJ (2005) The soils of Iran: new achievements in perception, management and use (in Persian). Soil and Water Research Institute, Karaj, Iran

Brevik EC, Hartemink AE (2010) Early soil knowledge and the birth and development of soil science. CATENA 83:23–33

Dewan ML, Famouri J (1964) The soils of Iran. FAO, Italy, Rome

Fallah Rad M (2012) Dovecote or Doocot (*Kabutarkhaneh*) unpublished report

Farhangi B (1993) A general view on Iranian dam: past, present and future. Ministry of Power, Tehran, Iran

Farshad A (1997) Analysis of integrated soil and water management practices within different agricultural systems under semi-arid conditions of Iran and evaluation of their sustainability. ITC publication 57, ITC Dissertation 45, Enscheda, Netherlands

Farshad A, Zinck JA (1998) Traditional irrigation water harvesting and management in semiarid western Iran-a case study of the Hamadan region. Water Int 23:146–154

Gomshadzehi A, Abraham TP (1968) Chemical fertilizer consumption in the main crops and estimation of required fertilizer of Iran. Soil and water research institute, technical bulletin No. 166

Khademi Z, Malakouti MJ, Mohajermilani P, Balali MR (2006) Optimizing fertilizer recommendation through computer model. J Soil and Water 20(2):191–202

Lightfoot DR (1996) Syrian Qanat Romani: history, ecology and abandonment. J Arid Environ 33:321–336

Mahler P (1979) Manual of land classification for irrigation (third ed). Soil and water research institute of Iran, Pub. No. 205

Mohammadi M, Balali MR (2002) Soil science society of Iran and its effects on promotion of science and technology in the country. Soil Science Society of Iran(SSSI),bulletin No. 14

Roozitalab MH (1987) National soil policies and its technical and organizational principles, Soil and water research institute, technical bulletin No 725
Sharif Rohani M, Aghaei Sarbarzeh M, Karimi Soghorlo M (2015) Agricultural research. AREEO Publication No, Education and extension organization at one glance, p 48242
Soil Science Society of Iran (2012) Review of soil science discipline in Iran, review report, publishes by the office of vice-president for science and technology, Tehran, Iran
Soil Science Society of Iran (2014) Review of scientific publication and scientific congress of soil science in Iran, Review report published by the office of the vice-president for science and technology

Author Biographies

Dr. Mohammad Reza Balali is an environmental philosopher working at Soil and Water Research Institute. He obtained MSc in soil science and MA in Islamic philosophy and theology emphasizing on environmental ethics. At the Applied Philosophy Chair group of Wageningen University, the Netherlands, he received his PhD entitled "*Reflexive land and water management in Iran: Linking technology, governance and culture.*" His interest is environmental and agricultural ethics and philosophy, sustainability, integration, and multidisciplinary research approach.

Dr. H. Siadat is a distinguished soil scientist and editor-in-chief of two scientific periodicals in Iran: Journal of Soil Research and Journal of Water Research in Agriculture. He is a retired research professor and has been an invited professor in The Academy of Sciences, Islamic Republic of Iran for more than 20 years and a member of the recently formed Council of Luminaries in Agricultural Research, Education and Extension Organization. He served as Deputy Director General of Soil and Water Research Institute from 1983 to 1995. His main field of expertise is soil–water–plant relationship.

Dr. M. H. Roozitalab is a distinguished soil scientist with wide range of scientific and working experiences in Iran and abroad, He received his MSc (1974) and PhD (1978) from Oklahoma State University, Stillwater, Oklahoma, USA. His major research and teaching activities at Tehran University have focused mainly on soil genesis and classification and clay mineralogy. He has contributed to the advancement of soil science in Iran as President of Soil Science Society for 17 years from 1992 to 2009 and Director General of Soil and Water Research Institute from 1983 to 1995. Dr. Roozitalab also served as Deputy Head of Agricultural Research, Education and Extension Organization (AREEO) in Iran from 1995 to 2006. He has contributed to strengthening the regional and global agricultural research collaboration as President of Association of Agricultural Research Institutions in the Near East and North Africa, AARINENA (1998–2000), Coordinator of Regional Highland Research Network of ICARDA (2008–2011) and Chairman of the Global Forum on Agricultural Research (GFAR) from 2002 to 2006.

3 Climate

Ali Khalili and Jaber Rahimi

Abstract

Iran is characterized by an extremely diversified climate, ranging from extra-arid to per-humid type. The main reasons are (a) vast extension of latitude, (b) stretching mountains and tremendous changes in altitude across the country, and finally (c) the position of the land considering seas and adjacent/far water. In this chapter, the effect of these factors on Iran's climate is explained, and main climatic parameters including precipitation, temperature, humidity, solar radiation, wind regime, and prevailing winds are classified, and their spatial/temporal distribution maps are presented. The climate-based zoning of Iran according to the modified de Martonne's classification system shows a rapid decrease from extra-arid toward per-humid in such a way that 64% of the whole country is arid and 20% semi-arid. Hence, Mediterranean to per- humid climate type is limited to 16%. The same study shows the future status of climatic condition during the 2050s and 2080s as becoming drier and warmer. Changes in the percentage of covered areas by different climates and sub-climates are presented quantitatively.

Keywords

Precipitation • Air temperature • Soil climate • Solar radiation • Winds • Humidity
Climate classification • Climate change

3.1 Factors Affecting the Climate of Iran at a Glance

Iran is geographically located in southern half of the temperate belt in northern hemisphere, within the arid and desert belt of the world. Major parts of this area coincide with air subsidence zones in general circulation model of the atmosphere. The deserts of Iran are categorized among the world's driest regions. The climate diversity is extremely high due to: (a) vast expanse of latitude being mountainous with tremendous changes in altitude across the country, and finally (c) the position of the land as related to the adjacent as well as the far water bodies.

A. Khalili (✉) · J. Rahimi
University of Tehran, Tehran, Iran
e-mail: akhalili@ut.ac.ir

3.1.1 Effect of Latitude

The vast expanse of latitude in Iran, from 25 to 40 °N, i.e., about 15°, directly affects solar radiation regime, air temperature, and other climate-related factors. For example, annual mean air temperature varies from 26 °C in Chabahar, at the Oman Sea, to 16 °C in Bandar-e Anzali, at the Caspian Sea borders, while both places nearly have the same elevation. This difference is mainly due to the difference in latitude.

3.1.2 Effect of Topography

Topography is known as the most important factor in climate diversity. The impact of mountains on climate can be outlined as follows:

M.H. Roozitalab et al. (eds.), *The Soils of Iran*, World Soils Book Series,
https://doi.org/10.1007/978-3-319-69048-3_3

- First, the great mountain chains, Alborz in the north and the western Zagros mountains, act as walls. Alborz causes the rising of humid air masses above the Caspian Sea or North Atlantic over the hillsides and extensive precipitation over the northern coastal area. Western Zagros range, on the other hand, obstructs the western humid air masses. This condition theoretically causes warm and dry winds on the lee sides of Zagros (Föhn phenomenon), and in further areas, it leads to arid condition, depending on the distance from the sea.
- Second, the mountainous heights, either as chains or as single, play a recognized role in reduction in temperature and increase in precipitation. Main points of these changes in mathematical equations of temperature and precipitation are reflected in Sects. 3.3.2 and 3.4.1.

The range of heights in Iran is from about 25 m under the sea level at the coastal areas of Caspian Sea to about 5600 m at Central Alborz chain. Figure 3.1 displays topography and location of meteorological stations referenced in the text.

3.1.3 Position of the Land as Related to Seas

The impact of great water bodies on the climate is twofold, including "the adjacent seas" and "the distant seas." The Caspian Sea, Persian Gulf, and Oman Sea are categorized in the first category, and the Mediterranean Sea, Black Sea, Indian Ocean, Atlantic Ocean, and the Red Sea in the second one. Prominent effect of the Caspian Sea is noticeable in a widespread region that stretches from Astara to the end of Gorgan Plain, which is situated between the foothills of Alborz mountain range and the sea. In this area, there is a type of humid to per-humid, moderate climate, and at higher areas cold humid to very cold humid climate. Reduction of temperature and a high amount of rainfall are known as traits of this kind of climate.

The role of the seas located in the south of the country, with respect to lower latitude and to the absence of rugged mountains that can act as a wall against wind flow, is not comparable with the situation in the Caspian Sea's coastal region, in terms of annual rainfall.

Its effect is dependent on the existence of special synoptic conditions. Under such conditions, relative intensity of showers and instant precipitation in southern regions are much higher than in central and northern regions. However, general climatic impact of the seas and the distance from the mean be explained in slight fluctuations in temperature and humidity during the year, in high vapor pressure throughout the coastal plains and in local wind regime.

The role of distant seas can mainly be summarized in supplying humidity of western moist flows, which are dependent on external low-pressure systems. Finally, the effect of Indian Ocean on the climate is dependent on monsoon's southeastern flows that reach Balochistan region and the southeastern part of the country in summers, with the frequency of four to five years in average. This causes heavy summer rainfalls in this region.

Effects of the monsoonal flows coming from the Indian Ocean into Iran are to notice in August at rainfall regime of southeastern stations, which is stretched to Fasa and Darab. Humidity resulting from the monsoon flows with a frequency of eight years, in August penetrates even to Tehran.

3.2 Air Masses

The types, sources, and tracks of air mass penetration to Iran during the warm and cold seasons of the year are shown in Fig. 3.2.

3.3 Precipitation

3.3.1 Source of Precipitation

General source of precipitation comes from the humid air flows moving across the country along with low-pressure centers, with a west to east direction, during a period of approximately seven months/year, from mid-October until mid-April. A major portion of these low-pressures directly come from the Mediterranean Sea (64%), another part from the North Atlantic Ocean, after crossing over the Caucasus or Black Sea (13%), and finally another portion that after crossing over the Red Sea and Persian Gulf reach Iran (23%).

In addition, in mid-scale, two other sources for regional precipitation in Iran are (1) advection of the polar air masses over the Caspian Sea that include main portion of the precipitation on the Caspian Sea coastal plains, especially in November and December (Khalili 1973), and (2) occasionally penetration of seasonal humid flows of the Indian Ocean in the summer from southeast of the country. The main location for generating rainy low pressures in Iran is mainly at central and eastern Mediterranean (the Aegean Sea). Some researchers point out that the existence of Zagros mountains is effective on low-pressure centers generation in the region (Alijani 1981).

3.3.2 Precipitation Variation with Altitude

Investigation into the variation of mean annual precipitation with altitude in order to draw isohyetal map was carried out based on a network consisting of 1441 stations (Khalili and Rahimi 2014). In this study, Iran is divided into 17 zones, 16 zones of which show positive linear altitudinal gradient (at

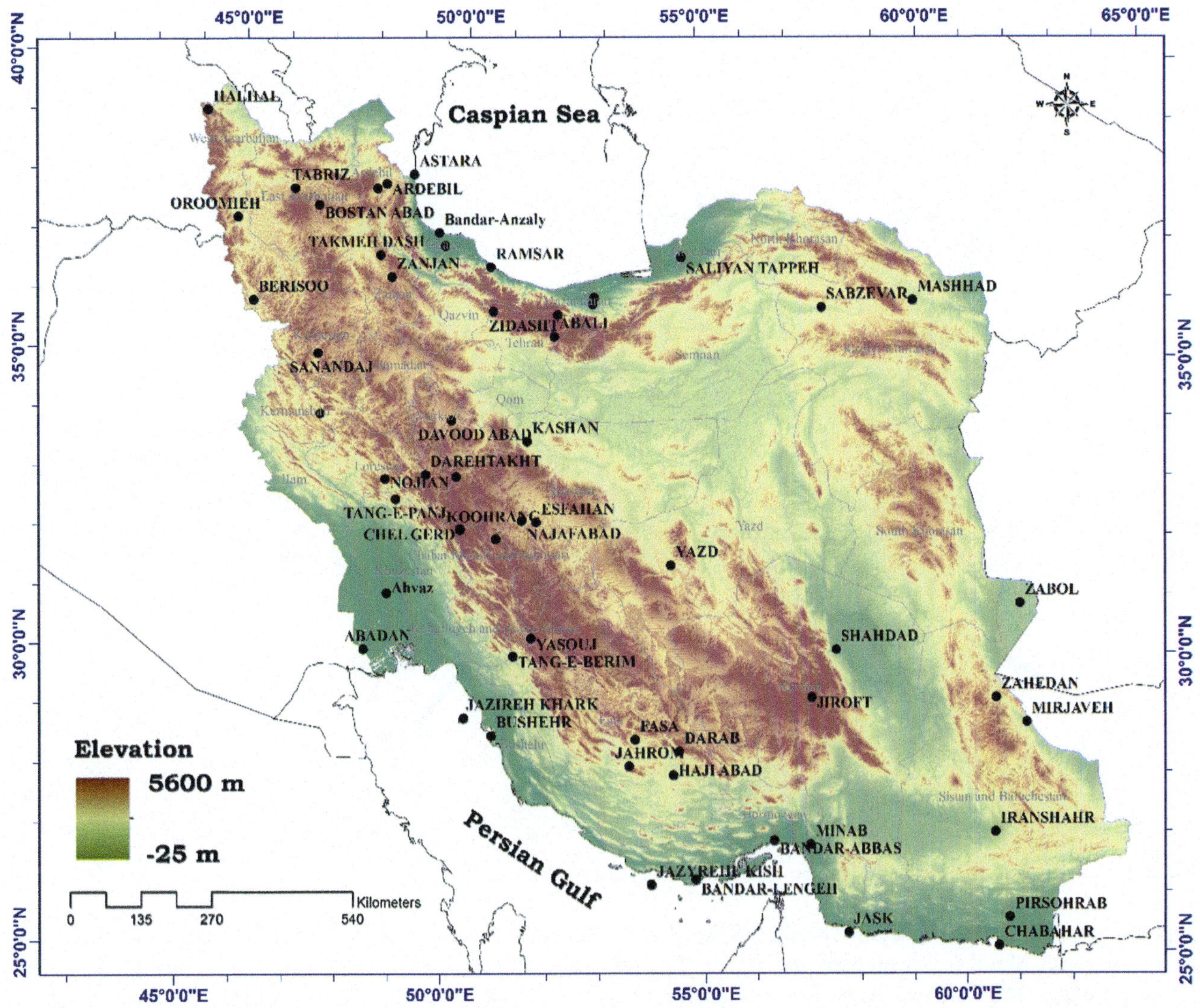

Fig. 3.1 Topography of Iran and location of typical meteorological stations cited in the text (Khalili 2005)

least significant at 5%). These gradients vary between 75 and 573 mm/km in various regions. In the coastal zone of the Caspian Sea to ridgeline of the Alborz chain, the correlation between precipitation and altitude is from second-degree type, which means that the precipitation reduces until a specific height (about 1000 m) and then increases again. This reduction is due to the dominant effect of proximity to the seas compared to altitude effect. The subsequent increase is due to the stronger influence of the altitude effect over the effect of distance from the sea. The annual isohyetal map of Iran with resolution of 1 × 1 km (Fig. 3.3) is the result of the study.

The amount of annual precipitation in the Iranian plateau ranges from 13 to 2003 mm per year. Lut Desert with <25 mm, southwest of Caspian Sea with 1500 mm, and Jazmurian pit and other extra-arid regions with <100 mm are a few cores to mention.

Areas of the country that annually receive precipitation between 100 and 200 mm are almost continuous areas stretching between mountains and coastal regions of the Persian Gulf and Oman Sea, or between central deserts and low-height or semi-high mountains of Alborz, Zagros, and central ranges. A major section of lands with good agricultural capability in Iran is located in the areas where annual precipitation is between 300 and 400 mm, which includes Azerbaijan Plain, northern Khorasan, slopes of Alborz, east/west of Zagros chain, and central mountains. At the group of areas with high level of precipitation in Iran, coasts of the Caspian Sea, heights of Alborz, branches of western river basins (e.g., Chelgard, 1400 mm), and some parts of Iran and Turkey border can be mentioned. The average of annual precipitation in Iran (Fig. 3.3) is 254 mm.

At a general view, precipitation in Iran decreases from the west to east and from north and south to the center. Figure 3.4 depicts precipitation height profile of Iran along the parallel 33°N, between border of Iraq and Afghanistan (A1A2), with resolution of 1 × 1 kilometer. This profile

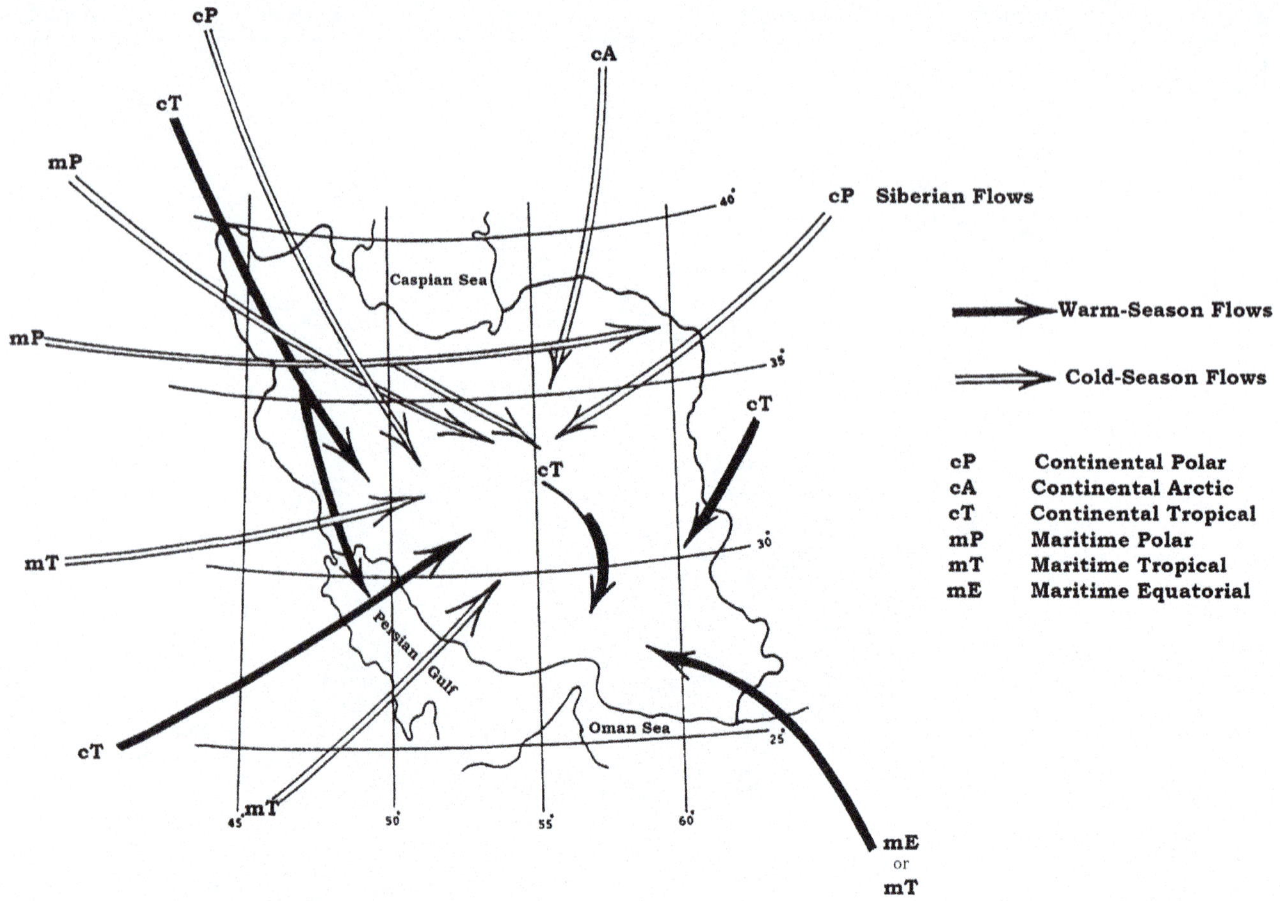

Fig. 3.2 Source regions and paths of various air masses over Iran (Khalili 2005)

appropriately displays the effective height of the Zagros chain influence. Moreover, it properly reflects the general effect of distance from the sea, unequal role of heights, and difference of altitudinal gradient of precipitation at different points of the country. In this Fig. 3.3b, another profile of precipitation height has been presented along meridian 49 ° E (B1B2), between the Caspian Sea and Persian Gulf, which especially reveals different roles of these two great water bodies' area.

Overall, the precipitation in Iran, except for the Caspian Sea coastal regions, is irregular. More irregularity is observed in the drier regions. The coefficient of variation (CV) for annual precipitation in Iran at coastal areas of the Caspian Sea is approximately 20%, while in the interior of the plateau, its values increase and in the Shahdad, located at Dasht-e Lut border, reach 80% (Khalili 1973).

3.3.3 Temporal Pattern of Precipitation

Trend of variation of precipitation in different months in Iran is mainly Mediterranean. Regardless of this general view, distribution of precipitation amount in different regions of the country is not uniform, and the occurrence time of the rainiest month as well as the gradient of precipitation versus time during the year and from a month to another month at different areas varies considerably.

Khalili and Rahimi (2014) classified the precipitation regime, divided into groups and subgroups, and in summary "representative types" (Table 3.1). Each type is the average of all studied stations in the region.

3.4 Air Temperature

Climatic parameters of air temperature in Iran are controlled firstly by altitude and then by latitude, and lightly by longitude. Mean annual air temperature network varies from 5.3 °C in Halhal in northwest of Azerbaijan at the height of 2050 m to 28.3 °C in Pir-e-Sohrab in southeast of the country at an altitude of 120 m above sea level.

3.4.1 Spatial Variation of Air Temperature

Studies based on the multivariate correlation analysis have shown that temperature parameters in Iran can be determined

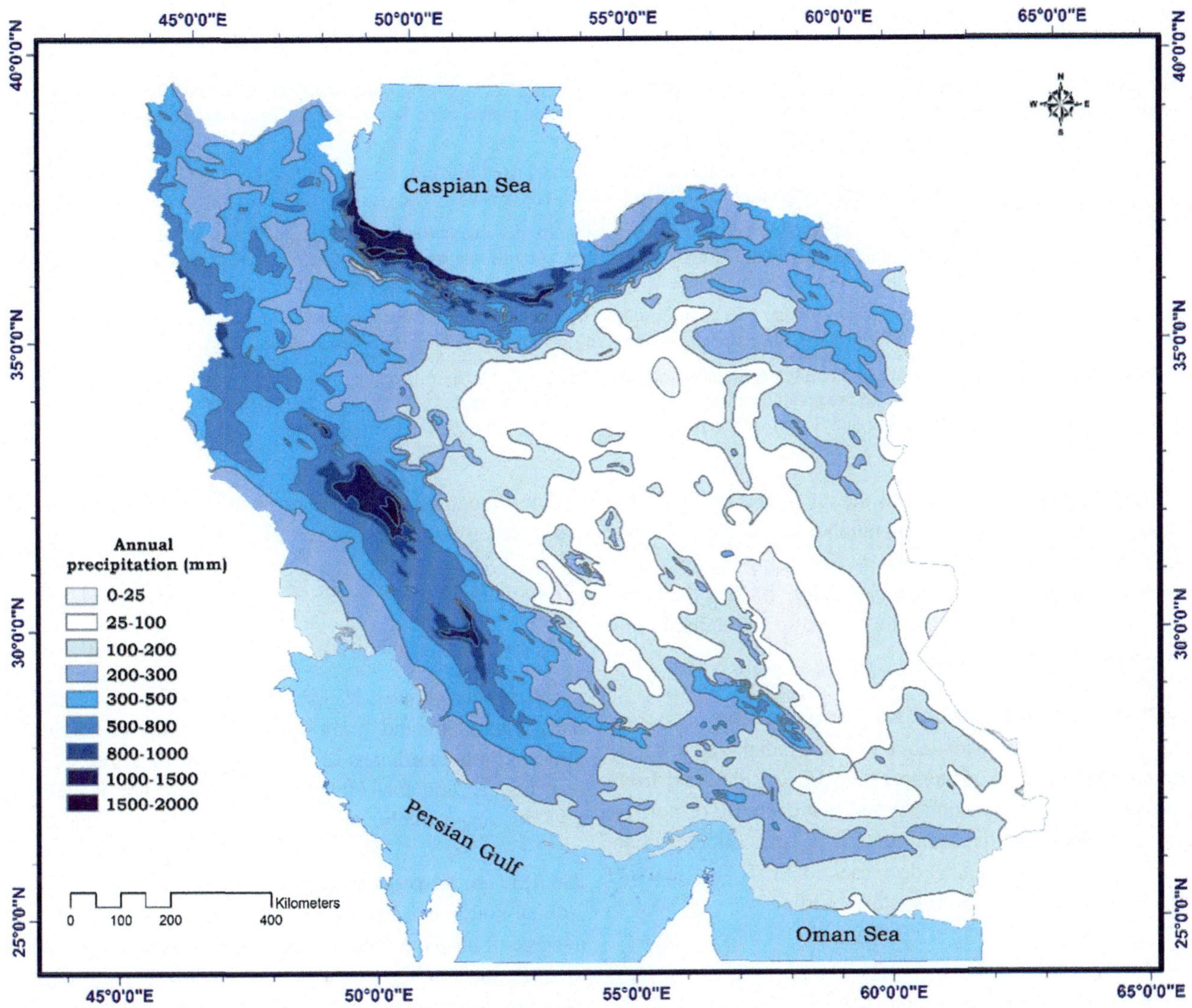

Fig. 3.3 Spatial distribution of precipitation in Iran (1961–2005); only major isohyets map are shown (Khalili and Rahimi 2014)

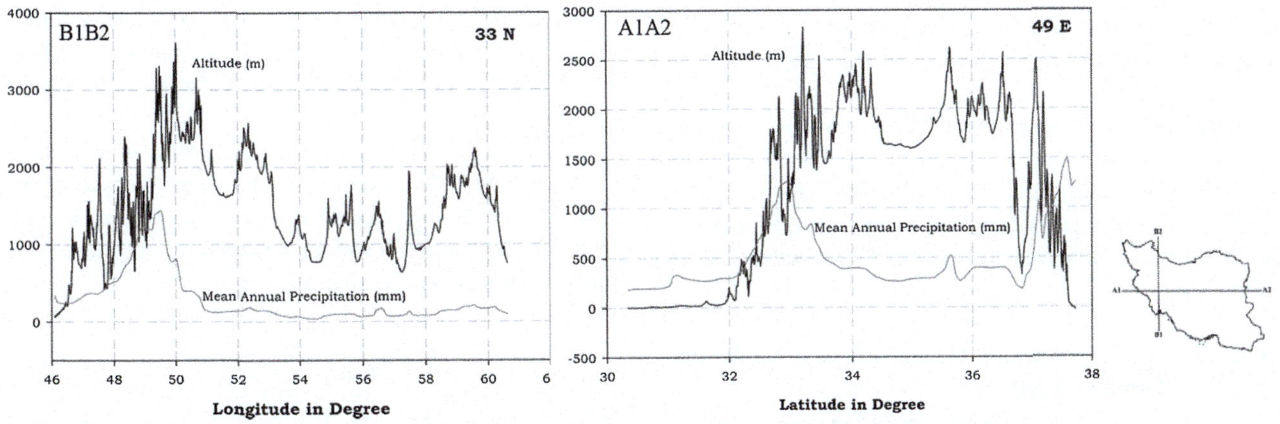

Fig. 3.4 Precipitation height profile of Iran along the latitude 33°N (A1A2) and meridian 49°E (B1B2) (Khalili 2005)

Table 3.1 Classes of precipitation regimes (Khalili and Rahimi 2014)

Types	Description
(a) Caspian sub-Mediterranean types	There is no dry season, at the driest months of the year, on average 25 mm precipitation. Since all months have precipitation, temporal gradient is small. The maximum precipitation in the west and in center occurs at early autumn and in the east at the winter and spring
(b) Spring types	The maximum amount of precipitation takes place in March to April. Azerbaijan, basins of the Lake Urmia and the Aras River, northern Khorasan, and Alborz heights fall under this type
(c) Winter types	Major portion and absolute maximum precipitation in this area occur in the winter, and in addition, precipitation gradient around the maximum is high. Conversely, summer season is completely dry, and no rain sometimes for six months. This regime is dominant type of desert basins and southwest of the country
(d) Mediterranean monsoon types	There is one weak summer maximum resulting from monsoon flows from the Indian Ocean. A prominent example of this regime is in the southeast of the country and mainly at southern Balochistan basins (e.g., Iranshahr)
(e) Winter–spring types	The majority of monthly precipitation occurs with an insignificant difference in one of the months January until April, and increase of precipitation from November until May is gradual. This regime includes mainly the western Zagros, Karkheh, Zab, and Sirvan river basins as well as some high regions of Fars Province

by linear equations that connect the air temperature to longitude, latitude, and height. The number of equations to take into account the three parameters, including mean maximum, mean minimum, and mean daily, is 39 equations. Here, only the equations that are related to the annual temperature are briefly described.

3.4.1.1 A. Mean Annual Air Temperature

The following regression equation of spatial variation of annual mean of daily air temperature (T_{mean}) obtained from 30 years data of 507 stations shows that decrease of normal annual temperature is equal to 5.7 °C per kilometer; in addition, air temperature decreases 0.93 °C as latitude increases each degree from south to north.

$$T_{mean} = 53.78 - 5.7 \times 10^{-3} Z + 5.9 \times 10^{-3} X - 9.3 \times 10^{-1} Y$$

(Significant at 0.01; R^2 = 0.914) where Z is height (m), X is longitude (arc degree), and Y is latitude (arc degree).

Although, the effect of longitude on temperature is negligible and is about 0.01 °C per each degree that can be due to the existence of desert ranges at the center to east and southeast of Iran.

Based on this equation, annual temperature at the heights of 3000 m in northwest of the country is about 1.6 °C, and at the heights of sea level on the southern shores is about 28.0 °C. These two values define the general range of annual long-term normal temperature fluctuations in Iran (Fig. 3.5).

3.4.1.2 b. Mean Annual, Maximum, and Minimum Temperature

The aforementioned method was applied in order to analyze annual normal maximum temperature (T_{max}) and minimum temperature (T_{min}). Three-dimensional regression equation of these parameters is significant at 0.01 and verified as follows (coefficients of determination for these equations are, respectively, 0.925 and 0.815).

$$T_{max} = 68.66 - 5.6 \times 10^{-3} Z - 1.94 \times 10^{-2} X - 1.119 Y$$

$$T_{min} = 38.90 - 5.9 \times 10^{-3} Z + 3.11 \times 10^{-2} X - 0.741 Y$$

These relationships also reveal that the annual average of daily maximum and minimum temperatures at heights of 3000 m in the northwest of the country is 10.4 and −5.3 °C, respectively, and at sea level in the southern Iran is 40.6 and 19.8 °C, respectively.

3.4.1.3 b. Annual Range of Temperature

Annual range of temperature is the difference between daily average of air temperature in the warmest and coolest month of the year. This range for the coastal regions with per-humid climate is much less than for arid areas of the central deserts, because of air humidity that causes adjustment in temperature fluctuations. The mean range of annual temperature (A) at southern coastal areas is low (Bandar Abbas 16.4 °C, Bandar-e Lengeh 16.1 °C), and gradually with going far from the sea, it increases (Hajjiabad 23.8 °C, Jiroft 21.9 °C). In the coastal areas of the north, this law is also observed.

Generally, the amount of A in the coastal areas is about 15 °C, and in cold areas of the northwest as well as the central desert of Iran reaches about 27 °C. Conrad's continentality coefficient (C) was used to measure the effect of air humidity on temperature regime.

$$C = \frac{1.7A}{\mathrm{Sin}(\varphi + 10)} - 14$$

where φ is latitude, and A is range in terms of degree centigrade. This coefficient in the most continental points of the world like Verkhoyansk, Siberia is about 100, and in the maritime tropical regions is around zero.

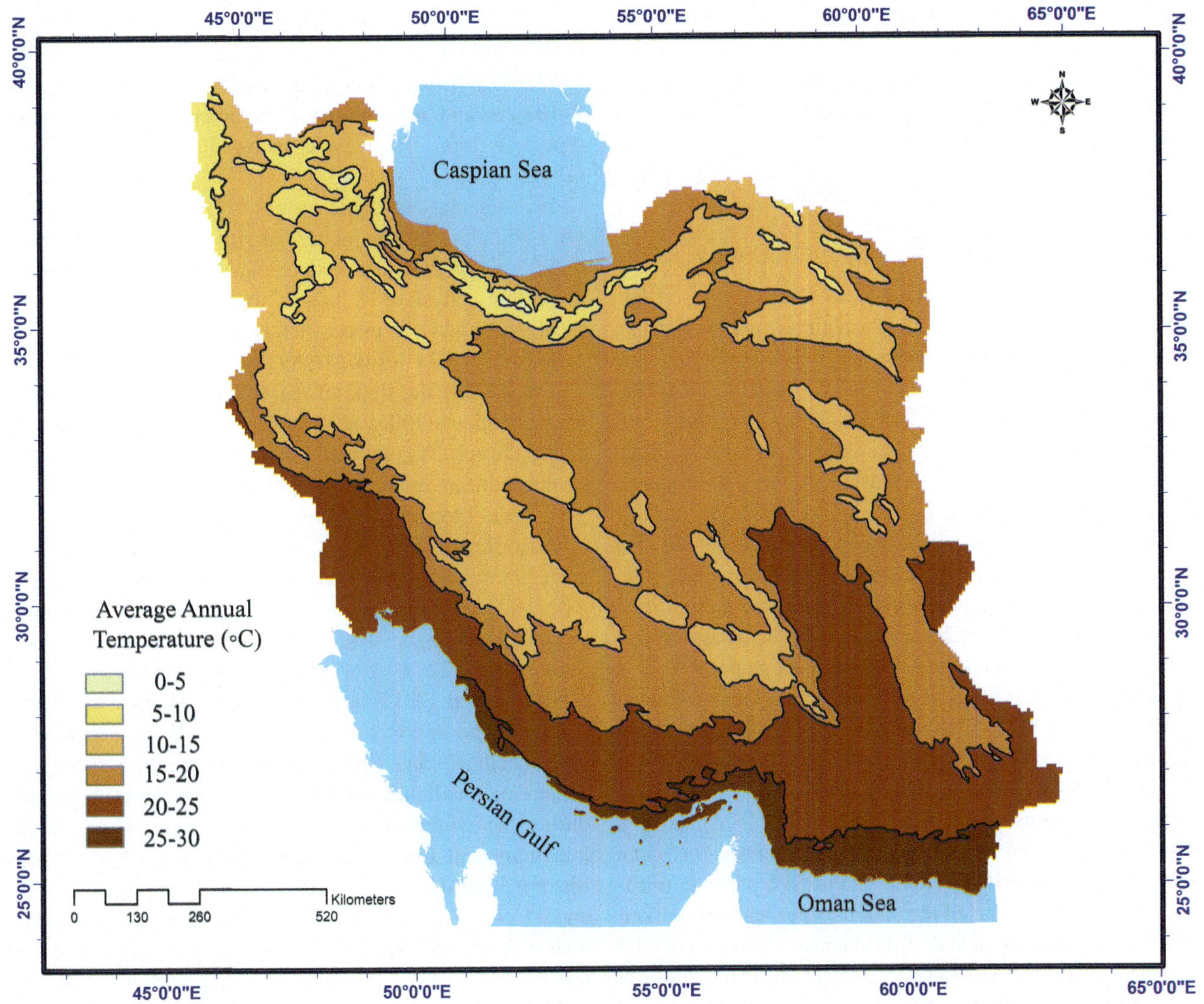

Fig. 3.5 Spatial distribution of mean annual air temperature over Iran (1961–2005) (original work by the authors for the chapter)

In Iran, the "C" value in the coastal zones near the Caspian Sea, Persian Gulf, and Oman Sea is about 24; in cold areas of the northwest, heights of the Zagros are about 50, and in the central desert and Dasht-e-Lut are about 60.

3.4.2 Frosts

In this context, the number of frost days, which are the days in which the daily minimum temperature reaches zero or below, was studied. The most suitable method to estimate the number of frost days in Iran is the relationship between annual mean temperature and the number of frost days. This analysis was performed, and its results are presented as follows:

$$N = 207.9 - 8.7T$$
$$R = 0.91$$
$$n = 249$$

where N is annual mean of the number of frost days; T is annual mean of daily temperature (normal temperature); n is the number of stations; and R is correlation coefficient.

From the 249 stations, Kish, Chabahar, Jask, Bandar-e Lengeh, and Minab are frost free; Bushehr is on average 0.1 frost day (once at 10 years); Bostan Abad (altitude = 1750 m) has the top number, i.e., 169.1 days/year, which is equal to 46% of time over a year. Above equations were used to convert the data to a three-dimensional spatial distribution, the basis for drawing the map of frost day's number considering longitude, latitude, and height or in order to point estimates (Khalili 2005). In general, one-degree increase of annual normal temperature would mean 8.7 days less frost, i.e., the number of annual frost days at the mentioned place minus 8.7.

3.5 Soil Climate

Climate is one of the major state factors controlling soil formation (Jenny 1941). Soil climate regimes, manifest in the combined factors of soil moisture and soil temperature, were defined and are used at different levels in various soil classification systems (e.g., USDA Soil Taxonomy and WRB). Soil moisture regime "refers to the presence or absence of either groundwater or water held at a tension <15 bars in the soil or in specific horizons (soil moisture control section) by periods of the year"; a soil characteristic that plays an important role in the analysis of the state of the soil water. On the other side, also soil temperature controls plant growth and is important for soil formation. Both characteristics can be measured and described, but there also some methods and computer models have been proposed to relate soil moisture to meteorological records (USDA 1975). The best-known computer program is that of Newhall (1972), written in COBOL language, where climatic data of several years are used to simulate the soil climatic conditions. In 1976, van Wambeke (1976) set up a research perspective, and while strictly followed the same principles proposed by Newhall, published a paper to introduce some subdivisions of the soil moisture regimes worked out in a software written in BASIC language, known under the name NSM. The BASIC software differs in its approach from the original Newhall model by using average years (http://www.css.cornell.edu/faculty/dgr2/research/nsm/nsm.html). NSM was used by Banaei (1987), in cooperation with Ghent University, Belgium to prepare the soil climate map of Iran at the scale of 1:2,500,000. The required climatic data of 300 meteorological stations were used to run the program; the results of which were combined in a GIS environment to produce the map.

In another attempt, Sabziparvar et al. (2010) used 10-year (1996–2005) daily meteorological data to investigate the relationship between soil temperatures at standard depths of 5–100 cm and meteorological parameters in four climate types of Iran. The results showed strong correlations (Table 3.2).

Investigating the table values shows that in arid and semi-arid climates, the higher the air temperatures, the more differences between air and soil temperatures are. Based on the regression analysis of 18 stations' data over Iran, the following relationship was obtained which is significant at 1%.

$\text{MAST} = 2.34 + 1.07\text{MAAT}$.

Considering the slope of regression line in this equation, abovementioned difference is a little more than 2 °C in Iran. Using the results combined with the isothermal map of the country, mean annual soil temperature map at depth of 50 cm was prepared (Fig. 3.6).

According to this map, 0.24, 13.44, 52.37, and 33.95% of the country area is covered with frigid (temperature range: 4–8 °C), mesic (temperature range: 8–15 °C), thermic (temperature range: 15–22 °C), and hyperthermic (temperature range: >22 °C) soil types, respectively.

Khalili et al. (2012) have also worked on "freezing depth" as one of the important climatic indices in the soil science and concluded that SHAW model is the best model for estimating maximum soil frost penetration depth, especially during the winter months. Based on their findings, mean maximum soil frost penetration depth at the four stations Shahr-e kord, Urumia, Sanandaj, and Yazd was 22, 9, 21, and 11 cm, respectively. Also, the absolute soil frost penetration depth was 36, 20, 39, and 21 cm, respectively. Soil texture at a depth of 20–50 cm in Shahrekord was silty clay and silty loam in others.

3.6 Air Humidity

Important indicators of air humidity mainly include the relative humidity, vapor pressure, and the amount of these two parameters during the year as well as their trends of

Table 3.2 Main parameters of soil temperature regime in four climate types of Iran (original work by the authors for the chapter)

Climate type	Typical station	Average MAAT	Soil temperature at 50 cm depth[a]				
			MAST	MSST	MWST	MSST-MWST	MAST-MAAT
Temperate–humid	Rasht–Sari	17.2	19.5	25.3	10.7	14.6	2.3
Cold semi-arid	Tabriz–Urmia	12.1	15.2	26	5.2	20.8	3.1
Warm semi-arid	Isfahan–Shiraz	17.1	21.6	30.7	13.2	17.5	3.8
Warm-arid	Yazd–Zahedan	18.8	24	33.2	14	19.2	5.1

[a]MAST (mean annual soil temperature in °C); MWST (mean soil temperature in winter °C); MSST (mean soil temperature in summer °C) all at a depth of 50 cm below soil surface

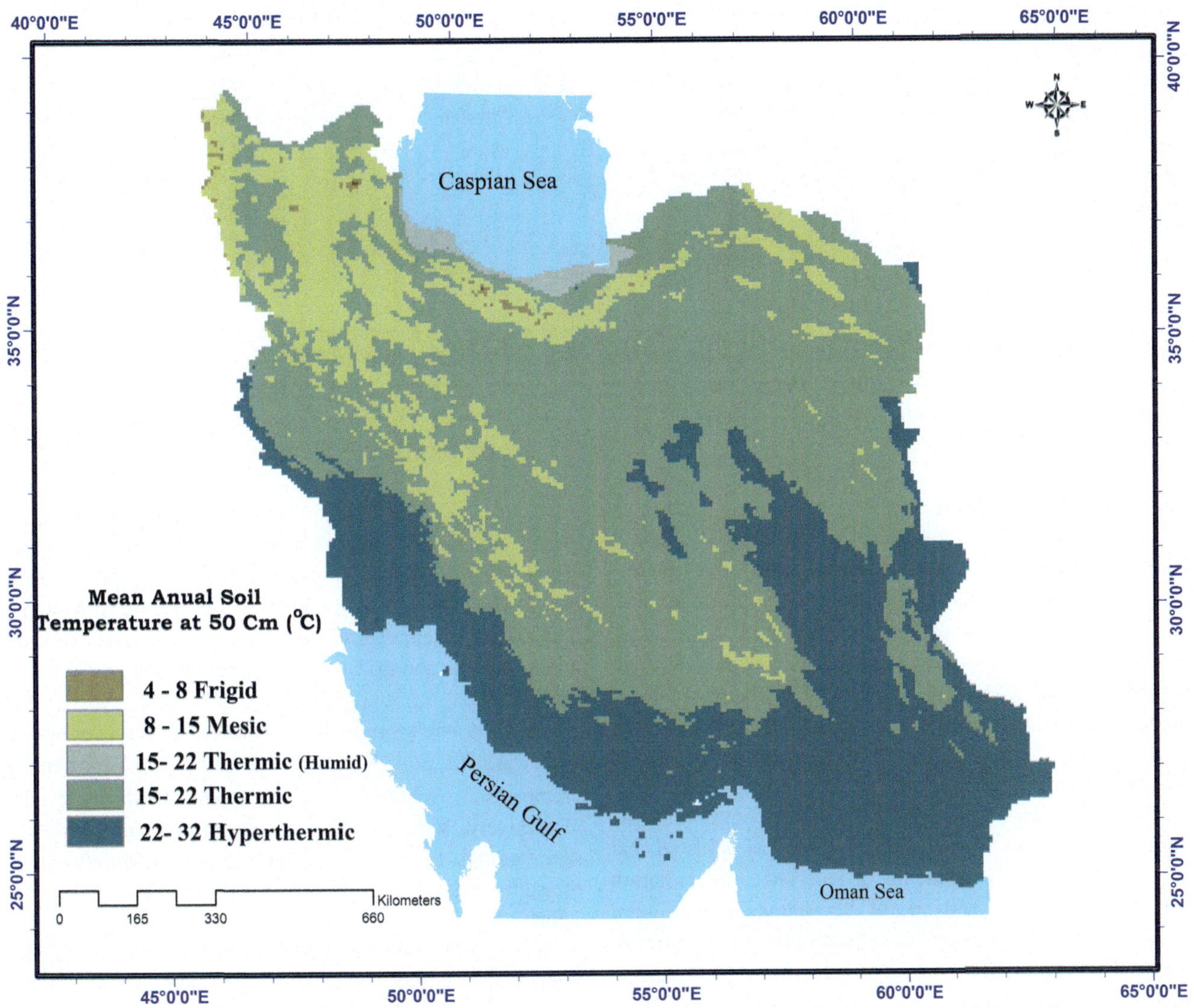

Fig. 3.6 Spatial distribution of soil temperature regime over Iran (original work by the authors for the chapter)

fluctuations, relative to the distance from the sea, proximity to mountains, and the presence/absence of vegetation coverage.

3.6.1 Relative Humidity

The annual average of relative humidity in Iran swings between 29% in Mirjaveh to 85% in Qaran talar at the coastal plains, near the Caspian Sea (Table 3.3).

Review of trends indicates that (1) in the southern coasts, the air relative humidity reflects two peaks, one in the winter (February) and another one in the summer (August). An appropriate sample of this regime is that of Bandar Abbas. And (2) in other areas of the country, the amount of relative humidity in the winter is maximum and in the summer is minimum, with high domain of humidity fluctuations. The monthly average of relative humidity in Yazd in the summer is about 17%, while in the semi-arid regions or the areas with Mediterranean climate, located in the west, such as Kermanshah, this minimum reaches about 20–25%. In the mountainous areas like Abali, the minimum amount extends more, even up to 30%.

3.6.2 Water Vapor Pressure

The amount of vapor pressure, similar to relative humidity, is dependent on the distance to seas and on vegetation coverage in the surroundings. The vapor pressure in different stations in winter is lower than in summer. Also, Persian Gulf's coast tends to have a higher vapor pressure than the Caspian Sea regions. For examples, variations in the monthly average of water vapor pressure in Bandar Abbas

Table 3.3 Variation of mean annual relative humidity (%) in Iran based on the data of 249 stations (Khalili 2005)

Relative humidity parameter	Morning (6:30 LT)		Noon (12:30 LT)		Afternoon (18:30 LT)		Mean	
	Station	Value	Station	Value	Station	Value	Station	Value
Min	Mirjaveh	39.3	Zahedan	22.1	Yazd	25.5	Mirjaveh	29.4
Max	Rasht	93.8	Chabahar	78.5	Ghoran talar	87.6	Ghoran talar	84.7
Mean	65.0		44.3		50.6		53.3	
SD	12.0		13.2		14.1		12.7	

(Persian Gulf coast) are 13–35 mbar, in Bandar-e Anzali (Caspian Sea coastal areas) 7–25 mbar, and in Yazd, Tehran, and Kermanshah (Central Iran Plateau) 4–10 mbar.

3.7 Solar Radiation

Direct measurement of solar radiation intensity is merely taken in few weather stations in Iran; among them, the data of 17 stations can be processed from climatic point of view. The lowest amount of received energy is related to Ramsar with an annual average of 283 $cal.cm^{-2}$ day^{-1}, and the highest value is associated with Yazd with an annual average of 507 $cal.cm^{-2}$ day^{-1}. In order to estimate the amount of received solar radiation (Rs) in other stations, angstrom relationship between Rs from one hand and sunshine duration/cloudiness from the other hand was adjusted and applied. Therefore, the amount of received solar radiation was prepared for a network of 85 stations (Khalili and Rezaee Sadr 1997). Figure 3.7 indicates the main isolines of annual radiation in Iran that have been drawn with above-mentioned considerations.

3.8 Winds

In order to give an overall view concerning the winds in Iran, the data from 75 synoptic stations were assessed that their statistics period were more than five years. Average speed of winds is tabulated (Table 3.4), with a remark that average speed in summers is two to four times greater than in winters.

Regarding prevailing wind, that is the direction of the wind with the highest frequency, 62% of the prevailing winds blow from western sector (including northwestern and southwestern winds). The northern prevailing winds have the lowest frequency and can be seen purely at 3% of stations.

3.9 The Climate of Iran

3.9.1 Iran's Climate Zones in Modified de Martonne System

Iran's climatical classification has been done so far at different scales and various systems. For example, Ganji (1954), Shafi Javadi (1966), Sabeti (1969), and Khalili et al. (1992) (Fig.3.8) can be mentioned. The last classification is done at the Integrated Water Plan of Iran using 1100 rain gauge stations and 507 stations for temperature as is shown in Fig. 3.8.

De Martonne's original classification is based on aridity index (A_i), ($A_i = \frac{P}{T+10}$), where P and T are mean annual precipitation (mm) and mean annual temperature °C), respectively. Values of A_i were categorized and resulted in seven groups, from arid to per-humid. Names and thresholds of aridity index classes are presented in the columns 1–3 of Table 3.5.

While maintaining the general skeleton of the de Martonne's aridity index, Khalili (1997) came up with a modification where more classes are foreseen. Table 3.5 illustrates the areas (in %) of the different climate and sub-climate types according to the proposed modified system. A quick review of this table shows that 36% of the country area is covered with extra-arid climate, and 29% is associated with the arid deserts; in general, 65% of whole country lands are under arid climate. Semi-arid climate covers about 20%. Hence, the portion Mediterranean climate to per-humid climate totally is limited to 16%.

Mentioned sample stations in Table 3.5 provide a clear picture of necessity for climate separation of climate types into thermal sub-climate types, the modification that is applied. For example, although the climate of both Koohrang (2650 m) and Bandar-e Anzali (−21 m and close to the Caspian Sea) are categorized as per-humid B (A7 class), their vegetation coverage and flora are different. While in the

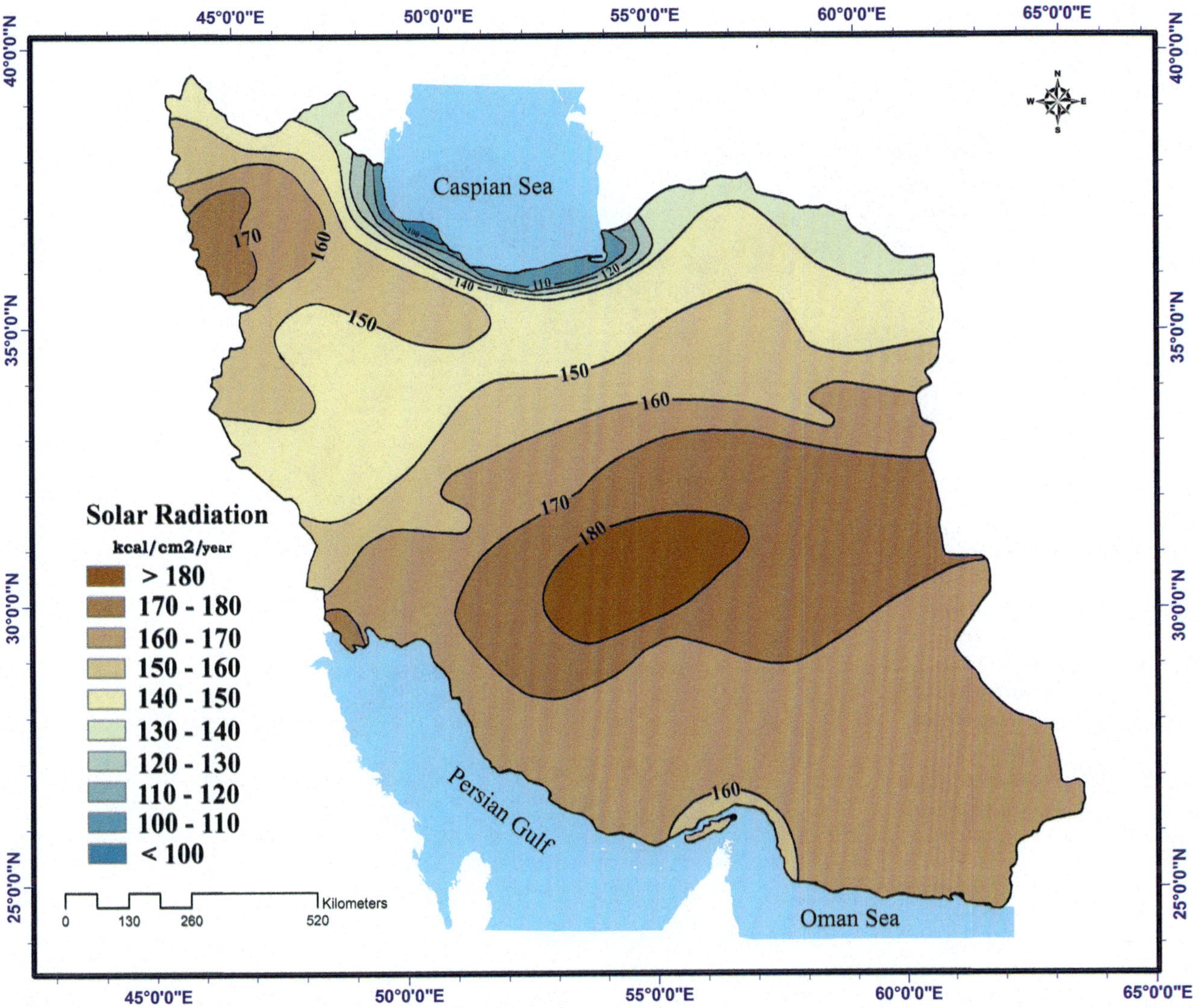

Fig. 3.7 Mean annual global radiation on horizontal surface (Khalili and Rezaee Sadr 1997)

Table 3.4 Range of variation in mean wind speed in Iran (m/s) (Khalili 2005)

Parameter	Annual mean		Mean of June		Mean of January	
	Station	Value	Station	Value	Station	Value
Min	Kashan	0.6	Kashan	0.8	Kashan	0.2
Max	Zabol	4.8	Zabol	9.1	Jazireh Khark	4.3
Mean	All stations	2.3	All stations	2.8	All stations	1.8
SD	All stations	0.9	All stations	1.8	All stations	0.9

modified classification, Koohrang and Bandar-e Anzali are categorized as per-humid B-very cold (A7m1) and per-humid B-moderate (A7m3) climate, respectively.

3.9.2 Climate Change in Iran

In Iran, alike many other countries, studies have been carried out on climate change, whether in form of trend analysis of

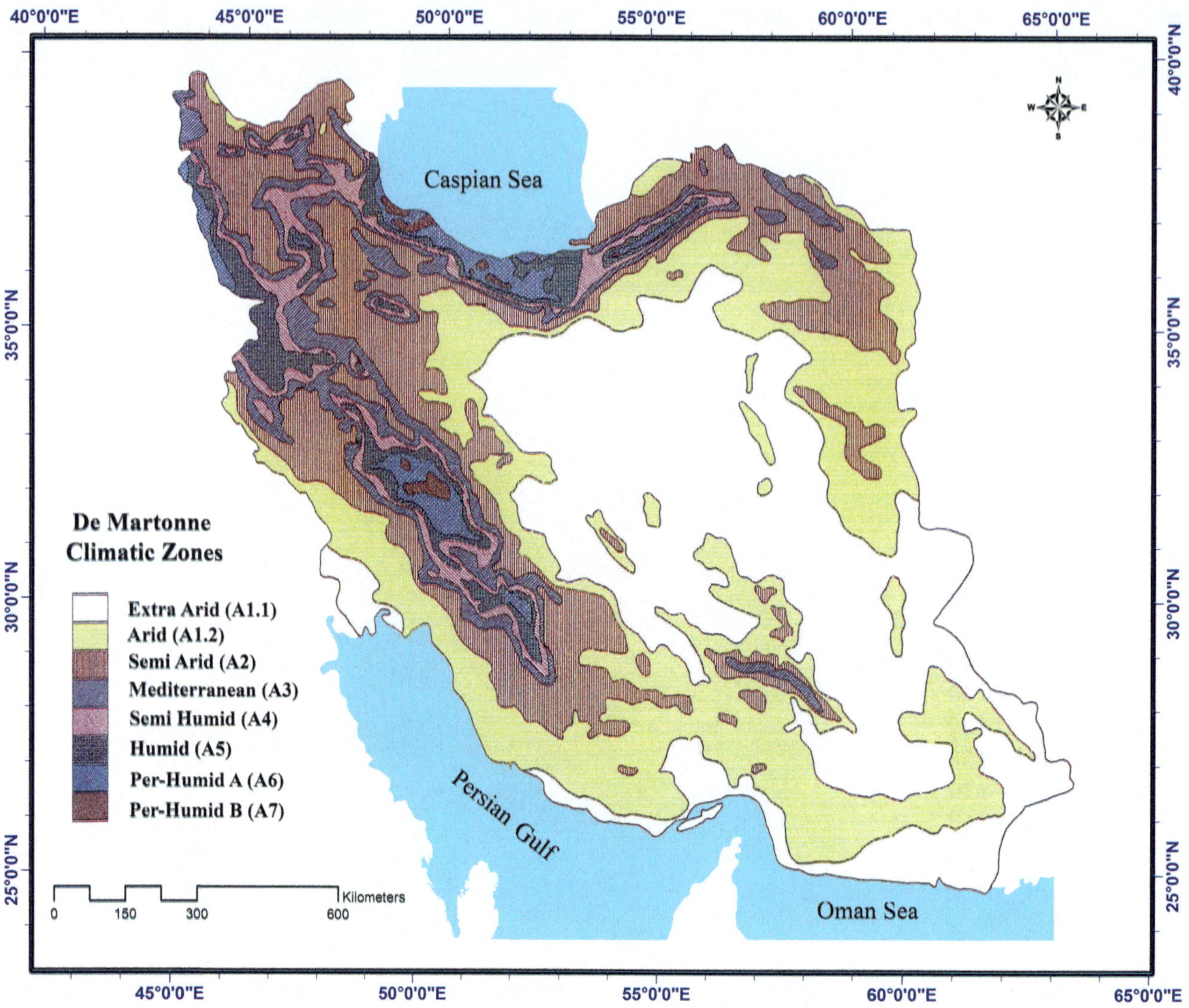

Fig. 3.8 Climate zones of Iran based on original de Martonne classification system (Khalili 1997)

historical climate parameters (Tabari et al. 2011; Kousari and Zarch 2011) or for searching in future climate under different scenarios (Rahimi et al. 2013). In the revision of the Integrated Water Plan of Iran (1997), climate change in terms of mean temperature was considered for 507 meteorological stations in a 30-year period (1965–1995) as baseline temperature; then, consecutive 10-year means of the same network were studied (Table 3.6).

Besides, Rahimi et al. (2013) evaluated the future status of climatic zones in the country during the 2050s and 2080s, using the modified de Martonne's classification system and concluded that the general tendency is toward drier and warmer climate in the future decade. In Table 3.7, changes in percentage of covered areas by different climate and sub-climate types have been presented during two decades, the 2050s and 2080s, and under the scenarios A1B and A2.k

Table 3.5 Areas of different climates and sub-climates of Iran according to modified de Martonne's classification system (% of total area[a]) (Khalili 2005)

Main climate	Thermal sub-climate	Climate type	Typical station	Areas (%)	
Extra-arid A1.1 (Aridity Index <5)	Very cold (m < −7)	A1.1m1	–	0.14	35.54
	Cold (−7 ≤ m < 0)	A1.1m2	Esfahan	15.1	
	Moderate (0 ≤ m < 5)	A1.1m3	Shahdad	13.9	
	Warm (m ≥ 5)	A1.1m4	Chabahar	6.4	
Arid desert A1.2 (5 ≤ Aridity Index <10)	Very cold (m < −7)	A1.2m1	Davood Abad	1.12	29.15
	Cold (−7 ≤ m < 0)	A1.2m2	Najaf Abad	16.1	
	Moderate (0 ≤ m < 5)	A1.2m3	Jahrom	4.76	
	Warm (m ≥ 5)	A1.2m4	Ahvaz	7.17	
Semi-arid A2 (10 ≤ Aridity Index <20)	Very cold (m < −7)	A2m1	Takmeh Dash	6.79	20.08
	Cold (−7 ≤ m < 0)	A2m2	Zanjan	10.55	
	Moderate (0 ≤ m < 5)	A2m3	Yazd	2.0	
	Warm (m ≥ 5)	A2m4	–	0.74	
Mediterranean A3 (20 ≤ Aridity Index <24)	Very cold (m < −7)	A3m1	Sarab Hendeh	3.08	4.903
	Moderate (0 ≤ m < 5)	A3m2	Ardebil	1.5	
	Warm (m ≥ 5)	A3m3	Tang-e-Berim	0.31	
	Very cold (m < −7)	A3m4	–	0.013	
Sub-humid A4 (24 ≤ Aridity Index <28)	Very cold (m < −7)	A4m1	Zidasht	2.08	3.375
	Cold (−7 ≤ m < 0)	A4m2	Berisso	0.95	
	Moderate (0 ≤ m < 5)	A4m3	–	0.34	
	Warm (m ≥ 5)	A4m4	–	0.005	
Humid A5 (28 ≤ Aridity Index <35)	Very cold (m < −7)	A5m1	Sareine Ardebi	2.01	3.562
	Cold (−7 ≤ m < 0)	A5m2	Yasouj	1.24	
	Moderate (0 ≤ m < 5)	A5m3	–	0.31	
	Warm (m ≥ 5)	A5m4	Tang-Panj	0.002	
Per-humid A A6 (35 ≤ Aridity Index <55)	Very cold (m < −7)	A6m1	Darehtakht	1.54	2.93
	Cold (−7 ≤ m < 0)	A6m2	Nojian	0.98	
	Moderate (0 ≤ m < 5)	A6m3	Astara	0.41	
	Warm (m ≥ 5)	A6m4	[b]	0.0	
Per-humid B A7 (Aridity Index ≥55)	Very cold (m < −7)	A7m1	Koohrang	0.26	0.46
	Cold (−7 ≤ m < 0)	A7m2	–	0.8	
	Moderate (0 ≤ m < 5)	A7m3	Bandar-Anzali	0.12	
	Warm (m ≥ 5)	A7m4	[b]	0.0	

[a]Iran's Area: 1,620,703 km^2, [b]Not to Exists—There Isn't Typical Station

Table 3.6 Trend of changes in the regional averages of daily temperature in the network of Iran's stations in 10-year periods based on the data of 507 stations (°C) (Khalili 1997)

Climatic period	1965–1975	1975–1985	1985–1995	1965–1995
Regional averages of daily Temperature	15.9	16.2	16.3	16.1

Table 3.7 Changes in areas (%) of climate zones of Iran during the twenty-first century under A1B and A2 scenarios (Rahimi et al. 2013)

Climates		Areas (%)							
		2050s				2080s			
		A1B		A2		A1B		A2	
A1.1	m1	+0.2	+10.5	−0.8	+10.1	0.0	+3.8	−0.8	+5.4
	m2	−4.1		−19.2		−4.0		−18.5	
	m3	+9.2		−2.7		+8.0		−2.2	
	m4	+5.2		+26.5		+6.1		+26.9	
A1.2	m1	+0.4	−0.7	−4.0	+1.6	−0.8	+1.4	−3.5	+2.6
	m2	−0.9		−6.3		+0.6		−5.7	
	m3	−0.2		+7.1		+0.3		+7.1	
	m4	0.0		+4.6		+1.5		+4.6	
A2	m1	−6.4	−5.5	−10.4	−6.8	−8.5	−1.5	−10.9	−4.0
	m2	+1.3		+0.1		+1.4		−1.1	
	m3	−0.4		+7.4		−0.2		+6.9	
	m4	0.0		+1.4		0.4		+1.1	
A3	m1	−1.7	−2.1	−2.2	−2.4	−2.0	−1.8	−2.3	−2.0
	m2	−0.5		−0.4		−0.6		−0.5	
	m3	+0.1		+0.6		0.0		+0.5	
	m4	0.0		+0.3		+0.1		+0.3	
A4	m1	−0.8	−1.0	−1.0	−1.2	−0.9	−0.7	−1.0	−0.8
	m2	−0.1		−0.1		−0.2		−0.1	
	m3	−0.1		+0.1		−0.1		+0.1	
	m4	+0.1		+0.2		+0.1		+0.2	
A5	m1	−0.8	−0.9	−0.8	−1.0	−0.8	−0.9	−0.8	−0.9
	m2	−0.1		−0.2		−0.2		−0.2	
	m3	−0.1		−0.1		−0.1		−0.1	
	m4	+0.1		+0.2		+0.1		+0.2	
A6	m1	−0.1	−0.2	−0.1	−0.3	−0.1	−0.3	−0.1	−0.3
	m2	−0.1		−0.1		−0.1		−0.1	
	m3	−0.1		−0.5		−0.2		−0.5	
	m4	0.0		+0.4		+0.1		+0.4	
A7	m1	0.0	0.0	0.0	0.0	0.0	0.0	0.0	0.0
	m2	0.0		0.0		0.0		0.0	
	m3	0.0		0.0		0.0		0.0	
	m4	0.0		0.0		0.0		0.0	

References

Alijani B (1981) Synoptic origin of precipitation in Iran. Ph.D. thesis, Michigan State University, p 328

Banaei MH (1987) Map of soil moisture and temperature regimes of Iran. Soil and Water Research Institute of Iran, Tehran, Iran

Ganji MH (1954) The climate of Iran. Bulletin de La Socie'te' de Ge'ographie d' Egypte 28:195–298

Shafi Javadi (1966) Distribution climatiques en Iran. Monographie de La Me'te'o, Nationale, Paris

Jenny H (1941) Factors of soil formation. McGraw-Hill, New York

Khalili A (1973) Precipitation patterns of central Elburz. Archiv für Meteorologie, Geophysik und Bioklimatologie, Serie B 21(2–3):215–232

Khalili A (1997) Synthesis proceeding of integrated water plan of Iran. Jamab Consulting Engineering Co., Climate Section, The Ministry of Energy, Tehran, Iran

Khalili A (2005) Climate of Iran. In: Banaei et al (eds) The soils of Iran. New achievements in perception, management and use (in Persian). Soil and Water Research Institute of Iran, Tehran, Iran, pp 24–71

Khalili A, Hajam S, Irannejad P (1992) Integrated water plan of Iran. In: Meteorological studies, climatical classification map 1964–1984, vol. 4. Jamab Consulting Engineering Co., The Minisitry of Energy, Tehran

Khalili A, Rahimi J (2014) High-resolution spatiotemporal distribution of precipitation in Iran: a comparative study with three global-precipitation data sets. Theoret Appl Climatol 118(1–2):211–221

Khalili A, Rezaee Sadr H (1997) Estimation of global solar radiation over Iran based on climatological data. Geographical Res 12(3):15–36

Khalili A, Rahimi H, Aghashariatmadary Z (2012) Validation of SHAW model in determination of maximum soil frost penetration depth in typical arid and semi-arid zones of Iran. J Agric Sci Technol 14:1185–1192

Kousari MR, Zarch MAA (2011) Minimum, maximum, and mean annual temperatures, relative humidity, and precipitation trends in arid and semi-arid regions of Iran. Arab J Geosci 4(5–6):907–914

Newhall F (1972) Calculation of soil moisture regimes from the climatic record, Revision 4. USDA Soil conservation service, Washington D.C., USA

Rahimi J, Ebrahimpour M, Khalili A (2013) spatial changes of extended de Martonne climatic zones affected by climate change in Iran. Theoret Appl Climatol 112(3–4):409–418

Sabeti H (1969) Bioclimatic Studies of Iran (in Persian). University of Tehran Press, p 1231

Sabziparvar AA, Tabari H, Aeini A (2010) Estimation of mean daily soil temperature by means of meteorological data in some selected climates of Iran. J Sci Technol Agric Nat Resour 14(52):125–138

Tabari H, Somee BS, Zadeh MR (2011) Testing for long-term trends in climatic variables in Iran. Atmos Res 100(1):132–140

USDA (1975) Soil taxonomy. A Basic system of soil classification for making and interpreting soil surveys. US Government Printing Office, Washington, D.C., USA

van Wambeke A (1976). NSM program to calculate soil moisture and temperature regimes using a mathematical model developed by Newhall F, New York State College of Agriculture and Life Sciences, Cornell University Soil Management Support Services, Soil Conservation Service, U.S. Dept. of Agriculture, Ithaca, N.Y. Washington, DC

Author Biographies

Dr. Ali Khalili is professor emeritus of the University of Tehran, Faculty of Agricultural Engineering and technology, Irrigation & Reclamation Engineering Dept. Meteorological Division. He is member of the Academy of Sciences of I.R.I., editor in chief of Journal of Agricultural Meteorology, member of working group on Climate Change and Global Warming in Science Foundation of I.R.I. His current research activities are centered on climate, climate change, extreme climate events risks and agro-climatology.

Dr. Jaber Rahimi is a graduate from Department of Irrigation & Reclamation Engineering (Meteorological Division), University of Tehran. His research interests focus on climate change, extreme climate events, risk assessment, agro climatology, and drought assessment.

Geology and Geomorphology

4

Mostafa Karimian Eghbal, Nikoo Hamzehpour, and Mohammad Hady Farpoor

Abstract

Soil formation is a complex phenomenon. The role of geology is mostly prominent through the influence of lithology in the production of parent material and tectonics in the geoform of structural origin. Geomorphology plays an important role in soil formation through relief, surface morphodynamics, morphoclimatic context, weathering and time (age). The geological zones and their related geomorphic features are dealt with as related to soil distribution. The Iranian Plateau is divided to eight structural zones. The Zagros mountain chain contains the most important sedimentary basins comprised mainly of calcareous detrital sediments. Karst features are the major expression of the Zagros landscape. The other mountain chain—Alborz zone—is composed of sedimentary rocks including limestone, dolomitic, and clastic rocks. Large areas in northern Iran are covered by loess deposits with a thickness of 30–70 m. Central Iran is composed of a series of inland basins and more than 60 playas. Makran zone, in east and southeast of Iran, is composed of uplands with varying width and altitude, irregular in trend and physiography. Various landforms such as fans, terraces, valleys, outwash deposits studied in several case studies from different regions of Iran are reviewed to show the close relationship between soils and geomorphology.

Keywords

Iran • Geology • Geomorphology • Structural zones • Landforms

4.1 Introduction

Soil formation is a complex phenomenon which is driven by various soil forming factors (see Chap. 6). The role of geology is mostly prominent through the influence of lithology in the production of parent material and tectonics in the geoform of structural origin (Zinck et al. 2016). The role of geomorphology in soil formation is crystallized through relief, the surface morphodynamics, the morphoclimatic context, weathered materials, and the factor of time and is also important in mapping soils in Iran. In the past few decades, only physical appearance of the earth surface has been used in soil surveys as a representative of topography. More and more, Iranian pedologists consider the role of geomorphology as one of the most important factors contributing in soil formation (e.g., Khademi et al. 1997a, b; Toomanian et al. 2001; Farpoor et al. 2004; Karimi et al. 2009; Khormali and Ajami 2011; Farpoor et al. 2012), but still this area of soil studies is relatively new and requires more consideration and work.

Internal and external processes leading to the formation of the Iranian plate have complicated outcomes through space and time. Such complicity has resulted in structural division of the plate into several zones, which will be

M.K. Eghbal (✉)
Tarbiat Modares University, Tehran, Iran
e-mail: mkeghbal@modares.ac.ir

N. Hamzehpour
University of Maragheh, Maragheh, Iran

M.H. Farpoor
Shahid Bahonar University of Kerman, Kerman, Iran

M.H. Roozitalab et al. (eds.), *The Soils of Iran*, World Soils Book Series,
https://doi.org/10.1007/978-3-319-69048-3_4

discussed in this chapter. Each zone is characterized by specific geological style, including unique records of stratigraphy, metamorphism, tectonism, or orogenic events. The style is often closely linked with the existing landforms. Geomorphology of the country will also be discussed in this chapter, focusing on the main geomorphic features and landforms in different geological structures. The chapter also covers soil formation and its relation to different geologic and geomorphic units, using several case studies lately published on this subject: e.g., soil–geomorphology relations in Sirjan playa (Farpoor et al. 2012); Gypsisols in Southeast (Farpoor et al. 2004) and Northwest Esfahan (Toomanian et al. 2001); paleosols in South (Kehl et al. 2005); Argids and Aridisols of central Iran (Khademi and Mermut 2003), among others.

4.2 Geology and Geomorphology of Iran

Iran is a part of the Alpine-Himalayan orogenic belt, which extends from Atlantic Ocean to Western Pacific. It is a high plateau with one of the world's most diverse geological and geomorphological settings. It includes the world's greatest number of playas, sand dunes, salt domes, longest playa called Dasht-e Kavir, the hottest desert called Dasht-e Lut, and longest salty cave called Namakdan diapir (Krinsley 1970).

The country is known as one of the most seismically active countries in the world, implying internal and external processes that lead to the formation of the Iranian plate. It has been likened to a bowl, with high mountains as its rim, surrounding the irregular and lower interior (Fisher 1968). Northern Alborz and western Zagros mountain ranges reach high elevations and cover extensive areas as compared to the mountains in east and south of the country, which are lower in elevation and are interrupted by lowland basins.

The north branch of the Alpine-Himalayan orogenic system in Iran is called Alborz range, of 960 km length, separating the Caspian Sea from central Iran. The Alborz range forms a relatively narrow area with high rainfall in its southern flank facing the Caspian lowland (see Chap. 3).

The southern part of the Alpine-Himalayan system is occupied by the Zagros and Makran ranges, which extend from northwest to southeast of the Iranian Plateau. Hence, the central part of the country is surrounded by the Alborz range in its north and the Zagros range in the west and southwest, including unique geomorphic features with extended deserts, depressions, playas, sand dune fields, and broad alluvial fans.

In 1968, tectonic map of Iran was released by Stocklin, who has divided the country to ten structural units (Fig. 4.1). Later on, several other interpretations and models have been published by some other authors (e.g., Nabavi 1976; Alavi 1994; Aghanabati 2004). A combined summary of different structural zones of Iran is presented in the economic geology of Iran (Ghorbani 2013), in which Iran is divided to eight structural zones, named Zagros, Alborz, central Iran, Sanandaj-Sirjan, Azerbaijan, eastern Iran, southeastern Iran (Makran), and Kopet Dagh. The summary of the major geological features and the related geologic times are presented in Table 4.1.

4.2.1 Zagros Structural Zone

Zagros system is the most developed of all the mountain ranges of Iran. It is a set of NW–SE trending parallel anticlines and synclines, which extends from Kermanshah in the northwest to Bandar Abbas in the south. It occupies the whole western part of the country, about one half of the total area. It is the result of the collision of the Arabian and Iranian plates sometime in the late cretaceous. In fact, it is considered the northeastern edge of the Arabian plate. Zagros mountains are the typical examples of a morphostructural fold style with breached antiform and synform valleys (Cucchi and Zini 2003).

According to Alavi (2004), there are four groups of rocks in the Zagros. The youngest group of rocks consists of *Surgah*, *Savak*, *Ilam*, *Pabdeh*, *Gurpi*, and *Asmari* formations. The second series of rocks includes Jurassic to Upper Cretaceous sedimentary deposits, e.g., *Fahliyan*, *Gadvan*, and *Dariyan* formations. The third oldest group of rocks includes Permian to Triassic rocks consisting of conglomerate, arkosic, and quartzose sandstones, evaporates and dolomites with interlayers of shallow-marine and low-energy limestone. The oldest group of rocks consists of rocks from Neoproterozoic to Devonian marine and non-marine deposits with *Hormoz* series and volcano sedimentary layers at the base (Zarasvandi et al. 2008).

The Zagros zone is characterized by some special features, which have made it outstanding among the other structural zones. Some of these important features (Ghorbani 2013) are:

- Absence of magmatic and metamorphic events after Triassic;
- Low abundance of outcrops of Paleozoic rocks;
- Structurally consists of large anticlines and small synclines;
- Continuous sedimentation from Triassic to Miocene.

The Zagros contains the most important sedimentary basins in Iran with sediments from great depths of more than 12 km, comprised mainly of carbonate and detrial sediments (Zarasvandi et al. 2008). Most of the oil and gas fields in Iran are located in the sedimentary basins of Zagros. In addition

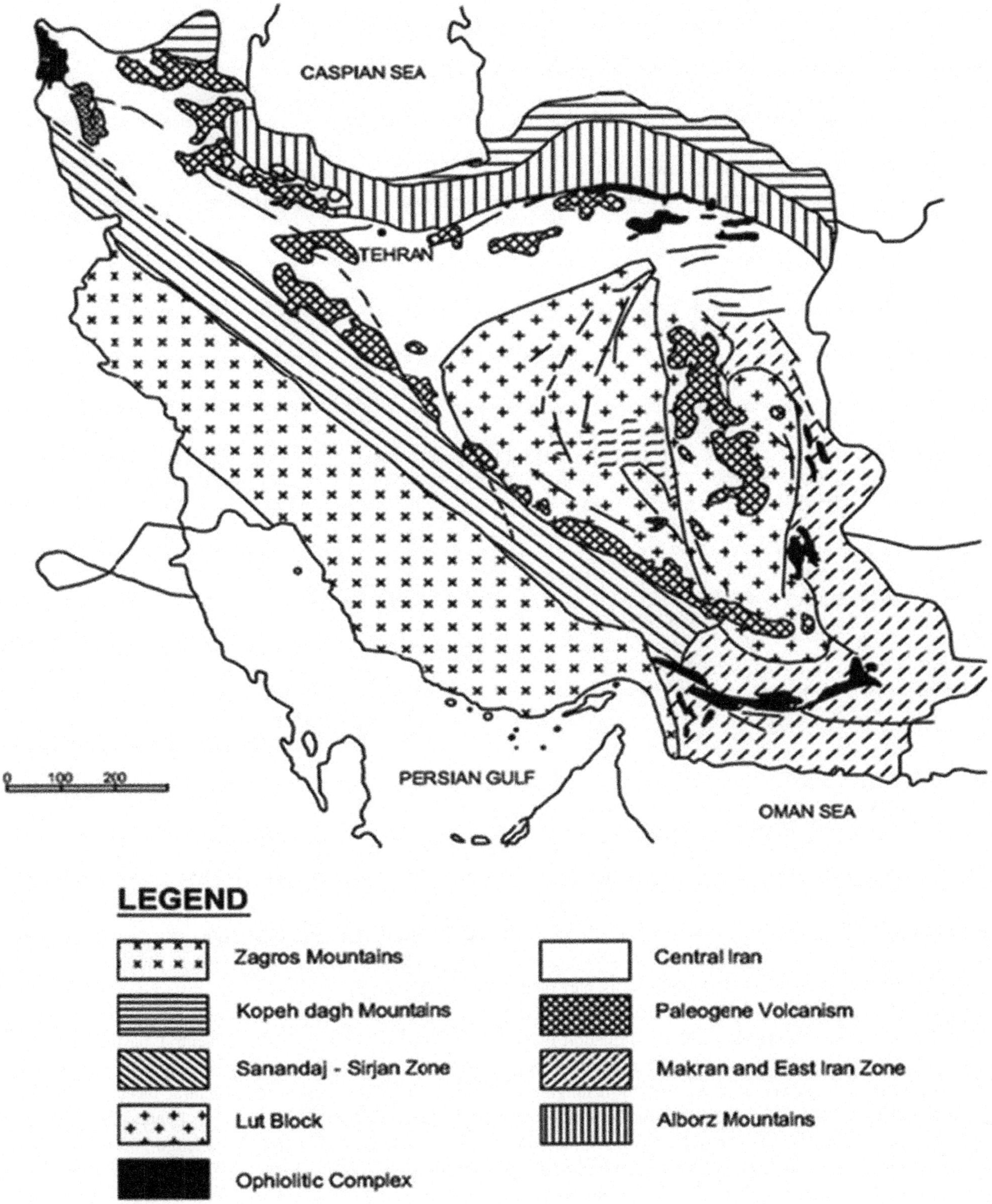

Fig. 4.1 Structural zones of Iran (Stocklin 1968)

to oil and gas, laterite-bauxite deposits are one of the most important non-hydrocarbon resources in Zagros (Zarasvandi et al. 2008) which are found in Savak and Ilam formations.

Allen et al. (2004) has divided Zagros structural zone to two main units, named high Zagros and folded Zagros (Fig. 4.2).

Table 4.1 Major geological subdivisions of Iran (Nezafati 2006)

Subdivision	Geologic feature	Geologic time
High Zagros	The same as Zagros fold belt plus ophiolite mélanges	Mesozoic
Folded Zagros	Mainly fold thick sediments (carbonate rocks and carbonates) overlying a high-grade metamorphic Precambrian basement salt diapirs in the south	Late Precambrian–Early Cambrian, Carboniferous to late Cretaceous, Cretaceous to recent
Alborz mountains	Volcanic and volcano-clastic rocks	Tertiary
Central Iran	Horsts of Precambrian crystalline basement, Paleozoic platform sediments, Cambrian to Triassic cover rocks, and magmatic rocks	Precambrian, Paleozoic, Triassic, Tertiary
Sanandaj-Sirjan Zone	Mainly metamorphic rocks along with intrusive bodies	Phanerozoic (mainly Mesozoic)
Eastern Iran	Flysh-molasse sediments	Post-Cretaceous
Kopet Dagh	Sediments (mostly carbonates)	Quaternary (Neogene)
Lut Block	Old stable platform covered by thick Mesozoic sediments and Eocene volcanic rocks	Mesozoic-Eocene
Makran	Colored mélange + thick flysh-molasse sediments	Upper Cretaceous-Paleocene for mélanges, post-Cretaceous for flysh sediments
Orumieh-Dokhtar Zone	Mainly volcanic (basalt to dacite), pyroclastic (i.e., tuff and ignimbrite) and plutonic (diorite to granite) rocks sporadically with numulitic limestone	Post-Upper Jurassic to Quaternary, peak of magmatic activities: Eocene

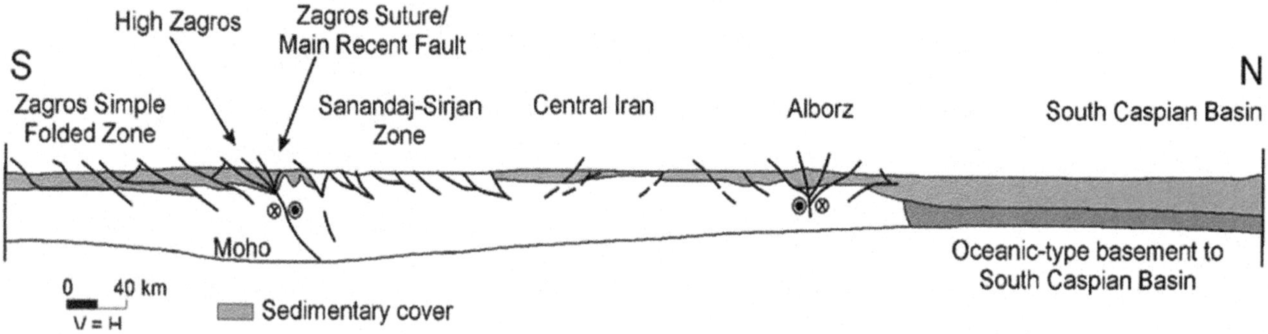

Fig. 4.2 High and folded structural units of Zagros range and their tectonic relations to central Iran and Alborz mountains (Allen et al. 2004)

High Zagros

High Zagros, with a thickness of 10–65 km, is part of the Zagros structural unit where the Arabian plate is subducted beneath the Iranian Plate, resulting in the formation of the central Iran's main stratovolcanoes and ore deposits. It consists of a series of massive structures, mostly belonging to Upper Cretaceous, Miocene, and Plio-Pleistocene which have been wrapped, altered, and fractured by folding. Fracturing and dissolution of the rock series have resulted in the rise of magma, forming extruded igneous material on the surface.

One of the main features of the high Zagros is 'stair topography,' led to the formation of a number of fault valleys and down-throw basins. Aras Valley is an example of the fault valleys that is a sequence of rifts and fault troughs. Because of the intense erosion in the high Zagros that is due to great elevation differences, gorge-like valleys have been cut by rivers in less resistant strata. As a result of erosion, rifts and fault troughs in the Aras Valley have been joined as a single valley.

Drainage pattern in the northwestern Zagros is complex, chiefly radial drainage with a central basin, called Uromiyeh[1] Lake, which is also an example of a down-throw basin. Uromiyeh Lake is a structural basin resulted from tectonic activities. It is now a hypersaline lake with approximately 5200 km^2, 140 km length, 55 km width, and only a few meters deep. The streams/rivers feeding it, especially those from the east pass through the formations mostly consisting of limestone and igneous intrusions, provide considerable amounts of salt that have now accumulated in the lake.

[1]Also written as Urmieh and Uromieh.

Another very significant and often a typic element of the Zagros is the superposition of large volcanic cones upon the high plateau surfaces, e.g., Sahand (3707 m) and Sabalan (4811 m) mountains.

Folded Zagros

The southeastern part of the Zagros consists of thousand meters of thick marine sediments dated to Mesozoic and Neogene ages. These sediments were folded and uplifted during Upper Miocene to Pleistocene. This part of the Zagros shows major differences from other parts. In this area, the general trend of NW–SE still remains and aligned ridges are composed of upper Mesozoic and lower Tertiary series. To the south, predominance of parallel ridges shifts to the widely separated uplands. Deposits of recent sediments and wind deposited sands begin to appear.

Perhaps, karst landforms are the major expression of the Zagros landscape (Heidari et al. 2004; Khanlari 2010), covering 102,000 km^2 (Heidari et al. 2011). Karst features have been reported by Bahrami (2013) around Danehkhoshk anticline, in Kermanshah province due to the outcropping of carbonate rocks (*Asmari* formation). In this area karrens, dolines and caves are the most common karstic features which are formed during colder glacial periods of Quaternary.

One of the outstanding works on karst landforms of the Zagros is done by Cucchi and Zini (2003) near the Karkheh River. They recognized three types of karst in *Gachsaran* formation.[2] Sinkholes, collapse dolines and caves were found to be the main karst landforms at the contact between limestone (*Asmari* formation) and gypsum (*Gachsaran* formation[3]). In this area, distribution of karst features seems to be controlled by a combination of stratigraphic and structural features. Sinkholes have a diameter of 0.5–5 m with a sub-vertical extension, occurring either in consolidated or nonconsolidated deposits. They are formed from the collapse of some underground cavities (Fig. 4.3).

Dolines are 5–250 m in size and similar in the formation of sinkholes. Two types of dolines were recognized, suffusion and collapse dolines. Suffusion dolines occur due to the dissolution of gypsum on the bottom of the valley in alluvial deposits (Figs. 4.4 and 4.5).

Although salt karsts are rare around the world due to the lack of salt outcrops, they are relatively dominant in southeast Zagros, mostly on the Iranian islands of Hormoz, Namakdan, and Larak in the Persian Gulf. These salt domes have been cropped out through overlying strata and have created well-developed exo- and endo-karst landforms (Bruthans et al. 2010) (Fig. 4.6). Some of the plugs are 600–1200 m above the ground with pinkish white to yellowish color. The world's largest salt cave (3 N Cave) in Namakdan diaper (Fig. 4.7) with a length of 1.9 km from the upper ponor to the outflow has also formed in this area (Bruthans et al. 2010).

Sometimes the rise of the salt domes coincides with the volcanic activity resulting in irregular veins of igneous and metamorphic material with darker color than the salt. Such materials might preserve the salt rocks from erosion through time (Fisher 1968).

Salt rock which is a part of the *Hormoz complex* consists of halite, gypsum, anhydrate, and fragments of sedimentary, igneous, and metamorphic rocks (Bosak et al. 1998). It has been deposited in late Precambrian to mid-Cambrian. In comparison to calcium carbonate karst, salt karst features need less time to form. e.g., in case of the *Namakdan* diaper, a 30,000 year period is enough for the evolution of large valleys, canyons, big dolines and caves (Bruthans et al. 2010).

Alluvial fans (although less common than karst landforms) are important in identifying the level of tectonic activity in the Zagros mountains. One of the few works on alluvial fans has been done by Bahrami (2013), who evaluated the relationship between quantitative characteristics of alluvial fans and active tectonics on 103 alluvial fans in the south of Sarpole-e-Zahab in Kermanshah province in the Zagros belt, southwest of Iran. According to his studies, climate of the Zagros was wetter and colder during glacial periods in Quaternary. For instance, crenulated and steep cliffs, bedrock slopes covered by taluses, and consecutive talus cones of the Zagros are related to the dominant processes in the glacial periods (Brooks 1982).

4.2.2 Alborz Structural Zone

The Alborz Mountain zone is located in the north, with a general trend of west–east from Azarbaijan to Khorasan. The peak of mountains is extremely high, especially in their northern border with the Caspian Sea due to its direct rise from the sea. The border between the Caspian Sea and the Alborz is distinct, but its southern limit is unclear and is a kind of transitional area passing through the central Iran (Alavi 1991). Tabriz fault (Alavi 1991), Garmsar fault, Semnan fault (Nabavi 1976), and Attari fault have been known as southern boundary of the Alborz. There are many similarities in terms of lithostratigraphy when comparing Alborz and central Iran. Alborz can indeed be described as marginal folds of the central Iran zone formed as a result of Turan and Iran plate's collision.

There have been two remarkable mountain building activities in the Alborz, one in late Cretaceous or early Tertiary and the other in late Miocene. In the Alborz

[2]Consists of medium-bedded marlstone and gypsum in a ratio of 1:2. This formation is strongly affected by karst phenomena.

[3]Consists of limestone and marly limestone locally dolomitized with a few marlstone interbeds with a thickness of 110 m.

Fig. 4.3 Sinkhole opening in residual soils near the Karkheh River (after Cucchi and Zini 2003)

Fig. 4.4 Alluvial doline near the Karkheh River (after Cucchi and Zini 2003)

structural zone, sedimentary rocks including limestone, dolomite, and clastic rocks are dominant (Ghorbani 2013). Several subzones are distinguished in the Alborz zone, considering the dissimilarity in structural stratigraphy in different parts of the zone (Nabavi 1976).

Studies revealed that northern and southern Alborz uplifts did not occur at the same time intervals, resulting in significant differences between these two areas of the Alborz zone. For instance, the Paleozoic deposits in northern side of the Alborz are thicker in places like in Amol and Kandovan.

Fig. 4.5 A collapsed doline in the Zagros mountains (after Cucchi and Zini 2003)

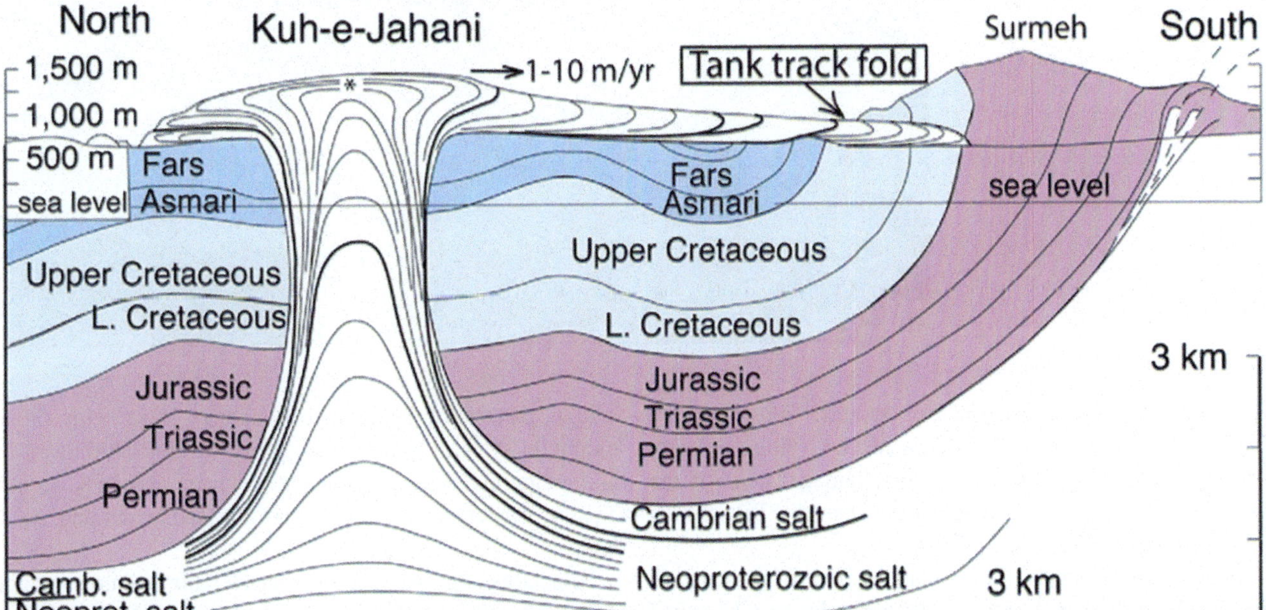

Fig. 4.6 N–S profile of the Hormoz salt (~500 Ma) that rose up an opening pulled apart along a NS transfer fault in the Zagros mountains to extrude the Kuh-e-Jahani, probably the largest current salt fountain in Iran (from Talbot et al. 2000). Star indicates flow separation point beneath summit dome (Talbot and Pohjola 2009)

Also, thickness of coal-bearing sediments in Upper Triassic–middle Jurassic is thicker in the northern side as compared to the southern Paleozoic–Mesozoic part. But in Cenozoic, when northern side of the Alborz uplifted, the southern side was under regressive sea with thick layers of pyroclastic-clastic sediments.

The Alborz zone can be studied as two different parts: Talesh hills in the west and the northwest and the main

Fig. 4.7 **a** Sub-horizontal old cave levels filled with sediments exposed in the side of 3 N Cave. Varicolored folds of rock salt in the cave ceiling; **b** blind valley with flat bottom above the Donkey Cave (after Bruthans et al. 2010)

Alborz massifs in the center and in the east. Talesh hills are composed of metamorphic rocks with a varying elevation between 250 and 3000 m. It is the drainage areas of the Gilan-Mazendran plain which lies between the Talesh hills and the Caspian Sea.

Sefid Rud River which is the largest river of the country with a drainage basin of about 61.600 mk^2 has deposited thick layers of sediments over the southern Caspian Sea lowland and has formed deep valleys, gorges, and a large delta (Kazanci et al. 2004; Kazanci and Gulbabazadeh 2013). It separates the Talesh hills from the main Alborz as it flows in a longitudinal structure through a syncline to the south. The strong erosive power of this river has produced a wide valley between the Talesh hills and the main Alborz. Moving to the east, geological features begin to change and simple pattern of ridge and trough replaces with broad mountain system, rock series change, overall altitude rises, and a significant development in volcanic activities occurs. Two of the largest volcanic cones are Damavand and Alam Kuh.

Glaciares are one of the considerable features of the Alborz. Small ice fields can be found around the summits of the Alam Kuh. To the northeast of the massif, three permanent tongues of ice can be seen which forms a glacier of nearly 8 km long.

The Damavand Mountain, which is another large volcanic cone of the Alborz zone, is located almost in the center of the entire system. Damavand peak is very recent, belonging to late Pliocene and early Quaternary. It is made up of andesite, intercalated bands of limestone, mudflows, breccia, and material exposed from the crater. Damavand glacier is about 2 km long. Unlike the glacier of the Alam Kuh, the ice surface is not covered by rock debris.

Large areas in northern Iran are covered by loess deposits. The thickness of loess is about 30 m in the northern footslopes of the Alborz and close to 70 m in the loess plateau of the city of Gonbad (Khormali and Kehl 2011). Kehl (2010) believes that the loess deposits in northeastern Iran are quite homogenous, and their most likely source are alluvial plains of the Atrak and the Gorgan rivers.

The loess cover in northeast Iran is less thick with patchy distribution as compared with the northern ones (Karimi et al. 2011). The loess deposit of the Binaloud Mountain range in south of Mashhad has a thickness of more than 12 m. They have been classified as recent and old Quaternary deposits. Their Aeolian nature and local origin is confirmed by grain size characteristics, microtexture, mineralogical composition, and geochemical signatures (Karimi et al. 2009).

Loess deposits have also been identified in other parts of Iran, such as in the Kopeh Dagh zone (Okhravi and Amini 2001); Presepolis Basin in southwest Iran (Kehl et al. 2005); Kashan, Yazd, and the central Iran (Darvishzadeh 2001).

4.2.3 Central Iran Structural Zone

The central Iran structural zone, which is located in the center of Iran, is bordered by the Great Kavir Fault in the north, the Nain-Bar Fault in the west and southwest and the Harirud fault in the east. It is one of the most complicated structural zones in Iran, comprising rocks from Precambrian to Quaternary with different stages of orogeny, metamorphism and magmatism (Ghorbani 2013).

Central Iran has three crustal domains, the Lut Block, the Kerman–Tabas Block, and the Yazd Block (Fig. 4.8). It was a stable platform during Paleozoic times but distinct horsts and grabens have formed due to Triassic movements. Kerman–Tabas region is an example of the grabens which is filled by thick Jurassic sediments.

Central Iran is composed of a series of inland basins and their related playas with an area of over 750,000 km^2, which despite their dissimilarities in size and shape, have one thing in common; they have no outlet to the sea. Altitude is a varying feature in the central Iran with mountain chains of up to 3000 m above the sea level to lowlands with altitude of less than 300 m. Much of the inlands of the central Iran once had been covered by large lakes which are now occupied with residuals of salt lakes.

More than 60 playas, ranging from 25 to 52,825 km^2, have been identified in the central Iran, and their surficial features have been studied by Krinsley (1970). Clay flats, salt crusts, wet zones, intermittent and perennial lakes, fan deltas, and swamps are the main surficial types in playas. A summary of playa types and their related areas are presented in Table 4.2.

The largest basin of the central Iran is called the Great Kavir (Kavir-e-Bozorg) lying immediately to the south of the Alborz mountains. It covers an area of over 800 km long and more than 320 km wide in an east–west direction, underlain by Miocene evaporates and mudstone. Its western, southern, and eastern boundaries are occupied by alluvial fans and dune fields and the northern boundary by an alluvial apron. The main surface types in the Great Kavir playa are wet zone, salt crust, clay flat, eroded surface of the Miocene rocks and dune fields (Krinsley 1970).

Along the south and southeastern margins of the Great Kavir, three sand dune fields are active. The biggest, Rig-i-Jin, occupies an area of 3855 km^2 and is considered as the second largest dune field in Iran, after the Lut Desert dune field. Bobek (1963) believes that alluvial fans to the north of the Great Kavir were the main source of the sand for all the sand dunes in the area. Krinsley (1970) believes the Miocene sediments are more likely to be the source of these sand dunes.

The Lut Desert is a hollow feature in the southeast of Iran with an area of about 80,000 km^2 (Fig. 4.9) (Ehsani and Quiel 2008). It is the lowest part of the central Iran (250 m asl), and some consider it as the hottest area in the world (Alavi Panah et al. 2007). Mildrexler et al. (2006) described the Lut Desert as the thermal pole of the earth. This area is surrounded by three mountain systems, the southern Zagros, the ring of uplands between Iran and Afghanistan and highlands east of Kerman. Northern and northwestern margins of the Lut are covered by alluvial fans.

The Lut Desert is filled by fine-grained sediments with a horizontally bedded silty clay and limey, gypsiferous sand which is called *Lut* formation. It has a thickness of 135–200 m. Silty beds within the *Lut* formation deposits suggest that deposition has occurred in a playa environment. Three dominant landforms of the Lut Desert are sand dunes, yardangs, and playas (Farpoor and Krouse 2008). Yardangs are elongated, typically with a length three times or more than their wide, separated by flat-bottomed furrows (Fig. 4.10). They vary in their scales, from micro-yardangs through meso-yardangs to mega-yardangs (Goudie 2007). They have ridges up to 80 m high and 120 km long with a NW–SE direction (Figs. 4.9c and 4.10) (Gabriel 1938). They are formed in the direction of the strong prevailing winds with an abrasive sediment load in the environments where water is scarce. However, the primary agent in the formation of yardangs seems to be the erosive power of water by cutting deep troughs and gorges. In some areas in the Lut Desert, with no prevailing wind, nonparallel and isolated ridges are formed, which are called Kaluts (Fig. 4.10).

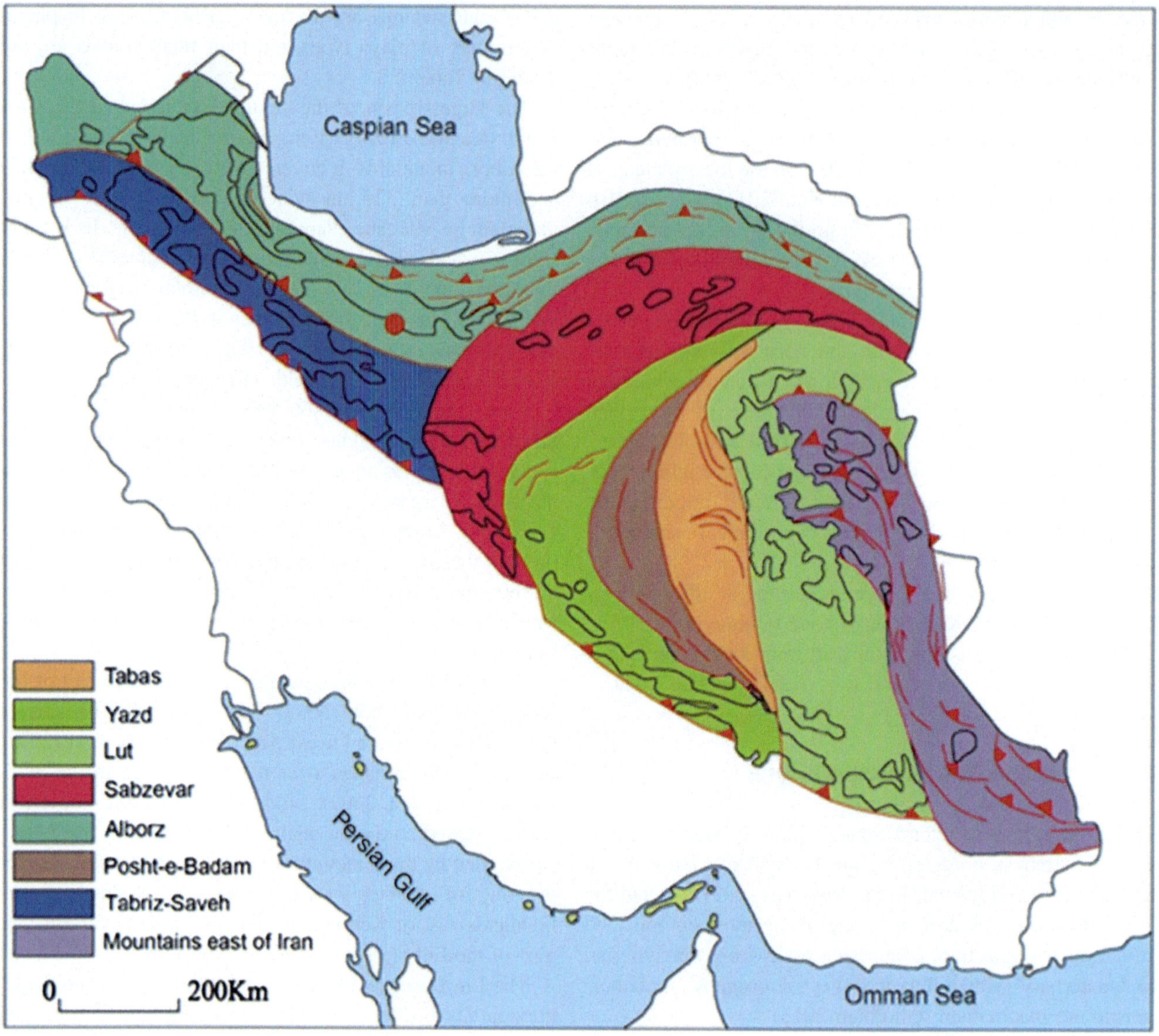

Fig. 4.8 Subzones in central Iran (Alavi 1991)

Table 4.2 Playas type, their area, and the percent of total area (Krinsley 1970)

Play type	Area (km^2)	Percent of total playa area
Wet clay flat–salt crust playa	38,364	57
Clay flat playa	9335	14
Intermittent lake playa	7422	11
Clay flat–salt crust playa	6934	10
Perennial lake playa	3089	5
Wet clay flat playa	1345	2
Salt crust playa	763	1
Total	62,252	100

According to Krinsley (1970), there are three fields of sand concentration in the Central Lut basin. First one is a series of sand ridges which extends from the southern end of the yardang area to an extensive second field of sand dunes at the eastern margin of the yardangs (Fig. 4.11). This area has a length of about 220 km which is the product of the erosion of the Central basin. The third one is a small dune field at the southern end of the Lut Plateau.

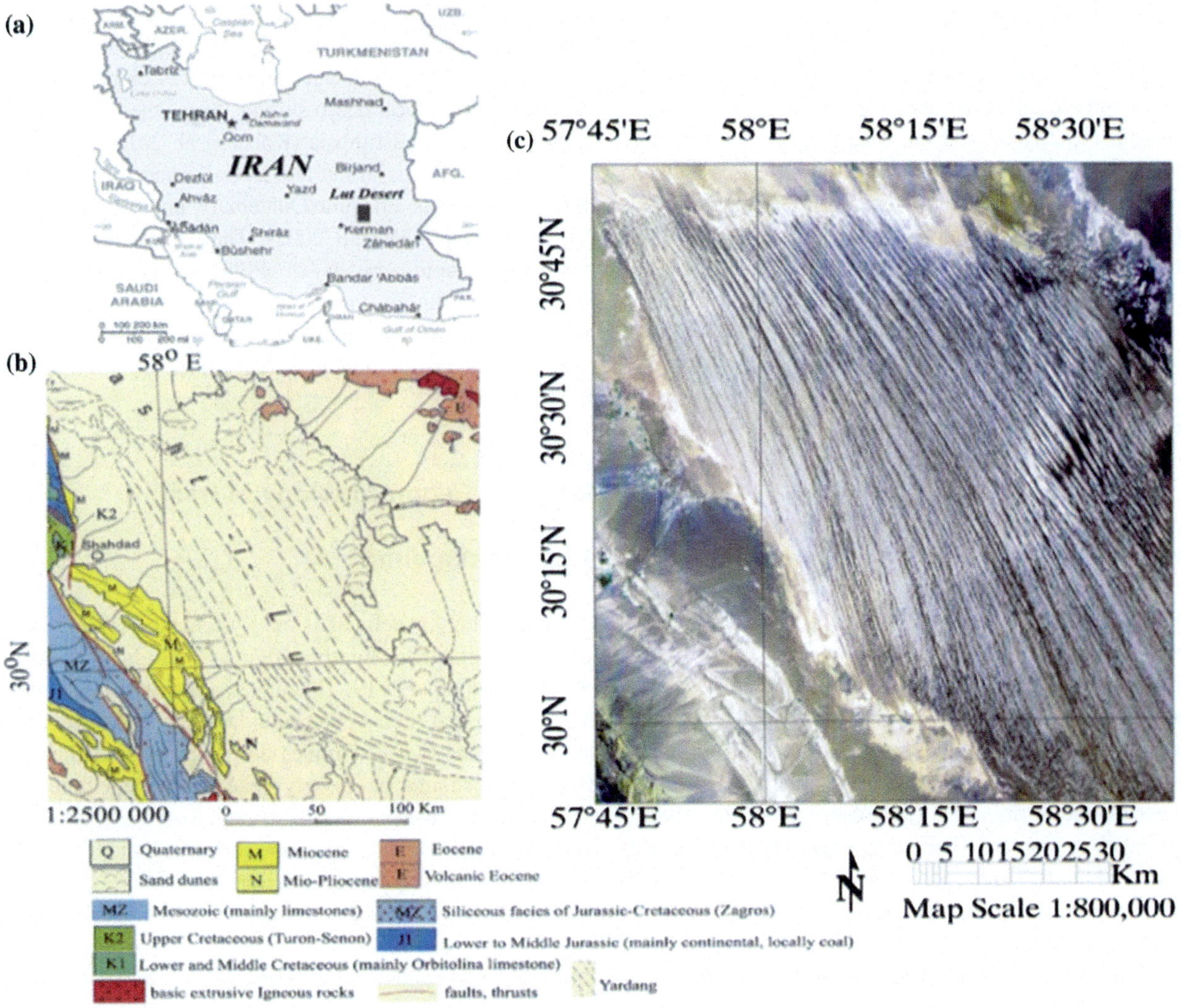

Fig. 4.9 The Lut Desert **a** location in the central Iran, **b** Geological map of the area and **c** NW–SE direction of Yardangs in the Lut Desert (after Ehsani and Quiel 2008)

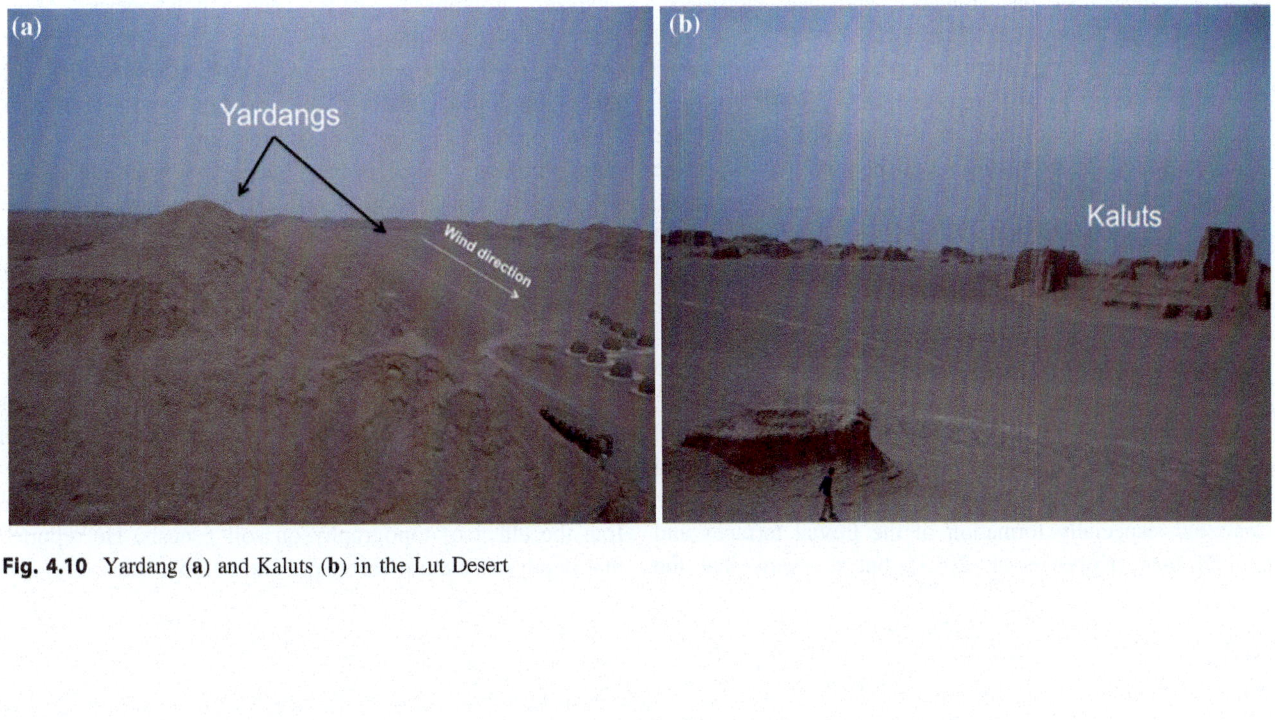

Fig. 4.10 Yardang (**a**) and Kaluts (**b**) in the Lut Desert

4.2.4 Sanandaj-Sirjan Structural Zone

It is a NW–SE linear structural element, parallel to the Zagros mountains in its western border. Despite the similarities with the Zagros, and the central Iran, there are significant differences in rock types, magmatism, metamorphism, and orogenic events (Ghorbani 2013). Complete lack of Tertiary formations and the presence of magmatic and metamorphic rocks of Paleozoic and Mesozoic have made the Sanandaj-Sirjan zone separable from the Zagros and the central Iran zones.

4.2.5 Eastern and Southeastern Iran (Makran) Structural Zone

The east and southeast zone is composed of uplands with varying width and altitude, irregular in trend and physiography, and separated by open or basin-like plains. The northern boundary of this zone lies to the south of the watershed that lies between the Gorgan and the Atrak rivers, and the southern border is limited to the Oman Sea. This area has a north–south spread of over 1300 km and east–west between 50 and 300 km. Wide varieties of landforms can be distinguished in this zone, but a rim of uplands is the predominant feature of the zone. The subregions of the eastern and southeastern Iran are Khurasan (in the north); Qa'in and Birjand highlands, and the lowland of Sistan; and Balochestan and Makran in the south.

Alluvial fans/Glacis are the most dominant landforms in Iran. Walker and Fattahi (2011) have reviewed the studies on alluvial fans of Sabzevar, Neyshabour (Hollingsworth et al. 2010), Kashmar (Fattahi et al. 2007), Ahar (Le Dortz et al. 2009), Bam (Fattahi et al. 2010; Talebian et al. 2010), and Minab (Regard et al. 2005) (Fig. 4.12). They have developed a scenario of the Holocene and late Pleistocene stratigraphy of eastern Iran (Fig. 4.13). According to their study, incision of rivers into the fan surfaces has occurred discretely in early to mid-Holocene during wetter periods, leading to the formation of river terraces. Possibly, this has occurred due to the decreased sediment supply and increase in precipitation. However, as within the late Holocene, no major geomorphic surfaces have been identified, and it seems that the increased aridity has slowed the evolution of the landscape (Walker and Fattahi 2011).

The Makran range (fold and trust) has an E–W trend, with a length of 900 km and width of 150 km. The oldest rocks in this zone are the ophiolites of late Cretaceous-Paleocene, overlain by a thick sequence of sandstone, shale, and marl (Ghorbani 2013). Although there are limited information about geomorphic evolution of the Makran range during late Cenozoic, especially formation of the fluvial terraces and alluvial fans (Kober et al. 2013), but it seems that the Cenozoic evolution of the Makran range and surface uplift processes have been controlled by the dominant SW-Asian monsoon, the Mediterranean winter rainfall climate, and the surface uplift processes resulting from the convergence between Arabia and Eurasia (Kober et al. 2013).

The Makran alluvial fan and fluvial terrace deposits have been classified as Pliocene/Quaternary units on 1:250,000 geological maps (McCall 2002). They are subdivided into bedded piedmont gravels of Pliocene/ Quaternary age: high-level Piedmont fans and valley terraces, low-level piedmont fans and valley terraces, silty outwash deposits and river alluvium (McCall and Eftekhar-Nezhad 1993; Kober et al. 2013).

4.3 Soil Geomorphology in Iran

4.3.1 Background

Geomorphology is defined as the 'systematic examination of landforms and their interpretation as records of geologic history …' (Howell 1957). On the other hand, pedology studies soils that their upper boundaries end up to the earth surface. As the two disciplines deal with parts of the same thing, they should be closely related. Graham and Buol (1990) believe genesis and distribution of soils are better understood if studied in landscape contexts, rather than at the level of individual pedons or taxonomic units.

In practice, every activity dealing with the distribution of soils on the earth's surface employs some geomorphic concepts. Soil and geomorphology are closely related. Pedology helps geomorphologists to better understand weathering and erosion processes. On the other hand, geomorphic properties that affect soil formation provide invaluable data for soil scientists. Geomorphology also helps analyze polyphase soils and polygenesis. Soil-geomorphology studies play also significant role in detailed soil mapping and classification. It means that with other conditions constant, one can find similar trends in soils located on similar geomorphic surfaces. Soil-geomorphology concepts help better understanding of distribution, formation, and spatial variability of soils (Walker 1986). Besides, soil science provides powerful data to investigate climate change and Quaternary sedimentation (Eghbal and Southard 1993).

The term 'geopedology' has been introduced to show the complex process of soil genesis controlled by geomorphology. Geopedology, also known as soil geomorphology, is defined by Zinck et al. (2016) as a systematic application of geomorphology for soil survey.

Abtahi (1980) was one of the first ones who studied in Iran the effect of topography on soil genesis. He reported that depth of calcium carbonate and soluble salts vary with

Fig. 4.11 Sand dune fields east of the Yardangs in the Lut basin

topography. Soil properties and erosion were studied by Rooyani and Southard (1985) in the Central Alborz foothills as related to topography. Many other pieces of research were also conducted by Soil and Water Research Institute of Iran (swri.ir) at different scales (detailed, semi detailed, and reconnaissance) soil surveys in Iran. It is important to be noted that the above-mentioned studies were conducted based on the physiographic units (Mahler 1979) that were initially described for land evaluation purposes.

Geomorphology provides a fundamental contribution not only to soil mapping but also to the interpretation of soil genesis (Zinck et al. 2016). The close relationship among geomorphology and clay mineralogy, micromorphology, erosion rate, genesis, and classification of soils was reported by Farpoor and Eghbal (1996) in different geomorphic positions of an area in Chaharmahal and Bakhtiari. Soil-geomorphology-based studies have been increasingly conducted since 90s. Like many other countries, use is made of the facilities offered by geographic information system (GIS), including modeling applying geostatistics, fuzzy theory, neural network. (Toomanian et al. 2001, 2006; Salehi 1994; Farshad et al. 2011).

Soil genesis and classification, clay mineralogy, micromorphology, and stable isotope geochemistry of soils related to landscape position will be discussed in the following section.

4.3.2 Soil Landscape Relationships

Many studies have been done in different parts of Iran, varying in geomorphology and soil climatic conditions (moisture and temperature regimes), on the soils-landscape relationships, a number of which will be shortly referred to, hereafter:

- Farpoor and Eghbal (1996) studied five transects with soils on different hillslope positions in the Imam Gheis area, Chaharmahal Bakhtiari province (Fig. 4.14). In the summit position, two subunits are distinguished, on the

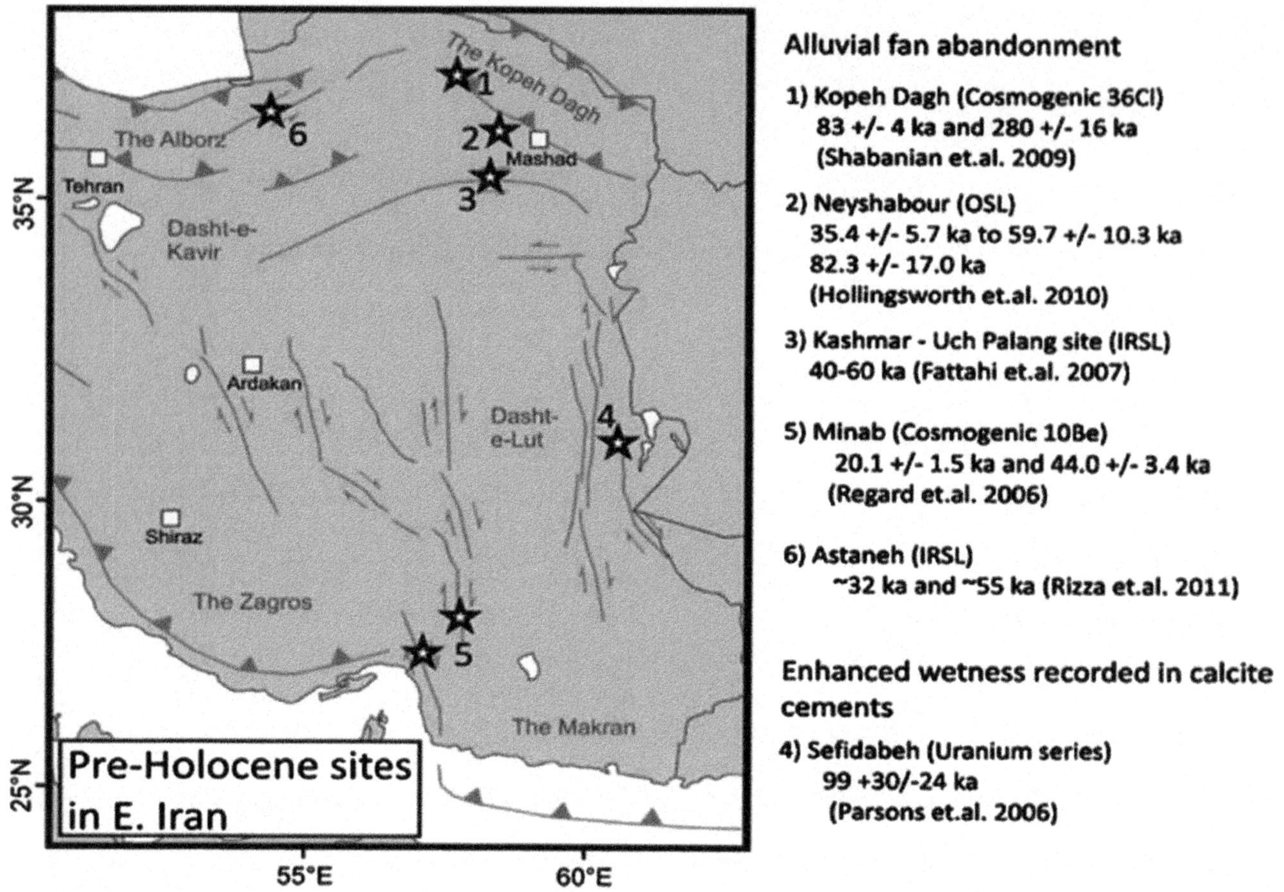

Fig. 4.12 Map showing the six sites (marked by stars) where there are constraints on the pre-Holocene history of alluvial fan deposition in eastern Iran. Cold and arid conditions during the last glacial period are indicated by luminescence dating of Aeolian deposits at Mashhad and Ardekan (Walker and Fattahi 2011)

basis of erosion intensity. In the part with less erosion, deep Calcic Haploxeralfs occur, whereas the soils of the eroded part are Typic Haploxerepts. Soils on the shoulder position in all transects under study were highly eroded and classified as Calcixerepts. The soils of the backslope position with slope gradient of around 18%, and convex in form, that is apparently less eroded are classified as Calcic Haploxeralfs. Footslope is reported to have young, less-developed soils, classified as Typic Haploxerepts. Soil texture is lighter in this position as compared to the other segments. This is reported to be due to the deposition of coarser sediments originating from the higher positions. Toeslope soils are dark gray with 4.3% organic matter content, slickenside and cracks, classified as Chromic Calcixererts. A close relationship between soil and geomorphology was found in this area. Three different soil orders are reported in the studied catena with an elevation difference of 25 m and in a distance of about 400 m.

- Farpoor et al. (2002) found a close relationship between soil genesis and landscape positions in gypsiferous Aridisols of Rafsanjan area, southeastern part of central Iran (Fig. 4.15). Moving down the slope from rock pediments toward playa landform, the amount of EC, silt, and especially clay content increased, but gypsum and the size of gypsum pendants decreased. The authors pointed out that coarse texture and high gravel content (up to 75%) of rock pediments play a role in the formation of gypsum pendants in this geomorphic position, while the gypsum content in fine-textured sediments of playas is disseminated.

The authors also reported on a close relationship between palygorskite morphology and the geomorphic positions. Upper geomorphic surfaces (rock pediments) included crystals greater as well in size as in number (Fig. 4.16). Conversely, the crystals are so small and few in the lower positions that could only be observed by electron microscopy. An increase in Mg/Ca ratio at the time of gypsum formation, high pH, and high soluble Si has made the palygorskite formation possible at the peripheries of the water bodies' interior in central Iran during Neogene

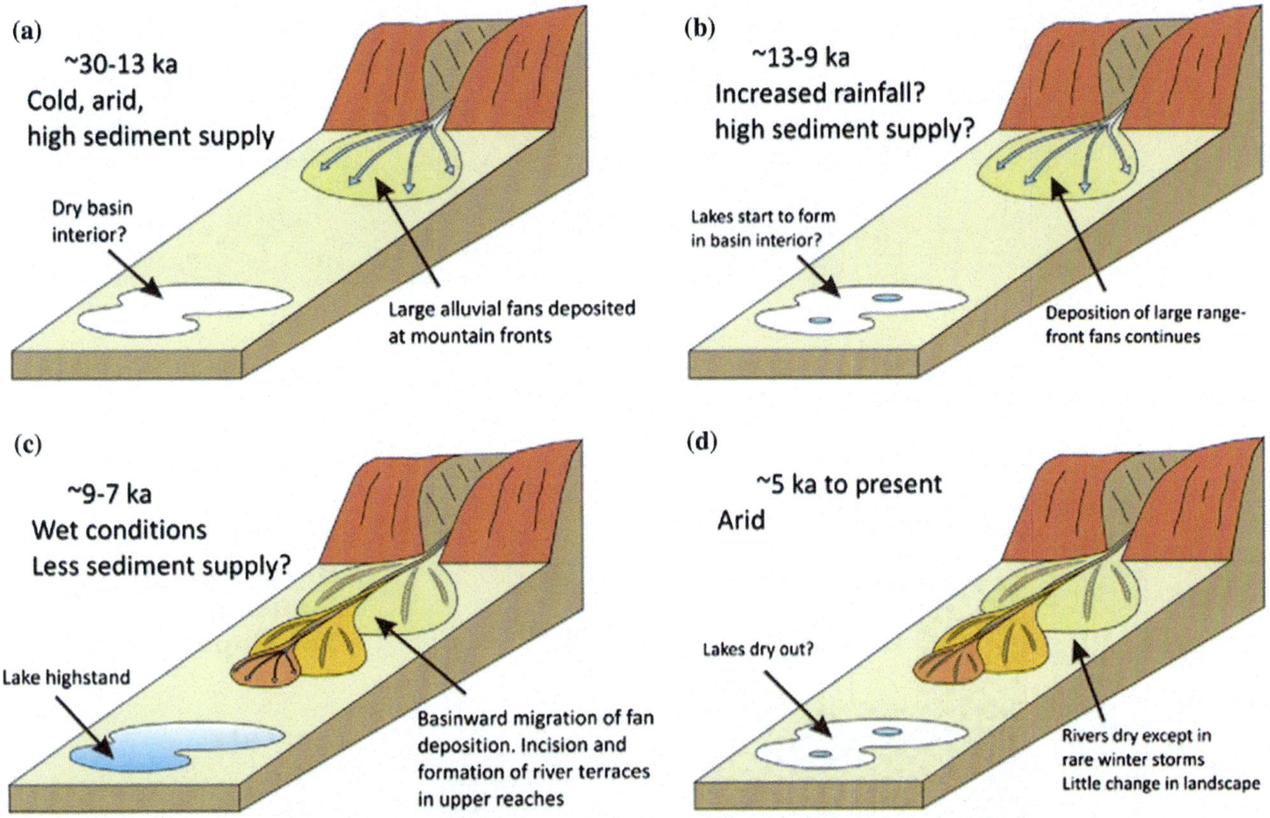

Fig. 4.13 Alluvial fans evolution as illustrated by Walker and Fattahi (2011) for eastern Iran

(Khademi and Mermut 1998). These peripheries are the present time rock pediment geomorphic surfaces (Farpoor et al. 2004). Palygorskite bundles have been preserved due to aridity since Neogene, as mentioned by Khademi and Mermut (1999) and Farpoor et al. (2002). Transformation of palygorskite to smectite was observed in the lower landscape positions by Farpoor et al. (2002). They suggested that this transformation has occurred as a result of less aridity and more water supplies by runoff.

The relationship between micro- and macro-morphology of gypsum crystals and geomorphic positions was also investigated by Farpoor et al. (2003) in the Rafsanjan study area, central Iran. Micromorphological observations together with other analyses showed that rock pediment geomorphic surfaces were the source of gypsum in the area. Large gypsum pendants and microforms of lenticular, vermiform, platy, and interlocked gypsum plates were reported as the dominant form in the rock pediments. Lenticular and vermiform gypsum crystals increased down the slope, but alabastrine gypsum was the most dominant form in the playa surface.
Studying the clay mineralogy of some soils and sediments in the Lut Desert, and in central Iran, Farpoor and Krouse (2008) found that presence of large amounts of palygorskite in different layers of yardangs might show that the old large Lut plain is formed under the post Tethyan shallow hypersaline water conditions. They concluded that clay mineralogy could shed some light on the formation of these huge landforms which are the remnants of the old valley.
Soil genesis on different geomorphic surfaces in the Mahan-Joupar area, southeastern part of central Iran, was studied by Nadimi and Farpoor (2013). The occurrence of argillic horizon on stable alluvial fan surface was attributed to the presence of a more humid paleoclimate in the history of the area. They reported that rock pediments of the area seem to be the source of gypsum for lower geomorphic positions. Due to the presence of argillic horizon in the stable alluvial fan surfaces and the lack of Bt in the soil of unstable alluvial fans, they concluded that stability of the geomorphic surfaces has been an important factor affecting the formation/preservation of the argillic horizon through time. The researchers reported that soil-geomorphology studies help to understand soil forming processes and soil management practices. In another study in an area in Karadj, Ghajar (1994) reports on the same conclusion, namely the close relationship between landforms and the soil.

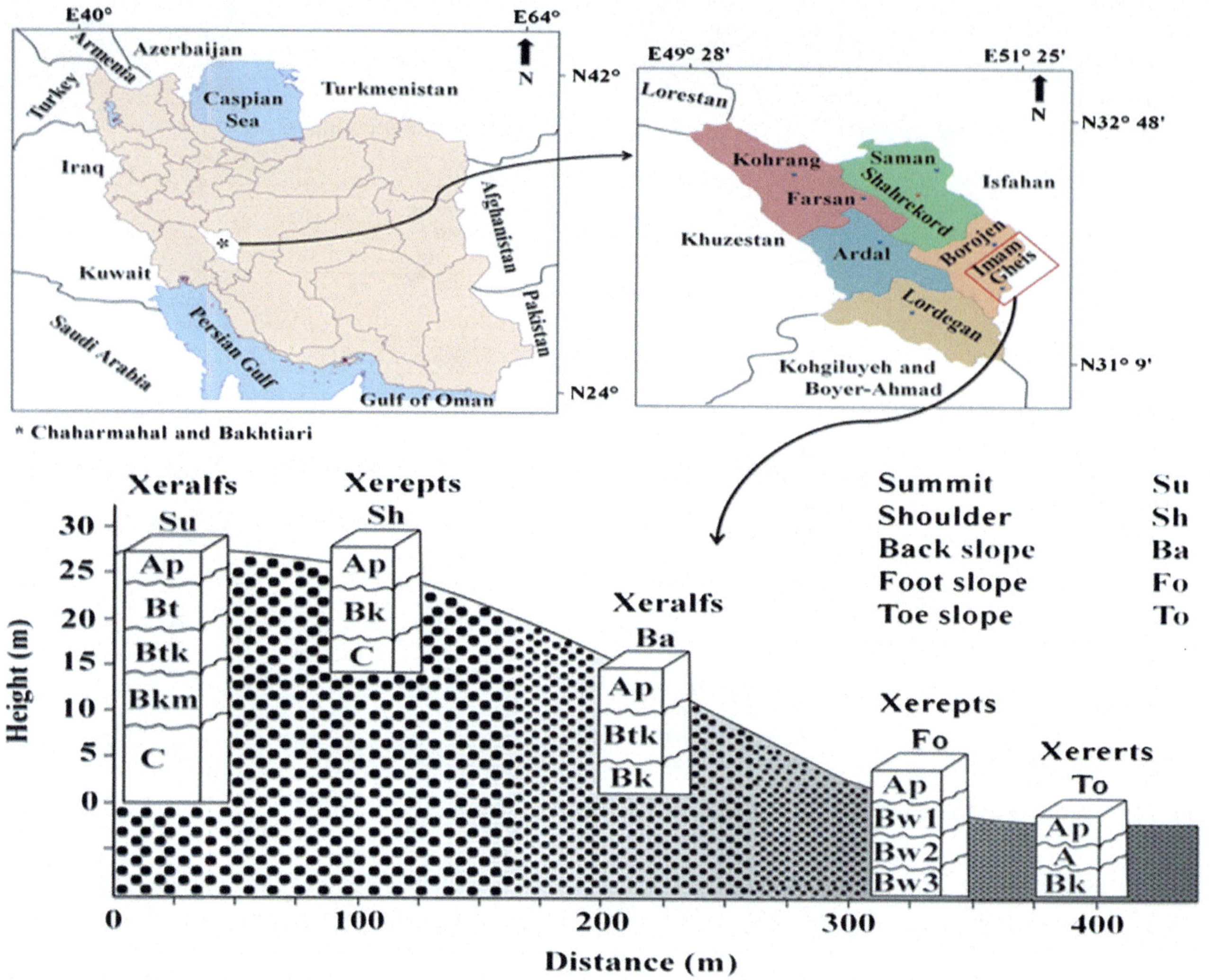

Fig. 4.14 Study area and location of representative pedons on hillslope positions in the study area (Farpoor and Eghbal 1996)

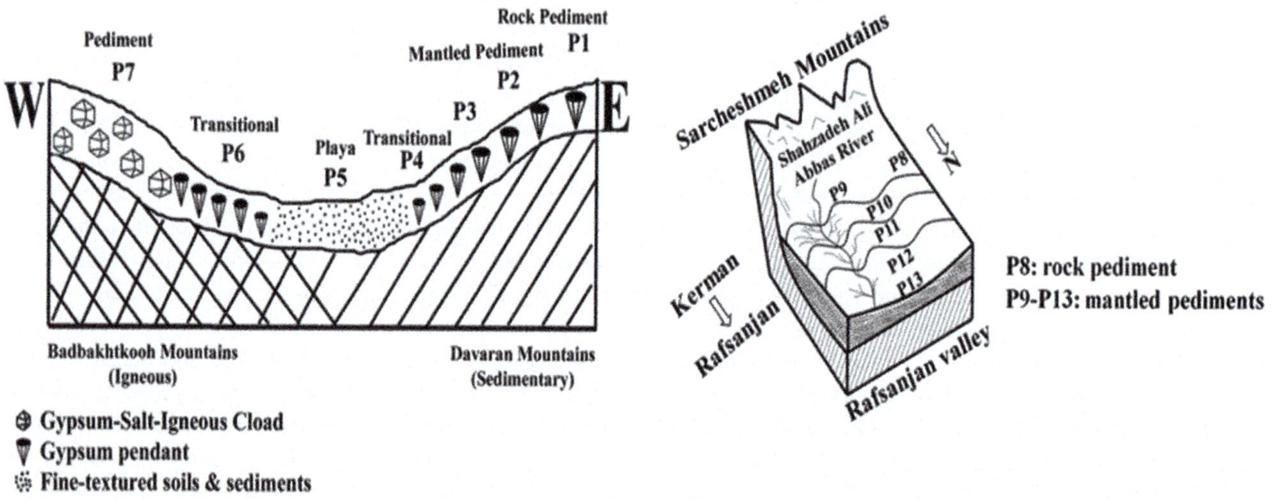

Fig. 4.15 Schematic diagrams of soils on two different transects in southeast of central Iran (adapted from Farpoor et al. 2002)

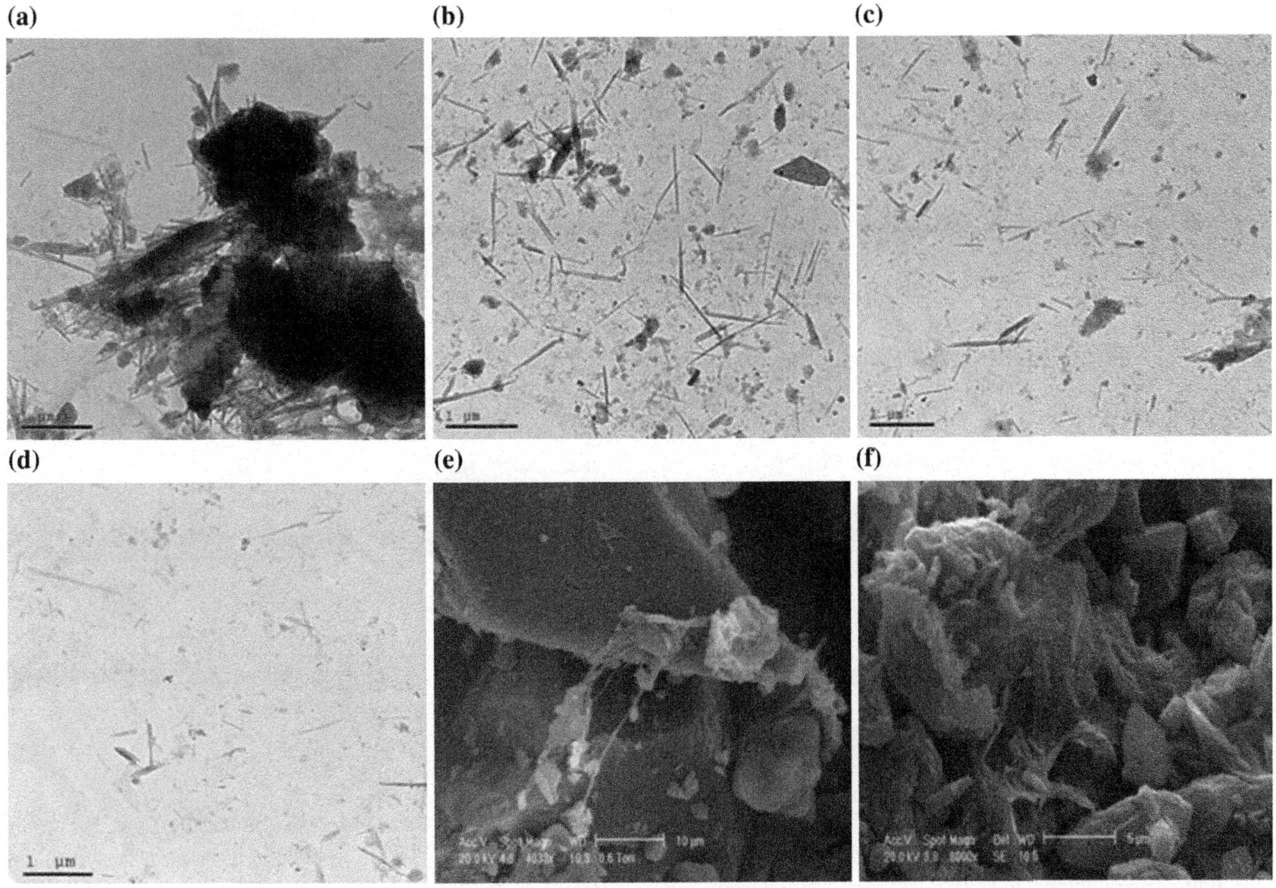

Fig. 4.16 Electron micrographs showing the distribution of palygorskite on different geomorphic positions in the Rafsanjan study area (Farpoor et al. 2002). **a** and **b**, TEM micrographs of *By1* and *By3* horizons from rock pediment, **c** and **d** TEM micrographs of *By1* and *Bz2* horizons from mantled pediment and the playa, respectively, **e** and **f** SEM micrographs showing palygorskite around gypsum and calcite crystals (*By2* and *Bk* horizons of mantled pediment), respectively

Developed soils with petrocalcic horizons were found on stable geomorphic surfaces, and less-developed soils such as Torrifluvents were formed in unstable flood plain surface. Farpoor and Irannejad (2013) compared stable and unstable geomorphic surfaces in an area in Rafsanjan as related to their soils and reasoned why stable surfaces of alluvial fans were more saline than the unstable surfaces.

Sanjari et al. (2011a) studied soil genesis in different geomorphic surfaces in an area in Jiroft. The identified landforms (alluvial fan, mantled pediment, intermediate surfaces, alluvial plain, and lowland) (Fig. 4.17) were divided to different surfaces on the basis of geomorphic stability and studied their soils in detail. Electrical conductivity, pH, and SAR values increased from mantled pediments toward lowland positions. They also reported on the correlation of chlorite, illite, palygorskite, smectite, and kaolinite clay minerals (Fig. 4.18) with the geomorphic surfaces. Moving from alluvial fan (higher position) toward alluvial plain (lower position, with high groundwater table), palygorskite stability decreases and smectite is reported as the dominant clay mineral. Inherited, pedogenic, and detrital origins of palygorskite were found in the mantled pediment, intermediate surfaces, and in the alluvial plain, respectively.

Different landforms including Kerman plain, rock pediment, piedmont plain, and lowlands of Lalehzar mountains were studied along a xeric-aridic climotoposequence in south central Iran by Moazallahi and Farpoor (2012). Soil parent material changed from recent Quaternary alluvium in north of the transect (Kerman plain) to the materials originated from weathering of granodiorites of the Lalehzar mountains in the south. Typic Haplocalcids and Gypsic Haplosalids are dominant soils of Kerman plain and rock pediment geomorphic surfaces, respectively. Typic Natrigypsids, Calcic Haplosalids, and Petrocalcic Calcixerepts occur in piedmont plains, and Fluvaquentic Epiaquolls and Typic Haplohemists are the soils formed in the lowland. Illite, smectite, palygorskite, chlorite, and kaolinite clay minerals are reported in almost all the soils studied. Smectite-vermiculite

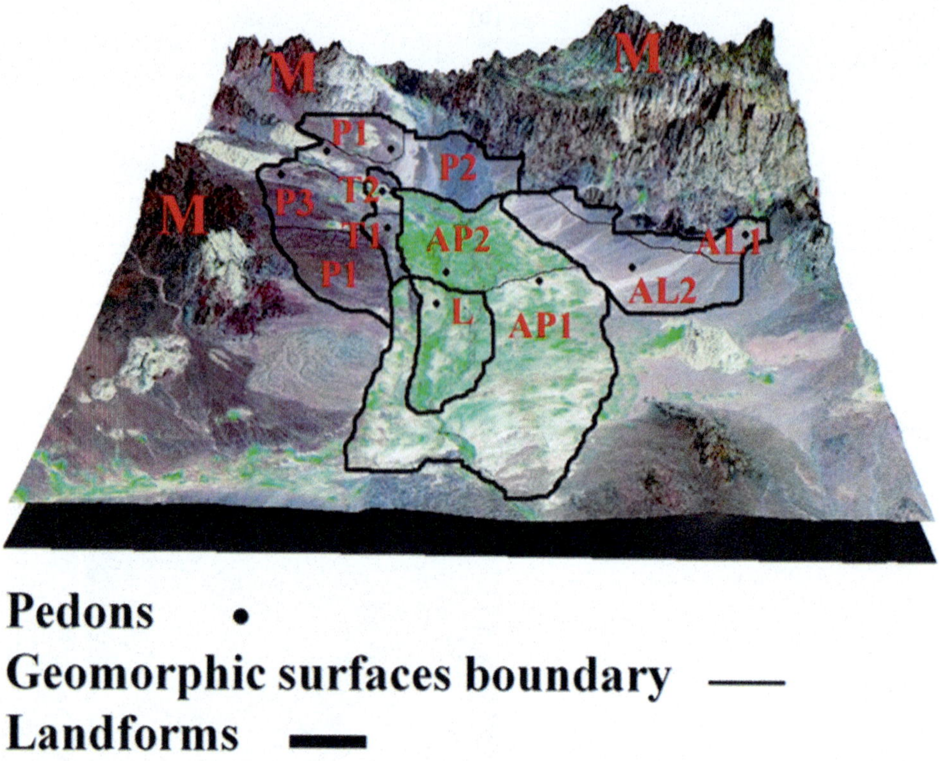

Fig. 4.17 Geomorphic positions and representative pedons as illustrated in a study by Sanjari et al. (2011a). *M* Mountain, *AL1* Upper unstable surface of alluvial fan, *AL2* Middle unstable surface of alluvial fan, *P1* Stable surface of mantled pediment, *P2* Relatively stable surface of mantled pediment, *P3* Unstable surface of mantled pediment, *T1* Intermediate stable surface of pediment and alluvial plain, *T2* Intermediate unstable surface of pediment and alluvial plain, *AP1* Stable surface of alluvial plain, *AP2* Unstable surface of alluvial plain, *L* Lowland surface

interstratified minerals are only found in Histosols in the lowland geomorphic position close to the Lalehzar mountain, which were attributed to the wetter soil moisture regime in this geomorphic surface. This was also considered to be the reason why palygorskite is absent in the lowland position. A close relation among soil properties including clay minerals, geomorphology, and climate was emphasized by authors of this research.

Farshad (2013) analyzed and cross-referenced data with some historic and archaeological information in the Hamadan-Komidjan area, northwest Iran, to reconstruct paleoecologic conditions in the past. He concluded that aridification has never been as degrading as it is today.

Borujeni et al. (2010) used geopedology (a systematic approach to map soilscapes) as a surveying method in Borujen Region. They hypothesized that the geopedology approach reduces the budget of accomplishing soil surveys and increases the purity of mapping units. Their objective was to validate the geopedological mapping methodology by statistical and geostatistical methods in the area under study. The results of the study showed that when comparing the same factors from sample and validation areas, the spatial distribution and dependency of soil attributes were different. They concluded that although the geopedological approach tries to distinguish more homogenous soil mapping units, but it cannot completely represent the real variability of soil properties. Further investigations on different techniques of stratifying the landscape were recommended by researchers for better analyzing and understanding soil forming processes and the nature of soil variability.

Genesis of gypsum-enriched soils in northwest Esfahan, central Iran was studied by Toomanian et al. (2001). They investigated the process of formation and alteration of gypsic horizons on a transect from mountain to piedmont plain and found out that most of gypsum-enriched soils in Esfahan and neighboring provinces occur on alluvial fans, dissected flood plains, and piedmont plains.

A geomorphic hierarchical downscaling method was used by Toomanian et al. (2006) in Zayandeh-rud Valley to determine the forms and processes of landscape formation and its subdivisions. The relationships between Shannon diversity index (K-entropy) and pedorichness versus increasing area were analyzed, and the effects of soil-landscape evolution on complexity of soil patterns in different geomorphic surfaces were investigated. The results of the study showed that soil diversity increased as geomorphic and taxonomic hierarchy levels decreased.

Clay mineralogy comparison of paleosols and modern soils located on different geomorphic surfaces in Jiroft area, Kerman province, showed that smectite, palygorskite, illite, chlorite, and kaolinite were present in these paleosols, but no chlorite was found in modern soils of the area (Sanjari et al. 2011b). Moreover, presence of palygorskite in soils of Jiroft area was highly affected by the stability of geomorphic

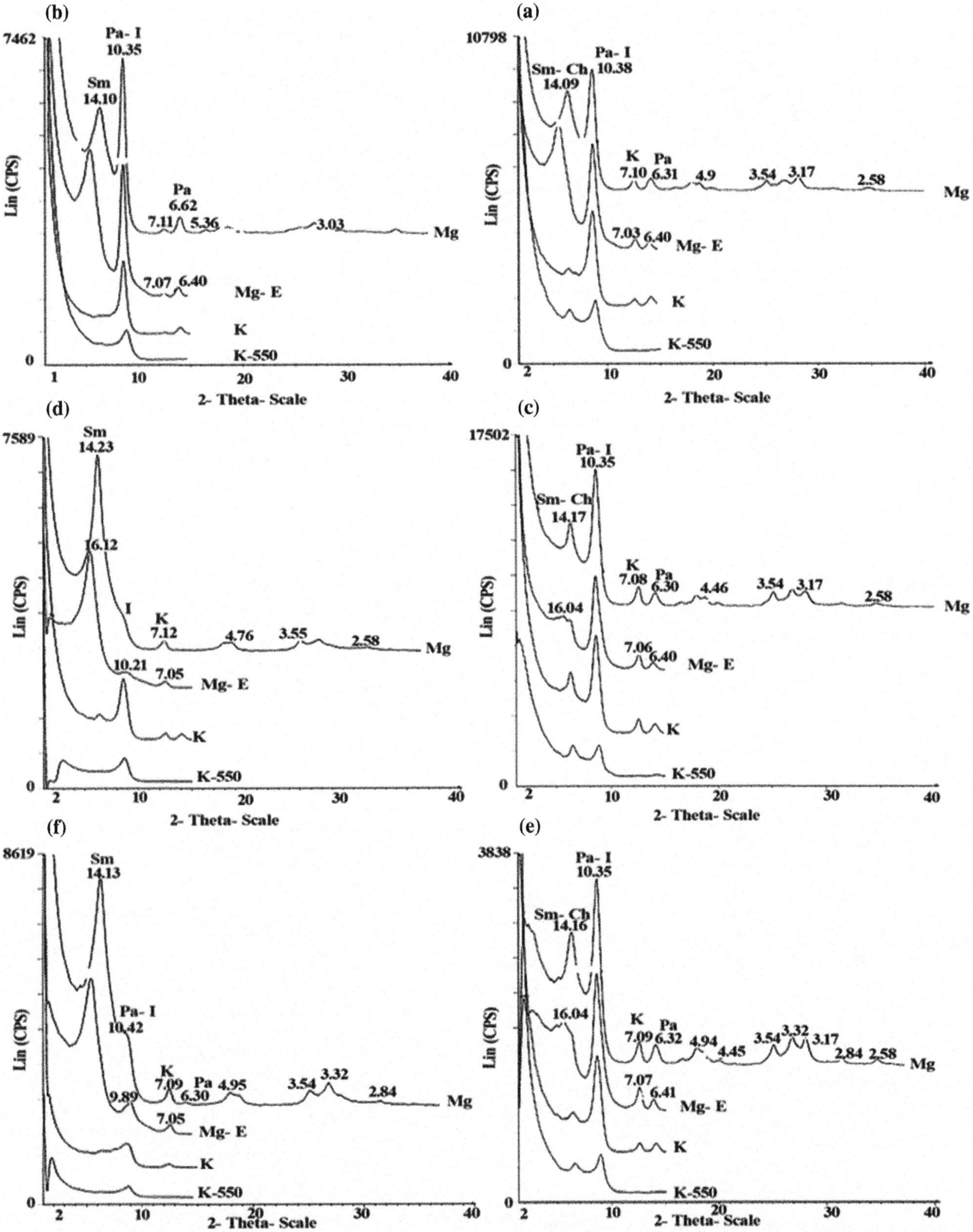

Fig. 4.18 X-Ray diffractograms of the clay fraction in soils of Jiroft area as reported by Sanjari et al. (2011a) **a** Btk horizon on stable mantled pediment, **b** By horizon on relatively stable mantled pediment, **c** Btk1 horizon on intermediate stable surface of pediment and alluvial plain, **d** Btn1 horizon on intermediate unstable surface of pediment and alluvial plain, **e** 2Btk horizon and **f** Btn2 horizon on stable alluvial plain. *Sm* Smectite, *I* Illite, *Pa* Palygorskite, *Ch* Chlorite, *Ka* Kaolinite. *Mg* Mg saturated; *Mg-Eg* Mg saturated with Ethylene glycol; *K* K saturated; *K-550* K saturated and heated to 550 °C

surfaces. Halil-rud River floods on unstable surfaces are reported as the main reason of palygorskite transformation to smectite. Sanjari et al. (2011b) concluded that while geomorphology helps to better understand soil processes and development, paleosols also play an important role in characterization and separation of different geomorphic surfaces of a given landscape.

Geochemistry of carbon, oxygen, and sulfur isotopes in soils along a climotoposequence in Kerman province was studied (Farpoor et al. 2011). Soils from 18 sampling sites from the Lalehzar mountains (4351 m asl) to Shahdad depression (250 m asl) were analyzed for $\delta^{13}C$ values of soil organic carbon and pedogenic carbonates, $\delta^{34}S$ values of sulfates and δD and $\delta^{18}O$ values of gypsum crystallization water. The $\delta^{13}C$ values of soil organic carbon and pedogenic carbonates and their trend along the transect showed that lower geomorphic positions have experienced higher proportion of C_3 plants. On the other hand, more C_4 plants in the history of upper landscape positions were concluded.

References

Abtahi A (1980) Soil genesis as affected by topography and time in highly calcareous parent materials under semiarid conditions of Iran. Soil Sci Soc Am J 44:329–336

Aghanabati A (2004) Geology of Iran. Geological Survey of Iran, Tehran

Alavi M (1991) Tectonic map of the Middle East. Geological Survey of Iran, Tehran

Alavi M (1994) Tectonics of the Zagros Orogenic belt of Iran: new data and interpretations. Tectonophysics 220:211–238

Alavi M (2004) Regional stratigraphy of the Zagros fold-thrust belt of Iran and its profogreland evolution. American J Sci 304:1–20

Alavi Panah SK, Komaki CB, Goorabi A, Matinfar HR (2007) Characterizing land cover types and surface condition of yardang region in Lut Desert (Iran) based upon Landsat satellite images. World App Sci J 2:212–228

Allen M, Jackson J, Walker R (2004) Lake Cenozoic reorganization of the Arabian Eurasia collision and the comparison of short-term and log-term deformation rates. Tectonics 23, TC2008. doi:10.1029/2003TC001530

Bahrami Sh (2013) Tectonic controls on the morphometry of alluvial fans around Danehkhoshk anticline, Zagros, Iran. Geomorphology 180–181:217–230

Bobek H (1963) Nature and implications of quaternary climatic changes in Iran. Symposium on changes of climate. Rome. Oct. 1961. UNESCO-WMO. Rome. Italy, pp 403–413

Borujeni IE, Mohammadi J, Salehi MH, Toomanian N, Poch RM (2010) Assessing geopedological soil mapping approach by statistical and geostatistical methods: a case study in the Borujen region, central Iran. Catena 82:1–14

Bosak P, Jaros J, Spudil J, Sulovsky P, Vaclavek V (1998) Salt plugs in the Eastern Zagros, Iran: results of regional reconnaissance. GeoLines 7:1–174

Brooks IA (1982) Geomorphological evidence for climatic change in Iran during the last 20000 years. P.P.H. communities in Eastern Mediterranean Region in late prehistory: British archeological reports. Int Series 133:191–230

Bruthans J, Filippi M, Zare M, Churackova Z, Asadi N (2010) Evaluation of salt diaper karst morphology during the last glacial cycle: effects of sea-level oscillation diaper and regional uplift, and erosion (Persian Gulf, Iran). Geomorphology 121:291–304

Cucchi F, Zini L (2003) Gypsum karst of Zagros mountains (I.R. Iran). Acta carsologica 32(1)

Darvishzadeh A (2001) Geology of Iran. Amir Kabir Press, Tehran (in Persian)

Eghbal MK, Southard RJ (1993) Stratigraphy and genesis of Durorthids and Haplargids on dissected alluvial fans, western Mojave Desert, California. Geoderma 59:151–174

Ehsani AH, Quiel F (2008) Application of self-organizing map and SRTM data to characterize Yardangs in the Lut desert, Iran. Remote Sens Environ 112:3284–3294

Farpoor MH, Eghbal MK (1996) Soil landscape relationships in Imam Gheis area, Charmahale Bakhtiari, Iran. Proc. of the 5th Soil Science Congress of Iran, Karadj, Iran

Farpoor MK, Eghbal MK, Khademi H (2003) Genesis and micromorphology of saline and gypsiferous Aridisols in different geomorphic surfaces of Rafsanjan area, Iran. J Sci Tech Agr Natur Resour 3:71–80

Farpoor MH, Irannejad M (2013) Soil genesis and clay mineralogy on Aliabbas River alluvial fan, Kerman Province. Arab J Geosci 6:921–928

Farpoor MH, Khademi H, Eghbal MK (2002) Genesis and distribution of palygorskite and associated clay minerals in Rafsanjan soils on different geomorphic surfaces. Iran Agr Res 21:39–60

Farpoor MH, Khademi H, Eghbal MK, Krouse HR (2004) Mode of gypsum deposition in southeastern Iranian soils as revealed by isotopic composition of crystallization water. Geoderma 121:233–242

Farpoor MH, Krouse HR (2008) Stable isotopic geochemistry of sulfur bearing minerals and clay mineralogy of some soils and sediments in Loot Desert, central Iran. Geoderma 146:283–290

Farpoor MH, Krouse HR, Mayer B (2011) Geochemistry of carbon, oxygen and sulfur isotopes in soils along a climotoposequence in Kerman Province, central Iran. J Agr Sci Tech 13:953–964

Farpoor MH, Neyestani M, Eghbal MK, Esfandiarpoor I (2012) Soil-geomorphology relationships in Sirjan Playa, south central Iran. Geomorphology 138:223–230

Farshad A (2013) Geopedology reports historical changes in climate and agroecology: a case study from northwestern Iran. Ecopersia 1 (2):145–158

Farshad A, Shrestha DP, Moonjun R (2011) Do the emerging methods of digital soil mapping have anything to learn from the geopedologic approach to soil mapping and vice versa? In: Shahid Sh (ed) Soil classification and reclamation of degraded lands in arid environments. ICSC 2010, Abu Dhabi, UAE

Fattahi M, Talebian M, Tabatabaei (2010) Applying OSL to determine the Bam—Baravat Fault Uplift Rate. UK TL/OSL/ESR Meeting 2010, 8–10th Sep, University of Oxford

Fattahi M, Walker RT, Khatib MM, Dolati A, Bahroudi A (2007) Slip-rate estimate and past earthquakes on the Doruneh fault, eastern Iran. Geophys J Int 168:691–709

Fisher WB The land of Iran (1968) In: Fisher WB (eds) Cambridge history of Iran. Cambridge University Press. ISBN:0-521-24693-8

Gabriel A (1938) The southern Lut and Iranian Baluchistan. Geogr J 92:193–208

Ghajar M (1994) Effect of landform in genesis, morphology, and soil physicochemical properties of Karadj plain. M.Sc. thesis. Tarbiat Modares Univ. of Tehran

Ghorbani M (2013) Economic geology of Iran: mineral deposits and natural resources. Springer, New York

Goudie AS (2007) Mega-Yardangs: a global analysis. Geogr Compass 1:65–81

Graham RC, Buol SW (1990) Soil-geomorphic relations on the Blue Ridge Front: II. Soil characteristics and pedogenesis. Soil Sci Soc Am J 54:1367–1377

Heidari M, Khanlari GR, Taleb Beydokhti AR (2004) Effective karstification carbonate rock acceleration occurrence sinkholes kabodarahang plain, Hamedan, Iran. Conference on karst problems of Iran (with special emphasis on karst areas of western part). Lorestan University and West Regional Water Company of Iran, Lorestan-Iran, pp 43–49

Heidari M, Khanlari GR, Taleb Beydokhti AR, Momeni AA (2011) The formation of cover collapse sinkholes in north of Hamedan, Iran. Geomorphology 132:76–86

Hollingsworth J, Fattahi M, Walker R, Talebian M, Bahroudi A, Bolourchi MJ, Jackson J, Copley A (2010) Oroclinal bending, distributed thrust and strike-slip faulting, and the accommodation of Arabia-Eurasia convergence in NE Iran since the Oligocene. Geophys J Int. doi:10.1111/j.1365-246X.2010.04591.x

Howell JV (1957) Glossary of geology and related sciences. Amer Geogr Inst

Karimi A, Frechen M, Khademi H, Kehl M, Jalalian A (2011) Chronostratigraphy of loess deposits in northeast Iran. Quat Int 234:124–132

Karimi A, Khademi H, Kehl M, Jalalian A (2009) Distribution, lithology and provenance of peridesert loess deposits in northeastern Iran. Geoderma 148:241–250

Kazanci N, Gulbabazadeh T (2013) Sefidrud delta and Quaternary evolution of the southern Caspian lowland, Iran. Mar Pet Geol 44:120–139

Kazanci N, Gulbabazadeh T, Leroy SAG, Ileri O (2004) Sedimentary and environmental characteristics of the Gilan-Mazenderan plain, northern Iran: influence of longe-and short-term Caspian water level fluctuations on geomorphology. J Mar Sys 46:145–168

Kehl M (2010) Late quaternary loesses, loess-like sediments, soils and climate change in Iran. Relief, Boden, Palaoklima, E. Schweizerbart Sceince Publishers, Stuttgart, Germany

Kehl M, Frechen M, Skowronek A (2005) Paleosols derived from loess and loess-like sediments in the Basin of Persepolis, Southern Iran. Quat Int 140–141:135–149

Khademi H, Mermut AR (1998) Source of palygorskite in gypsiferous Aridisols and associated sediments from central Iran. Clay Miner 33:561–575

Khademi H, Mermut AR (1999) Submicroscopy and stable isotope geochemistry of carbonates and associated Palygorskite in Iranian Aridisols. Eur J Soil Sci 50:207–216

Khademi H, Mermut AR (2003) Micromorphology and classification of Argids and associated gypsiferous Aridisols from central Iran. Catena 54:439–455

Khademi H, Mermut AR, Krouse HR (1997a) Sulfur isotope geochemistry of gypsiferous Aridisols from central Iran. Geoderma 80:195–209

Khademi H, Mermut AR, Krouse HR (1997b) Isotopic composition of gypsum hydration water in selected landforms from central Iran. Chem Geol 138:245–255

Khanlari GR (2010) The necessity of Hamedan sinkholes study from an environmental point of view. Proceedings of the 2nd congress of Water Industry Researches in Hamedan province, Hamedan, Iran, pp 162–173

Khormali F, Ajami M (2011) Pedogenetic investigation of soil degradation on a deforested loess hillslope of Golestan Province, Northern Iran. Geoderma 167–168:274–283

Khormali F, Kehl M (2011) Micromorphology and development of loess-derived surface and buried soils along a precipitation gradient in Northern Iran. Quat Int 234:109–123

Kober F, Zeilinger G, Ivy-Ochs S, Dolati A, Smit J, Kubik PW (2013) Climatic and tectonic control on fluvial and alluvial fan sequence formation in the Central Makran Range, SE-Iran. Global Planet Change 111:133–149

Krinsley DB (1970) A geomorphological and paleoclimatological study of the playas of Iran. U.S. Geological Survey. U.S. Government Printing Offic, contract. PROCP 700–800. US Air Force Cambridge Research. 329 pp

Le Dortz K, Meyer B, Sebrier M, Nazari H, Braucher R, Fattahi M, Benedetti L, Foroutan M, Siame L, Bourles D, Talebian M, Bateman MD, Ghoraishi M (2009) Holocene right-slip rate detemrined by cosmogenic and OSL dating on the Anar fault, Central Iran. Geophys J Int 179:700–710

Mahler PJ (1979) Manual of land classification for irrigation. Soil Institute of Iran. Pub. No. 205

McCall GJH (2002) A summary of the geology of the Iranian Makran. In: Clift PD, Kroon D, Gaedicke C, Craig J (eds) The tectonic and climatic evolution of the Arabian Sea region. Geolog Soc London, London, pp 147–204

McCall CJH, Eftekhar-Nezhad J (1993) Explanatory text of the Nikshahr Quadrangle Map, 1:250000-L14. In: Iran GSO (ed) Geological Survey of Iran, Tehran

Mildrexler DJ, Zhao M, Running SW (2006) Where are the hottest spots on earth? Eos, transactions. American Geophys Union: 87–461

Moazallahi M, Farpoor MH (2012) Soil genesis and clay mineralogy along the xeric-aridic climotoposequence, south central Iran. J Agr Sci Tech 14:683–696

Nabavi MH (1976) An introduction to the geology of Iran. Geological Survey of Iran. 109 pp (In Persian)

Nadimi M, Farpoor MH (2013) Genesis and clay mineralogy of soils on different geomorphic surfaces in Mahan-Joupar area, central Iran. Arab J Geosci 6:825–833

Nezafati N (2006) Au-Sn-W-Cu Mineralization in the Astaneh-Sarband Area, West Central Iran. Ph.D. thesis, University of Tübingen (Germany), 116 pp

Okhravi R, Amini A (2001) Characteristics and provenance of the loess deposits of the Gharatikan watershed in Northeast Iran. Glob Planet Change 28:11–22

Regard V, Bellier O, Thomas J, Bourles D, Bonnet S, Abbassi MR, Braucher R, Mercier J, Shabanian E, Soleymani Sh, Feghhi Kh (2005) Cumulative right-lateral fault slip-rate across the Zagros-Makran transfer zone:role of the Minab-Zendan fault system in accommodating Arabia–Eurasia convergence in southeast Iran. Geophys J Int 162:177–203

Rooyani F, Southard AR (1985) Characteristics and genesis of some soils in the southern foothills of central Alborz, Iran. Soil Sci 140:45–54

Salehi J (1994) Application of RS and GIS for evaluation of soil and water resources for development planning in the Hamadan area, Iran. Unpublished MSc thesis, ITC, Enschede, The Netherlands

Sanjari S, Farpoor MH, Eghbal MK, Esfandiarpoor I (2011a) Genesis, micromorphology and clay mineralogy of soils located on different geomorphic surfaces in Jiroft area. J Water Soil 25:411–425

Sanjari S, Farpoor MH, Esfandiarpoor I, Eghbal MK (2011b) Micromorphology and clay mineralogy comparison of paleosols and modern soils in Jiroft area. J Sci Technol Agric Natur Resour, Water Soil Sci 58:153–185

Stocklin J (1968) Structural history and tectonics of Iran, a review. AAPG Bulletin 527:1229–1258

Talbot CJ, Medvedev S, Alavi M, Shahrivar H, Heidari E (2000) Salt extrusion rates at Kuh-e-Jahani, Iran: june 1994 to november 1997. Geological 174:93–110 Society London Special Publications

Talbot ChJ, Pohjola V (2009) Subaerial salt extrusions in Iran as analogues of ice sheet, streams and glaciers. Earth-Sci Rev 97: 155–183

Talebian M, Tabatabaei SH, Fattahi M, Ghorashi M, Beitollahi A, Ghalandarzadeh A, Riahi MA (2010) Estimating slip rates of faults around Bam and their application in evaluation of earthquake hazard. Geosciences 19(74):149–156 (in Persian with English abstract)

Toomanian N, Jalalian A, Eghbal MK (2001) Genesis of gypsum enriched soils in north-west Isfahan, Iran. Geoderma 99:199–224

Toomanian N, Jalalian A, Khademi H, Eghbal MK, Rapritz A (2006) Pedodiversity and pedogenesis in Zayandeh-rud Valley, central Iran. Geomorphology 81:376–393

Walker AS (1986) Chapter 8: Plate E-19. In: Nicholas M, Short S, Robert J, Blair W (eds) Geomorphology from space: NASA http://daac.gsfc.nasa.gov/geomorphology/index.shtml

Walker RT, Fattahi M (2011) A framework of Holocene and Late Pleistocene environmental change in eastern Iran inferred from the dating of periods of alluvial fan abandonment, river terracing, and lake deposition. Quat Sci Rev 30:1256–1271

Zarasvandi A, Charchi A, Carranza EJM, Alizadeh B (2008) Karst bauxite deposits in the Zagros mountain belt, Iran. Ore Geol Rev 34:521–532

Zinck JA, Metternicht G, Bocco G, Del Valle HF (2016) Geopedology; An Integration of Geomorphology and Pedology for Soil and Landscape Studies. ISBN 978-3-319-19158-4 Springer (www.Springer.com)

Author Biographies

Dr. Mostafa Karimian Eghbal is Associate Professor at Tarbiat Modares University in Iran since 2004. Prior to that, he was in Isfahan University of Technology. He received his MS and Ph.D. from University of California at Davis and BS from Cal Poly San Luis Obispo. In addition of being active in his field of soils science and soil geomorphology, Dr. Eghbal has been one of the pioneers for establishment of science and technology parks in Iran.

Dr. Nikoo Hamzehpour is Assistant Professor in pedology at the University of Maragheh in Iran. Her field of interest is saline soils formation, development, and mapping especially in Urmia Lake Basin. She is also interested in dealing with uncertainties involved with field measured data and their error quantification.

Dr. Mohammad Hady Farpoor is Professor at Department of Soil Science, Faculty of Agriculture, Shahid Bahonar University of Kerman, Iran, and a member of Iranian Association of Official Experts (IAOE). His current research interests are soil–geomorphology relationships, clay mineralogy with environmental applications, soil micromorphology, stable isotope geochemistry, and desert soils and landforms.

Vegetation and Land Cover

5

Ali Farzaneh and Morteza Ebrahimi Rastaghi

Abstract

Four main factors, namely *climatic situation*, *phytogeographical regions*, *varied topography*, and *the impact of human activity* are considered responsible for the changes in distribution and character of the natural vegetation. The extensive Iranian plateau displays quite a variation in such a way that 167 families of vascular plants and 1215 genera, some of which include only one species and some (e.g. *Astragalus*) include about 800 species. The first and very small-scale map depicting major types of vegetation in Iran, Afghanistan, and Pakistan has been available since 1960s. In 2005, Forest, Range and Watershed Organization (FRWO) published the first land cover map using remote sensing and GIS techniques. This map shows the divisions and subdivisions of the vegetation cover according to location, quantity, and quality of the natural vegetation cover. In this chapter, we summarize the multipurpose functions of the vegetation cover in the country according to their status in different regions.

Keywords

Land cover map • Iran's floristic • Landsat -5 TM • GIS environment

5.1 Introduction

Climatic condition, physiographical regions, varied topography, and the impact of human activity play a vital role in the diversity of vegetation in Iran. According to some authors (Mobayen and Tregubov 1980), the existing vegetation maps were prepared in two ways. One is general land cover map based on climate and physiography, and the other one is based on physiography combined with the occurrence of such features such as climatic regions in the land cover. There was no accurate statistics on the types of vegetation cover because of the diffuse boundaries between vegetation and cultivated land (mixed units). However, this problem has been resolved in the recent land cover map in terms of species and the most dominant vegetation type.

A. Farzaneh (✉) · M.E. Rastaghi
Forest, Range and Watershed Organization (FRWO), Tehran, Iran
e-mail: alifarzaneh@hotmail.com

M.H. Roozitalab et al. (eds.), *The Soils of Iran*, World Soils Book Series,
https://doi.org/10.1007/978-3-319-69048-3_5

5.1.1 Definitions

The definition of the land cover is often confused with that of the land use. The following definitions are used in the lately published maps:

Land cover describes the most fundamental features of the land. In other words, land cover is an observed (bio) physical cover on the earth's surface. When considering land cover in a very pure and strict sense, it should be confined to describe both the vegetation and man-made features. In this way, areas of bare rock and bare soil are not considered land cover, but the land itself. It is still disputable whether water bodies should be labeled as land cover (Di Gregorio 2000).

Land use is characterized by the arrangements, activities, and inputs people undertake in a certain land cover type to produce, change, or maintain it. Definition of land use in this way establishes a direct link between land cover and the actions of people in their environment. For instance, "grassland" is a cover term, while "rangeland" or "tennis court" refers to the use of a grass cover; "recreation area" is a land use term that may be applicable to different land cover types, for instance, sandy surfaces like a beach, a built-up area like a pleasure park, woodlands (Di Gregorio 2000)." In Iran, Javanshir (1975) studied the vegetation applying "geographic structure" and "climatic zoning" approaches. The results of his studies have been used to produce a map that is called "land cover map" by some authors and "land use map" by some others.

5.1.2 Climatic Zones as Related to Vegetation Cover

A very small-scale map depicting major types of vegetation in Iran, Afghanistan, and Pakistan (Fig. 6.1) has been available since 1960. Zohari (1963) was probably the first researcher who published his study on woody vegetation. In 1973, he published his studies in more detail, where he considered five main zones as (1) Caspian zone, (2) Zagros zone, (3) Irano-Touranian zone, (4) Persian Gulf zone, and (5) Arasbarani zone. In the same period, Sabeti (1969) published his book titled "Les Etudes Bioclimatique de l'Iran," where he has included two valuable bioclimatic maps, prepared based on the Goosen method and also according to the Emberger method, at scale of 1:40,00,000. More than 10,000 species have been identified, classified, and listed under different zones. A much-generalized division might be the one given in the Cambridge History of Iran (Fisher 1968), shown in Fig. 5.1

5.1.3 Soils and Vegetation Cover

With reference to the five soil forming factors "ClORPT statement[1]" (Jenny 1941), vegetation is one of the important soil forming factors, symbolized as "O" in the statement. Generally, there is a close relationship between native vegetation and kinds of soil, yet there are important exceptions (USDA Soil Survey Manual 1993), in particular in regions with aridic climatic conditions. Soil surveyors take benefit of vegetation types as indicators of soil types, for instance, for understanding soil genesis and recognizing soil boundaries. Examples of such indicators are *Hydrophile* group (*Fraxinus Sp., Alnus glutinosa*); *Mesophiles* (*Fagus orientalis, Carpinus betulus, Quercus Sp.*); *Xerophiles* (*Glycyrhiza glabra Artemisia Spp.*); *Psamophile* (*Haloxylon Spp., Calligonum Spp.*); and *Hallophites* (*Salsola Spp.,Salicornia Spp., Tamarix Spp.*) which are used, respectively, as indicators for soils of humid (wet), semi humid, arid regions, sandy soils (mainly in active sand dunes), and for recognition of salt-affected soils. Farmers often know the relationships between the crops and different kinds of soil. A simple example is to make a choice between growing wheat or barley, when farmer knows that his/her soils are saline, or choosing between loamy-textured and clayey soils when the crop is sugar beet.

5.2 A General Overview of the Vegetation Cover

As reported in the Fourth National Report to the Convention on Biological Diversity (Department of Environment 2010), Iran is located in the Palearctic realm and is considered to be the origin of many genetic resources of the world vegetation including many of the original strain of commercially valuable plant species, such as wheat, medicinal, or aromatic species.

The southwest has some Afro-tropical features, while the southeast has some species from the Indo-Malayan subtropical realm. There are 12.4 million hectares of woodland and more than 10,000 ha of Mangroves along the Persian Gulf and the Sea of Oman coast. Ecosystem diversity of marine and coastal zones in the north and the south of the country consists of 25 ecological types and units, most importantly coral reefs, bays, and small islands. Because of its large size and varied ecosystems, Iran is one of the most important countries for conservation of biological diversity in the Middle East and West Asia. More basic technical works and field surveys in terms of eco-biological analyses are needed in order to fully understand the country's biodiversity characteristics. As reported in the 4th National Report, the country topography has given rise to four floristic zones, namely (1) Irano-Touranian arid and semiarid mountainous, (2) Zagrosian, (3) Hyrcanian, and (4) Khalij Fars-Omanian. Most of Iran's forest cover is located in Hyrcanian, Arasbaranian, and Zagrosian zones. Table 5.1 shows the general view on Iran's biomes and their status (Fig. 5.2).

5.3 Land Cover Map

In 2005, Forest, Range and Watershed Organization (FRWO) published the first land cover map using remote sensing and GIS at scale of 1:2,50,000 (Fig. 5.3). Landsat-5 TM images were visually interpreted on screen (Fig. 5.2) followed by a dense field check operation. The map includes climatic and physiographic regions and divisions or subdivisions of the vegetation cover identified according to the

[1]Cl (climate), O (organism: flora and fauna), R (relief/ topography), P (parent material/ lithology), T (time/age).

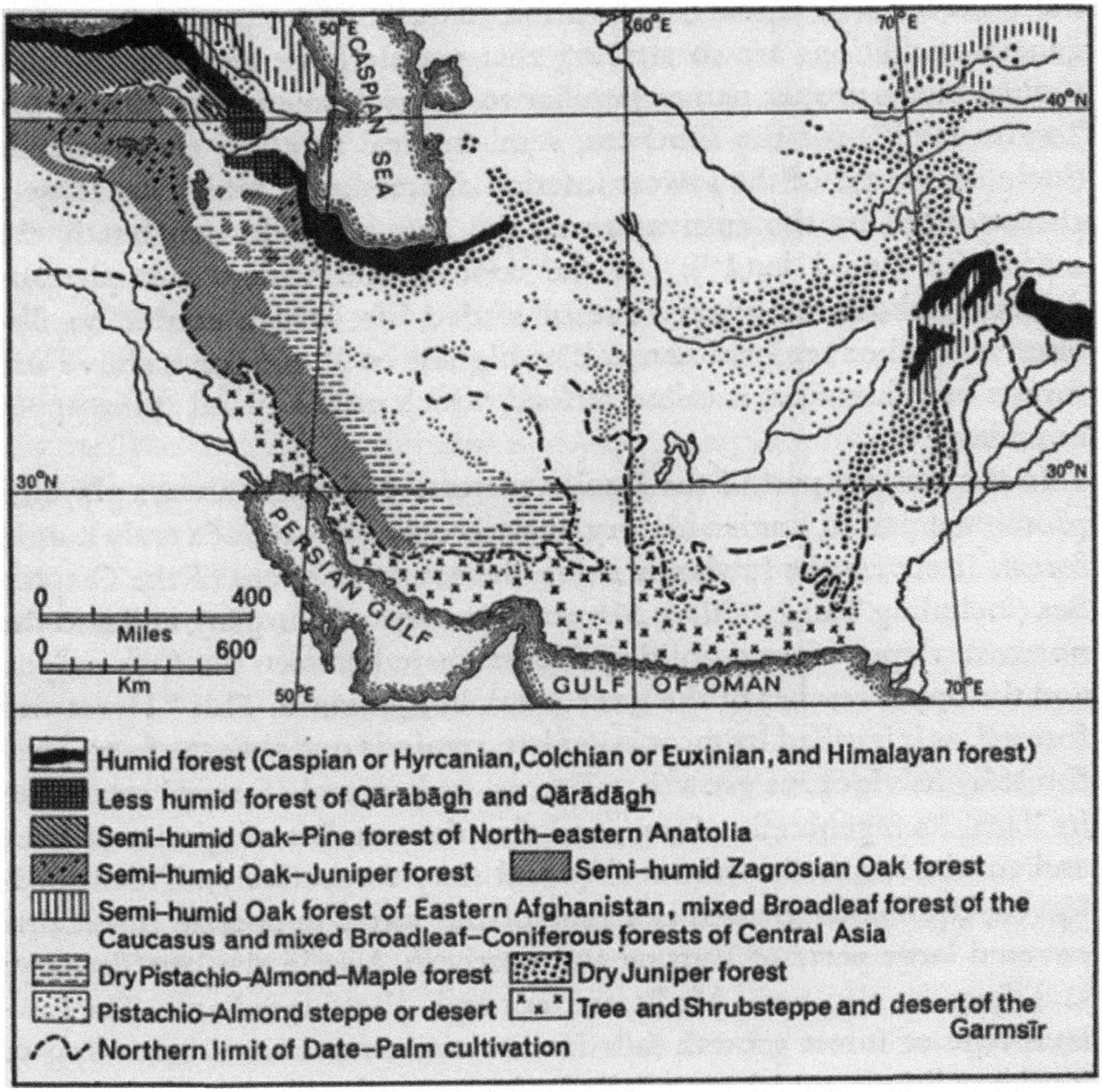

Fig. 5.1 Types of vegetation based on climatic zone (Fisher 1968)

Table 5.1 Area of Iran's biomes and their status area/hectares (DE 2010)

Irano-touranian plain and mountainous area	3,452,775
Zagrosian	4,749,000
Hyrcanian zone + Semi humid	2,800,000
Khalij Fars—Omanian	2,130,000

location, quantity, and quality of the coverage. The legend of the land cover map consists of 22 distinct land cover types, including residential areas (cities and rural), agricultural land (irrigated and dry-land farming), forests in six main classes distinguished according to density (dense to sparse), range and shrub land classified according to density, and desert areas that are classified in 6 classes. The features include lakes, water bodies, and infrastructure. Presently, this map

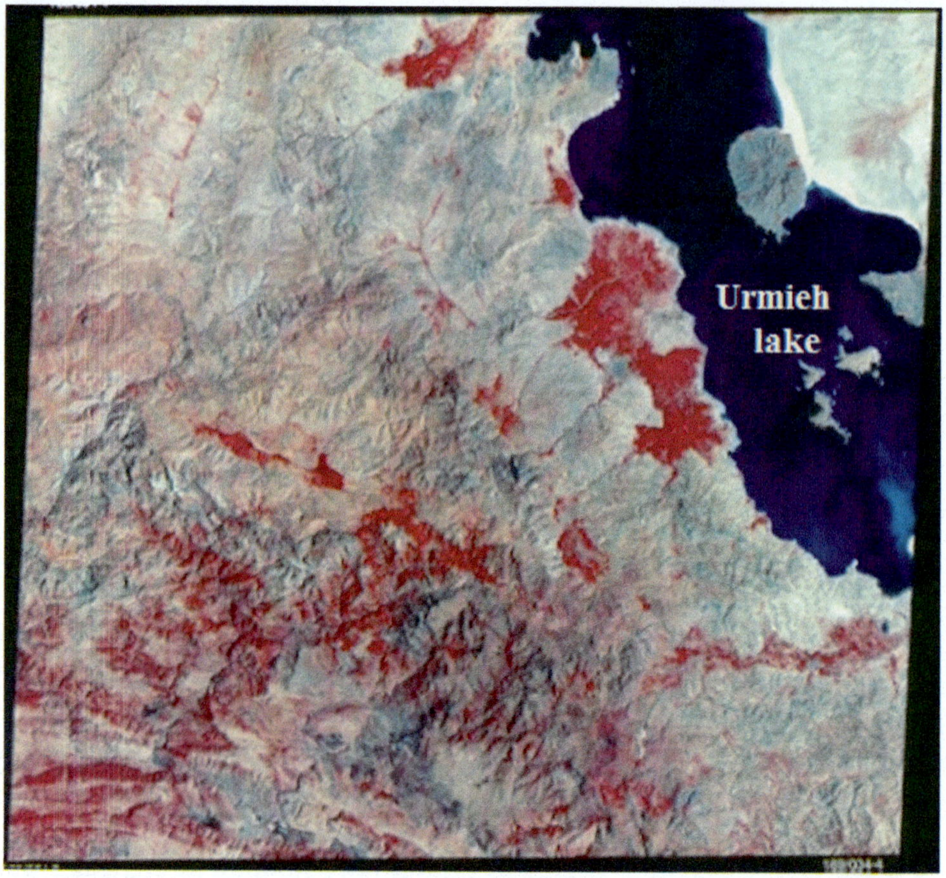

Fig. 5.2 A Landsat 5 TM 140 image showing land cover map of Iran (Urmieh Lake is in upper right)

and its accompanied statistics are one of the most reliable sources. The provincial agencies, public and private stakeholders also use this map for different purposes including development and conservation projects, where changes in land cover is a key issue. The latest version of this map shows that Iran consists of six big watershed units. The information is based on the land cover map prepared in 2004, which was slightly modified to align more clearly with the new 2010 version. This alignment process is described in more detail below.

5.3.1 Mapping Procedures

For more than a decade, FRWO Program focused considerable effort on mapping land cover to assist in the modeling of flora and biodiversity for large geographic areas of Iran. The program had been traditionally provinces-centered, each province having the responsibility of implementing a project design for the geographic area within their state boundaries. This was first formal project designed at a regional, multi-state at scale of 1:2,50,000. The project area comprised the whole country and lasted approximately 7 years: beginning in 1998 and ending in January 2005. The land cover map was generated using regionally consistent geospatial data: 140 Landsat TM5 images (Fig. 5.2) within GIS environment. This protocol with standardized land cover legend in 22 classes was carried out in 130 individual map sheets for the whole country at scale of 1:2,50,000 in a common modeling approach.

5.3.2 Iran's Floristic Zones

Different statistics are reported on vegetation, in terms of botanical classification. Some authors reported 167 families of vascular plants and 1215 genera, some of which include only one species and some others such as *Astragalus* include about 800 species. Mozafarian (2004) reported 146 families and vascular plants, and 1324 genera. Ghahreman and Attar (2000) reported "a total of 8000 taxa, including 6417 species, 611 subspecies, 465 varieties, and 83 hybrids. Of these, about 1810 are endemic to Iran." In this chapter, we have adopted the reported statistics by the Central Herbarium of University of Tehran.

Fig. 5.3 Land cover map of Iran at scale of 1:1,000,000 (FRWO 2005)

Hyrcanian zone (Humid region)

According to the four floristic zones division, Hyrcanian zone consists of broad-leaved forest at three levels of altitude:
A1—Three communities in lowland:

1. Oak-box community
2. Querceto-carpinetum community
3. Parrotio-carpinetum community

The most dominant species at this level are given in Table 5.2

A2—In midland Hyrcanian zone (humid region), the most important community is Fagetum hyrcanum with the most species shown in Table 5.3.

A3—Forest vegetation in upland Hyrcanian zone (humid region), summarized in two communities (Table 5.4) (Figs. 5.4 and 5.5).

1. Carpinetum Orientale
2. Juniperetum excels

Table 5.5 shows the correlation between the soil types and the Hyrcanian vegetation types. The soils are very diverse at different stage of development depending on the

Table 5.2 Dominant plant species in lowland

A1. In lowland	
Scientific name	English name
Quercus castaneaefolia	*chestnut/leaved oak*
Buxus hyrcana	*box/box tree*
Acer spp.	*Maple*
Alnus glutinosa	*European alder*
Albizzia julibrissin.	*Silk tree*
Ptrocarya fraxinifolia	*Caucasian wing nut*
Gleditschia caspica	*Caspian honey locust*
Populus caspica	*Eastern poplar*

Table 5.3 Dominant plant species in midland

A2. In midland	
Scientific name	English name
Fagus orientalis	*Oriental beech*
Carpinus betulus	*European hornbeam*
Ulmus glabra	*Scotch elm*
Quercus castaneaefolia	*Chestnut/leaved oak*
Sorbus torminalis	*mountain ash*
Alnus subcordata	*Caucasian alder*

Table 5.4 Dominant plant species in upland

A3. In upland	
Scientific name	English name
Carpinus orientalis	*Eastern hornbeam*
Quercus macranthera	*Persian oak*
Carpinus schuschensis	*hornbeam*
Juniperus excels	*Greek juniper*
J.communis	*common juniper*
J.Sabina	*common savin*

elevation, landscapes, amount of precipitation, and types of vegetation (see Chaps. 6 and 7)

Forest in semi humid, Arasbaranian zone

Arasbaranian zone is divided into two parts: forest above 1000 m altitude and forest below 1000 m above sea level. The most important species are listed in Table 5.6 (Fig. 5.6 and Table 5.7).

Zagrosian zone

With an area of about 4,749,000 ha covers the Zagros mountain ranges. This ecological zone extends throughout the western flank of the Zagros mountain range, parallel with the Persian Gulf coastline, and consists of numerous parallel ridges, with the highest peaks exceeding 4000 m. The forest and steppe forest areas of the Zagros Mountain ranges have a semiarid temperate climate, with annual precipitation ranging from 400 to 800 mm, falling mostly in winter and spring. Winters are severe, with winter minima often below 25 °C, and extreme summer aridity prevails in the region. The Zagros zone consists of two parts. The most important species in the northern and southern parts are shown in Tables 5.8 and 5.9, respectively (Fig. 5.7).

The correlation between soil and vegetation cover in Arasbaranian (semi humid forest) and semiarid Zagrosian zones shown in Table 5.10. The soils are generally calcareous and are mostly classified in Inceptisols and Entisols orders (see Chap. 7).

Irano-Touranian zone

Irano-Touranian covers an area of about 3,452,775 ha including arid to semiarid deserts and the plains of Central Iran. Considering topographical conditions and diversity of

Fig. 5.4 A view of Hyrcanian forest in lowland (Courtesy by FRWO)

species, the region is divided into plain and mountainous subregions. The most important plant species are shown in Table 5.11 (Fig. 5.8 and Table 5.12).

Table 5.13 shows a general correlation between vegetation cover types and soil in the Irano-Touranian zone.

Khalij-e Fars-Omanian

Khalij-e Fars-Omanian encompasses southern coastal plains of Iran with hot temperature and high humidity. The region, with an area of 2,130,000 ha, extends throughout the southern part of the country in provinces of Khuzestan, Bushehr, Hormozgan, and Sistan-Baluchistan. A sub-equatorial climate is prevalent in the region, and soils are mostly classified in Aridisols and Entisols (see Chap. 7) (Figs. 5.9 and 5.10).

5.4 Land Cover Classes and Statistics

The classification system that is presented in the land cover map was developed to meet the required needs of different sectors, in carrying out a reconnaissance survey of the natural resources (e.g. forest, range, desert, water bodies, and…) and to obtain the accurate statistics and data for different land units with different coverage (FRWO's report 2005). The proposed system uses the features of existing

Fig. 5.5 A view of Hyrcanian forest in upland (Courtesy by FRWO)

Table 5.5 Correlation of soil and vegetation in Hyrcanian zone

Vegetation cover zones		Soils (generalized)
Hyrcanian (humid Alborz along the Caspian sea coast)	- A1, Lowland: three communities	Entisols (Xerorthents), Inceptisols (Xerepts), Alfisols (Xeralfs), Mollisols (Xerolls)
	- A2, Midland Hyrcanian zone	Alfisols (Haplxeralfs, Hapludalfs, Paleudalfs), Inceptisols (Xerepts and Udepts), Mollisols (Udolls and Xerolls)
	- A3, Upland Hyrcanian zone	Entisols (Xeorthents and Udorthents) Entisols(Xerorthents),Inceptisols (Xerepts), Mollisols (Xerolls)

Table 5.6 Arasbaranian plant species above 1000 msl

Above 1000msl.	
Scientific name	English name
Quercus macranthera	*Persian oak*
Quercus petraea	*durmast oak*
Carpinus betulus	*European hornbeam*
Taxus baccata	*English yew*
Ulmus glabra	*scotch elm*
Fraxinus excelsior	*European ash*
Cotinus coggygeria	*common smoke tree*
Juniperus foetidissima	*Juniper*
Acer spp	*Maple*
Sorbus torminalis	*Checker mountain ash*
Cornus mas	*cornelian cherry*
Corylus avellana	*European filbert/filbert/hazel*

Fig. 5.6 Views of semi humid Arsbaranian Forest (left) below 1000 m and (right) above 1000 m

widely used classification systems, that are amenable to data derived from remote sensing sources. Despite the various and huge existing features, only 22 major classes were realized in the legend of the land cover map. The regional offices were responsible to create more detailed divisions and subclasses of land cover for their own purposes. The forests include woodland, shrublands, and forest-plantation cover about 14,907,355 ha. They are categorized into six classes according to density, quantity, and quality of forest (Tables 5.14, 5.15 and 5.16).

Table 5.7 Arasbaranian plant species below 1000 msl

Below 1000 msl.	
Scientific name	English name
Spiera crenata	*snow spirea*
Rhamnus spp	*Buckthorn*
Paliurus spina-christii	*christ thorn*
Punica granatum	*Pomegranate*
Rhus coriaria	*sisilian sumac*
Pistacia atlantica	*Mount Atlas pistache*
Celtis caucasica	*caucasian hackberry*
Ficus carica.	*common fig*

Table 5.8 Dominant plant species in northern Zagros

Northern Semidry forest	
Scientific name	English name
Quercus libani	*Lebanon oak*
Quercus infectoria	*Aleppo oak*
Quercuse brantii,	*Oak manna*
Quercus robur	*English oak*
Betula pendula,	*European white birch*
Cornus australis	*Cornel*
Sambacus nigra,	*European elder*
Sorbus graeca,	*White beam mountain*
Pistacia atlantica	*Mountain Atlas pistache*
Acer monspessulanum	*Montpellier maple*
Crataegus spp.	*Howthorn*

Table 5.9 Plant species in southern Zagros

Southern semidry forest	
Scientific name	English name
Quercus brantii	*Oak manna*
Amygdalus arabica	*Almond*
Cercis griffithii	*Judas tree*
Cupressus sempervirens var horizontalis.	*Horizontal cypress*
Juniperus excelsa	*Greek juniper*
Acer monspessulanum	*Montpellier maple*
Pistacia atlantica	*Mountain Atlas pistache*
Fraxinus rotundifolia	*Ash*
Crataegus spp.	*Howthorn*
Lonicera nummulari folia,	*Honey suckle*
Pyrus spp	*Pear*
Cerasus microcarpa	*Cherry*

Fig. 5.7 A view of Zagros forest mostly in west of Iran

Table 5.10 Soil–vegetation correlation

Vegetation cover zones	Soils (generalized)	
Forest in semi humid to semiarid	Arasbaranian	Dominantly Inceptisols (Xerepts including Calcixerepts and Haploxerepts), Entisols (Xerorthents), Mollisols (Haploxerolls)
	Zagrosian	Inceptisols(Calcixerepts), Entisols (Xerorthents), and Aridisols (Calcids)

Table 5.11 Dominant plant species in lowland of Irano-Touranian zone

Lowland of Irano-Touranian zone	
Scientific Name	English name
Tamarix Spp.	*Tamarisk*
Haloxylon Ammodendron,	*sacsaoul, saxaoul*
Calligonum spp,	*calligonum*
Popolus euphratica,...	*willows*

Fig. 5.8 Two views of Irano-Touranian forest (left) Kerman Province southeast of Iran and (right) Semnan Province in central part

Table 5.12 Dominant plant species in Mountainous Irano-Touranian zone

Mountainous Irano-Touranian zone	
Scientific name	English name
Juniperus excelsa	*Greek juniper*
Pistacia atlantica	*Mountain Atlas pistache*
Amygdalus Scoparia	*mountain almond*
A.arabica,	*almond*
Acer monspessulanum	*Montpellier maple*
Amygdalus spp.	*almond*
Celtis caucasica	*caucasian hackberry*
Berberis integerrima	*berberry*
Cotoneaster spp.	*cotoneaster /quince-leaved medlar*
Lonicera nummularifolia,	*honey suckle*

Table 5.13 Types of soils in the Irano-Touranian zone

Vegetation cover (sub-) zones	Types of soils
Irano-Touranian (plain) areas	Ardisols (Calcids, Gypsids, and Salids), Entisols (Torriorthents, Ustorthents), and Inceptisols(Calcixerpts and Calciustepts)
Irano-Touranian (mountainous) areas	Xerepts (Calcixerepts, Haploxerepts and Calciustepts), Entisols (Xerorthents and Ustorthents)

Fig. 5.9 Two views of vegetation of Khalij-e Fars-Omanian zone

5.5 Rangeland

According to the density of vegetation, rangeland is classified in three main classes as defined in Table 5.17. General properties and classification of the soils developed in the arid and semiarid regions of the country are reported in Chap. 7. The soils are mostly calcareous with an ochric epipedon and a calcic, gypsic, or cambic subsurface horizon that are generally classified in three orders of Entisol, Aridisol and Inceptisol (Figs. 5.11 and 5.12).

Table 5.18 shows the important rangeland's vegetation types.

5.6 DesertLand

Deserts classified into six classes according to the FAO/UNEP standard definitions. These areas are usually barren with no vegetation, except for some parts which are under *Haloxylon spp.*, *Tamarix spp.*, and other dry species artificially planted (Fig. 5.13 and Table 5.19).

Fig. 5.10 Mangrove Forest is a unique vegetation in Iran and is located in Khalij Fars-Omanian zone (left & right) (Courtesy by FRWO)

Table 5.14 Dominant plant species in the Khalij-e Fars-Omanian

Khalij-e Fars- Omanian zone	
Scientific name	English name
Ziziphus Spina-christii,	*christ thorn/nubk tree*
Acacia tortilis	*Acacia/binding bean tree/Egyptian thorn/wattle*
A.ehrenbergiana	"
A.oerfota	"
Salvadora persica	*kikuel oil plant/Mustard-seed/tooth brush tree*
Prosopis cineraria	*Cashew/Mesquite bean*
Tamarix aphylla	*athel tamarisk/desert athel*
Dalbergi asisso	*sissoo/sissu/rose wood black wood*
Dodonea viscosa	*clammy hop seed brush /switch sorrel*
Euphorbia larica	*euphorbia*
Capparis decidua	*caper berry/kureel tree/timbuktu caper brush*

Table 5.15 Dominant Soils in Khalij-e Fars-Omanian zone

Vegetation zone	Soils
Khalij-e Fars- Omanian (Southern coastal area)	Entisols (Torriorthents and, Ustorthents, Aridisols (Calcids, Gypsids, and Salids), Inceptisols (Calciustepts and Haplustepts)

Table 5.16 Total forest and shrub land areas in 2005(FRWO)

Description	Area/ha
Forest with more than 50% canopy cover	755,777
Forest with 25–50% canopy cover	2,806,846
Forest with 5–25% canopy cover	7,842,183
Mangrove forest	30,400
Shrublands with more than 10% Canopy cover	2,552,682
Plantation forest	919,468
Total forest and shrublands	14,907,355

Table 5.17 Total area of the rangeland as reported in the land cover map in 2005 (FRWO)

Legend	Description	Area/ha
R1	Rangeland with more than 50% canopy cover	6,345,923
R2	Rangeland with 25–50% canopy cover	20,694,347
R3	Rangeland with 5–25% canopy cover	56,095,560
	Total range land	83,099,831

Fig. 5.11 Dense vegetation of rangeland in south of Damavand Mountain (Courtesy by FRWO)

Fig. 5.12 Dense rangeland in southeast of Damavand Mountain (Courtesy by FRWO)

Table 5.18 The types of rangeland vegetation

Rangeland vegetation	
Scientific name	English name
Acanthophyllum spp.	*Acanthophyllum*
Acantholimon spp.	*prickly thrift*
Astragalus spp.	*locoweed, milk vetch, astragale, crow-toe*
Perennial & Annual grasses	*Perennial& Annual grasses*
Artemisia spp.	*Sage brush, wormwood, sage*

Fig. 5.13 Plantation in desert area (Courtesy by FRWO)

Table 5.19 Total desert areas (FRWO 2005)

Description	Area/hectares (ha)
Kavir (lowlands without canopy cover	5,839,891
Different type of sand dune	1,762,538
Smooth sand surfaces	613,684
Smooth clay surfaces in margins of Kavir	463,175
Salty lands	6,558,460
Rangelands with less than 5% canopy cover and out crop	17,368,883
Total desertland	32,579,631

5.7 Other Classes

Other classes like agricultural land, lakes, water bodies, cities, villages, and infrastructure, though reasonable in size, are not covered in this text. Existence of various cultures and ethnicities, varied climatic conditions, and long history of agriculture has contributed to the richness and uniqueness of biodiversity in the country.

5.8 Conclusion

Iran consists of 167 families of vascular plants and 1215 genera, some of them include only one species and some of them about 800 species (*Astragalus*). Due to the diversity of climate, topography, and edaphic conditions, limited areas are covered with vegetation, but in a very different and heterogeneous condition. Approximately, 8000 plant species of 167 families and 1200 genera have been recorded in Iran, nearly 20% of these species are considered endemic.

There are 12.4 million hectares of woodland and more than 10,000 ha of mangroves along the Persian Gulf and the Sea of Oman coast. Ecosystem diversity of marine and coastal zones in the north and the south of the country consists of 25 ecological types and units, most importantly coral reefs, bays, and small islands. Because of its large size and varied ecosystems, Iran is one of the most important countries for conservation of biological diversity in the Middle East and West Asia. Iran's topography has given rise to four floristic zones, namely Irano-Touranian arid and semiarid deserts and Irano-Touranian semiarid mountains, Zagrosian, Hyrcanian, and Khalij-e Fars-Omanian. Most of Iran's forest covers are located in Hyrcanian, Arasbaranian, and Zagrosian zones. The most important classes in land cover map are forest with about 14,907,355 ha; rangeland of about 83,099,830 ha; and desertland of about 32,579,630 ha.

References

Department of Environment, Islamic Republic of Iran (2010) Fourth national report to the convention on biological diversity, Oct 2010

Di Gregorio A (2000) Land covers classification system (LCCS) classification and user manual, FAO Rome

Fisher WB (1968) The Cambridge History of Iran. Vol. 1 The land of Iran. Cambridge at the University Press

FRWO—Forest, Range and Watershed Organization (2005) Report of land cover map of Iran. Engineering Bureau, Tehran, Iran

Ghahreman A, Attar F (2000) Biodiversity of plant species in Iran, Vol. 1. University of Tehran, Iran

Javanshir K (1975) Iran's wooden vegetation atlas. Protection of natural resources society

Jenny H (1941) Factor of soil formation a system of quantitative pedology. Dover Publication, INC, New York

Mobayen M, Tregubov (1980) Iran's vegetation map 1:2,500,000 Tehran University, Iran

Mozafarian V (2004) Kimaye sabz. FRWO's Publication, Tehran, Iran

Sabeti H (1969) Les Etudes Bioclimatique de l'Iran University of Tehran publication No. 1231

Zohari M (1963) Iran's geographic vegetation. Iran environment protection organization Tehran. Iran

Author Biographies

Dr. Ali Farzaneh is associated professor at Forest, Range and Watershed Organization (FRWO), senior expert of Remote Sensing/GIS, official expert of Agricultural and Natural Resources Engineering Organization (ANREO), general director of Land Evaluation Engineering Co. (LEE CO.), Tehran, Iran.

Morteza Ebrahimi Rastaghi is a senior expert of Forest, Range and Watershed Organization (FRWO), member of High Consultant of Forest, Range and Watershed Organization (FRWO). He is also a member of executive committee of International Beech and a member of scientific committee of international convention of LFCC, Tehran, Iran.

Soil-Forming Factors and Processes

6

Farhad Khormali and Norair Toomanian

Abstract

Soil-forming processes and the formation of diagnostic horizons in different regions of Iran are mainly influenced by the overall semiarid and arid climate and high calcareous conditions of the soils. Calcium carbonate redistribution, gypsum accumulation, soil salinization, and alkalization are therefore the major processes in these climatic and soil conditions. Clay migration is also a possible process occurring in these highly calcareous conditions either through the decalcification of the upper sola and the subsequent clay migration as a commonly accepted model or by the effect of previous high exchangeable Na in soils. Clay-accumulated horizons of the extreme arid regions could be regarded, as paleo-features. Calcitic features are mainly associated with the soil water availability and the vegetation. These features vary from pendants in the skeletal soils, nodules, cemented petrocalcic, and mycelium mainly in the arid zones. In areas with higher vegetative growth and favorable precipitation and temperature, needle-shaped calcite and cytomorphic type, which are associated with the biological activity, are also observed. Formation of gypsum accumulation horizons is the result of dissolution of gypsum and formation of gypsic and hypergypsic horizons. Two sources are mainly responsible for the soils high in gypsum which are geologic formations outcropped in some areas and saline lakes high in sulfate. Salt-affected soils are widespread in Iran. The evolutionary sequence suggested for these soils is as follows: salinization and alkalization; desalinization; solonetzation; and dealkalization. Vertic features, organic matter accumulation, gleization, etc., are among other processes happening in different climate zones.

Keywords

Soil formation • Arid and Semi-arid region • Carbonates • Salinization • Gypsum

6.1 Introduction

Climate, topography, parent material, organism, and time as soil-forming factors determine the dominant soil-forming processes and consequently the type of soils in a region. Considering the fact that climate of Iran is dominantly of arid and semiarid type, and the parent material is mainly containing high calcium carbonate, the major soil-forming processes are closely related to different climatic zones of Iran.

Arid zones: Areas with arid zones include mainly the central and eastern arid zones, coastal Persian Gulf, Oman Sea, and eastern Caspian Sea regions. Due to the extreme aridity, soluble salt and gypsum accumulation are dominant in the soils of these regions. Salt-affected soils are distributed throughout arid to semiarid regions of Iran (see Chap. 7). They occupy about 15% of the country's surface area, which is almost equal to the total arable lands. These soils are formed in nearly level to level surfaces including depressions under the

F. Khormali (✉)
Gorgan University of Agricultural Sciences and Natural Resources, Gorgan, Iran
e-mail: fkhormali@gau.ac.ir

N. Toomanian
Soil and Water Research Department, Isfahan Agricultural and Natural Resources Research and Education Center, AREEO, Isfahan, Iran
e-mail: norairtoomanian@gmail.com

M.H. Roozitalab et al. (eds.), *The Soils of Iran*, World Soils Book Series,
https://doi.org/10.1007/978-3-319-69048-3_6

influence of a fluctuating ground water table (Mahjoori 1975). The origin of salts in many regions is related to the parent materials that formed from saline and gypsiferous marl deposits and carried by water and redeposited. Among the countries in arid and semiarid regions, Iran seems to have the largest area of gypsiferous soils in the world (see Chap. 7), covering about 27 to 28 mha (Mahmoodi 1994; Roozitalab 1994). Saline soils comprise around 16 to 23 mha of arable lands in Iran (Siadat et al. 1997).

The major clay minerals in these arid saline conditions are palygorskite, smectite, chlorite, illite, and kaolinite (see Chap. 10). Chlorite and illite (micaceous minerals) are commonly believed to be inherited largely from parent rocks (Wilson 1999; Mahjoory 1975). Environmental conditions are suggested to lead to smectite formation in Aridisols. Smectite in arid soils has been noted in Iran (Abtahi 1977). Neoformation of smectite was reported by Gharaee and Mahjoory (1984) and by Givi and Abtahi (1985) under saline and alkaline conditions with high concentrations of Si, Mg, and Al in southern Iran. Very few occurrences of vermiculite have been reported in soils of arid and semiarid regions (Abtahi 1977, 1985; Khademi and Mermut 1998; Khormali and Abtahi 2001).

Most of the gypsum-enriched soils in Esfahan and neighboring Provinces occur on alluvial fans, dissected floodplains (old dissected alluvium), and piedmonts (Toomanian et al. 2001). The physical environment might play a great role in the formation of different gypsic pedofeatures. The form of gypsum crystallization changes intricately in space and is often ephemeral in time, making it difficult to draw firm conclusions linking crystal shape to particular environments or soil types (Jafarzadeh and Burnham 1992). Stable isotope geochemistry of gypsum deposits in central and south central regions provided useful information on the major source of gypsum, which is believed to be of evaporate associated with lower Cretaceous marine sulfate, which shows how far the shallow water bodies of post-Tethyan era extended in central and southern Iran (Khademi and Mermut 1998).

Coastal plains of Persian Gulf and Oman Sea also show aridic soil moisture regime (SMR) and mostly of hyperthermic temperature regime (see Chap. 7). Aridisols and Entisols are dominant, and their properties are closely associated with the parent material (Abbaslou et al. 2013).

The south-eastern Caspian Sea region is characterized by a pronounced precipitation gradient of about 800 mm year^{-1} over about 80 km distance from north (arid) to south (humid). In the lowlands, salt-affected soils are dominant. The main factor responsible for the formation of large area affected by salinity and sodicity is the low-lying topography and arid climate. The soils are mainly classified as Aquisalids with simultaneous salinization and solonetzation processes. Gleization and formation of iron/manganese oxides are also visible in the soil profile.

Semiarid zones: Mediterranean type climate or xeric soil moisture regime (SMR) comprises the semiarid regions of Zagros and Alborz Mountains and valleys and major south and south-eastern parts of the Caspian Sea lowlands. These areas are characterized by cold and rainy winters and hot dry summers. The Zagros topographic features extend from northwest of Iran to the southeast. The entire area shows mountains and intermountain valley regions. The calcareous parent rocks are dominant bedrocks, and therefore, calcium carbonate dissolution/precipitation plays a major role in the formation of the soils and the processes dominant in soils. Formation of secondary calcium carbonate, decalcification and following clay illuviation, and accumulation of gypsum are the major processes in these regions (see Chap. 7). Combined intermountain valley topography and the alternate dry and moist soils help develop pedoturbation features and the vertic properties. The Vertisols are typical for the western Iran, and the present studies have confirmed that the formation of these soils is closely related to the accumulation of clay and the occurrence of smectite (Heidari et al. 2005a).

Major parts of Khuzestan Plain and southern parts of Zagros Mountain Ranges are comprised of very hot semiarid climate (ustic SMR).The soils of Khuzestan Province, a primary crop production area in southwestern Iran, are extremely carbonated (see Chap. 7), with more than 40 percent carbonate content in most parts. The carbonatic nature of the parent material, in addition to the low precipitation of the region, has enhanced accumulation of secondary lime minerals in the soil mineralogy control section so that the carbonatic or mixed (calcareous) mineralogy classes are common in the region. Soils are mainly classified as orders of the Entisols and Inceptisols. The young (recently formed) Entisols of the region belong to Fluvents of the Holocene age. Redoximorphic features (brown mottling in a pale green soil matrix) in Fluvaquents represent hydric conditions (seasonally saturated) in young, flood plain soils.

In the Caspian Sea regions, the extensive loess deposits in the area reach a thickness of about 70 m in the loess plateau of Golestan Province. Organic matter accumulation and the formation of mollic epipedon are the dominant processes occur in the steppe (rainfall of 400–600 mm). Calcification and formation of calcic horizons are typical for loess-derived soils. In the north-facing slopes of the Alborz mountain ranges with the forest vegetation and higher rainfall, deeper leaching of calcite and formation of the argillic horizon are dominant. Soils are mainly Alfisols (see Chap. 7). The loess deposits of the study area contain several paleosols showing different degrees of weathering probably related of paleo-climatic conditions and duration of soil development (see Chap. 8). As based on pedostratigraphy and luminescence dating, the paleosols have been correlated with interglacial and interstadial phases of the Middle to Upper Quaternary (Frechen et al. 2009).

South and southwestern Caspian Sea areas are composed of lowlands and floodplains in the north adjacent to the Caspian Sea and piedmonts, hills, and mountains. The precipitation increases from east (Gorgan 600 mm) to west (Anzali 1500 mm). Aquic conditions in soils of lowlands and floodplains are often associated with redoximorphic features. The main soil types are Haplaquepts (Gleysols, WRB 2014), and the soils are showing either episaturation or endosaturation. Aquic moisture regime is associated here with the seawater intrusion resulting in the formation of saline and alkaline soils. The soils are mainly Aquisalids. In the mountainous and hilly areas, the Entisols and Alfisols are dominant. Gleization, salinization, and clay illuviation are the major processes in these regions.

Humid and sub-humid zones: Highlands of Alborz Mountain Ranges in the north-facing slopes are comprised of udic soil moisture regime up to about 1500 m asl. Role of climate is far important than the other soil-forming factors leading to intensive leaching of carbonates and in parts soil acidity. Emadi et al. (2012) studied the soils of Mazandaran Province along an east-west transect in xeric, ustic, and udic regions. Soils of the ustic and udic soil moisture regime were classified as Calcic Haplustalfs and Mollic Halpudalfs (Soil Survey Staff 2014) or Calcic Luvisol and Haplic Luvisol (WRB 2014), respectively, indicating the conditions were favorable for the downward decalcification and the subsequent clay illuviation and formation of moderately to strongly developed argillic horizons. Gleization is another main feature occurring in soils of udic regions mainly due to paddy rice cultivation.

6.2 Soil-Forming Processes

Soil genesis deals with the factors and processes of soil formation. The formation of soil is the result of the interaction of five soil-forming factors: parent material, climate, organisms, topographic position or slope, and time. As mentioned earlier, soil-forming processes are mainly directed by the dominant arid and semiarid climate and highly calcareous conditions. The dominant soil-forming processes are as follows: (1) translocation of silicate clays, (2) calcium carbonate redistribution, (3) gypsum accumulation, (4) salinization, (5) hydromorphism, (6) accumulation and humification of organic matter, and (7) swelling and shrinkage. The net effect of these processes is the development of soil horizons, that is, the genesis of soil.

6.2.1 Translocation of Silicate Clays

Mechanical migration of clay particles down the pores and their subsequent illuviation in the deeper horizons is the commonly accepted major process in the formation of Bt or argillic horizons. Considering the soil-forming factors, majority of the soils in Iran are highly calcareous formed mainly from limestone parent rocks (see Chap. 10). This includes aridic, xeric, and ustic SMR regions of almost all ecological zones (see Chap. 8). In humid regions of northern Iran, the leaching is more intense. Argillic horizons are of different development stages.

In order to quantitatively compare the different Bt horizons, a Micromorphological Index of Soil Evolution in Highly Calcareous Arid to Semiarid Conditions (MISECA) was developed (Khormali et al. 2003). Some main micromorphological features such as clay coating, decalcified zones, Fe/Mn oxides, microstructure, etc., are considered as factors of soil development in MISECA index. Table 6.1 shows classes of degree of soil development according to MISECA. (Khormali et al. 2003). According to the calculated MISECA values, the argillic horizons of the soils in the studied Fars Province differ in their degree of development. Soil-forming factors in this province are very variable, and therefore, the findings could be generalized to other parts of Iran since they are mainly derived from limestone.

Occurrence of clay-enriched horizons or Bt (argillic) has been studied in semiarid and arid regions of Iran (Khademi and Mermut 2003; Khormali et al. 2003). Khormali et al. (2003) studied the ten representative pedons with argillic horizons in Fars Province along with different climates and precipitations gradient of 200 mm to 600 mm. All the studied soils were formed in highly calcareous parent material. Based on the genesis and degree of development three types of argillic horizons could be distinguished:

- argillic horizons formed on older landscape units, showing few to highly developed calcite depletion pedofeatures in the groundmass (MISECA = 9–19) (semiarid to regions, precipitation of 400–600 mm)
- argillic horizons with little or no calcite depletion pedofeatures in soils that previously had a high Na content (MISECA = 7 and 9) (semiarid regions, precipitation of about 350 mm)
- argillic horizons with little or no calcite depletion pedofeatures formed in the past wetter conditions (MISECA = 7). (Arid regions, present precipitation of 200–300 mm).

6.2.1.1 Clay Illuviation in Highly Calcareous Conditions

Evidences show that these soils have experienced the decalcification of the upper horizons and subsequent clay migration from topsoil and accumulation in the subsoil. Micromorphology can be a good evidence by the fact that the upper layers and Bt horizons show striated and speckled b-fabric, confirming the decalcification of the

Table 6.1 Classes of degree of soil development according to the MISECA values, based on field and laboratory analyses

MISECA	Degree of soil development
0–8	Weakly developed
9–16	Moderately developed
17–24	Well developed

topsoil while in the Btk horizons combination of the calcitic crystallitic and speckled b-fabric are visible, indicating the simultaneous recalcification and partial dissolution of carbonates. In the Bk horizons, large accumulation of the carbonate as pedogenic features and presence of the matrix carbonate have resulted in the dominance of calcitic crystallitic b-fabric(Figs. 6.1 and 6.2).

In the Btk horizon, coatings of secondary carbonate along channels and other voids have been subsequently covered by coatings of strongly oriented reddish yellow clay, indicating that decalcification of the upper layers (Ap and Bt) and downward movement of the carbonates occurred prior to the formation of clay coatings. The presence of some calcite coatings overlying clay coatings suggests a later recarbonation followed by movement of carbonate (Fig. 6.2).

Most micromorphological studies of argillic horizons in arid, semiarid, and Mediterranean regions have shown few to no clay coatings present. This could be related to the strong shrink/swell properties of soils with large clay contents and periodic moist and dry phases (Kemp and Zarate 2000).

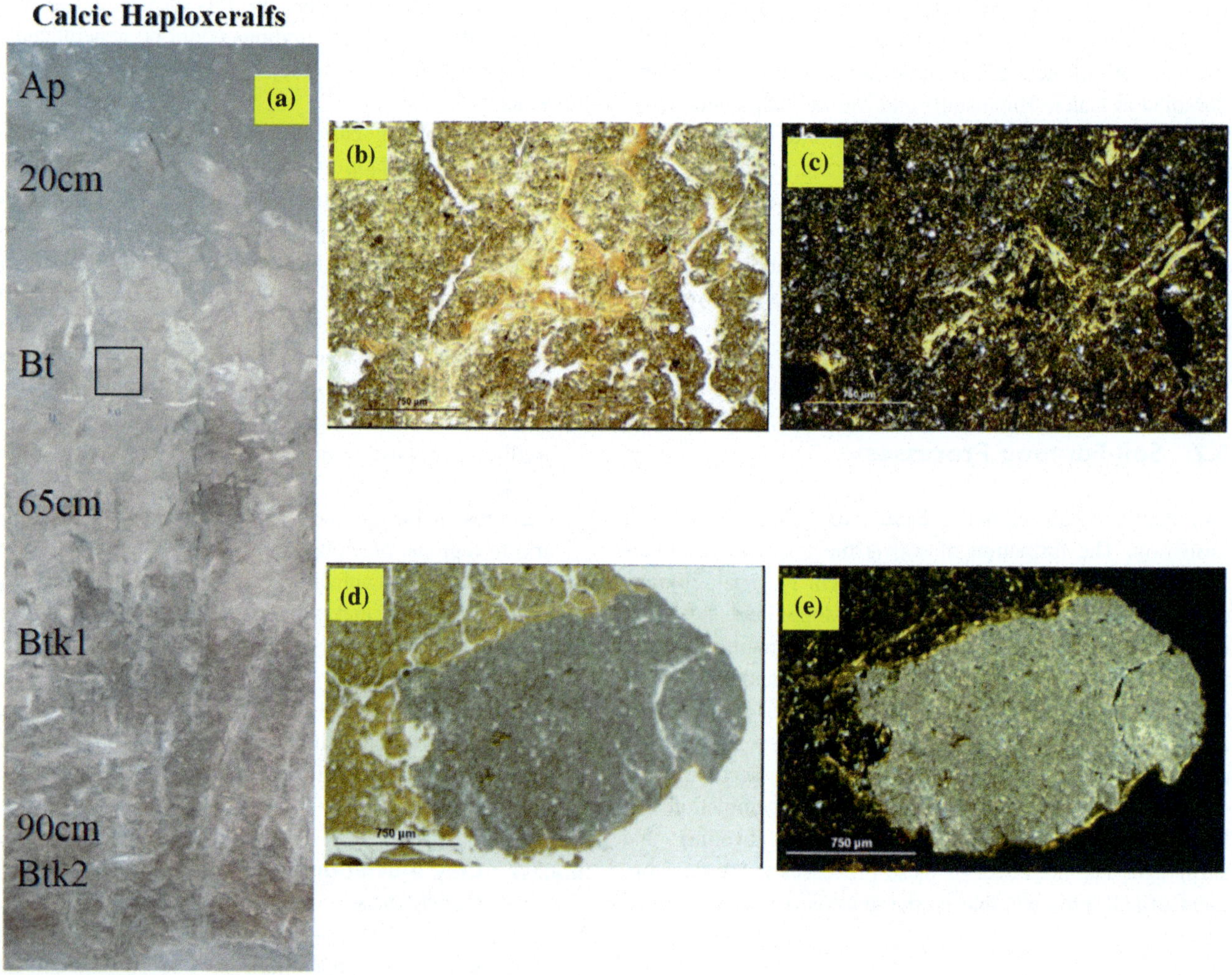

Fig. 6.1 **a** A Calcic Haploxeralfs in Fars Province; (**b** and **c**): Plain polarized light (PPL) and Crossed-polarized light (XPL) images showing speckled b-fabric and clay coatings in Bt horizon; (**d** and **e**): PPL and XPL of a typic calcite nodule coated by illuvial clay in the Btk horizon (Khormali 2003)

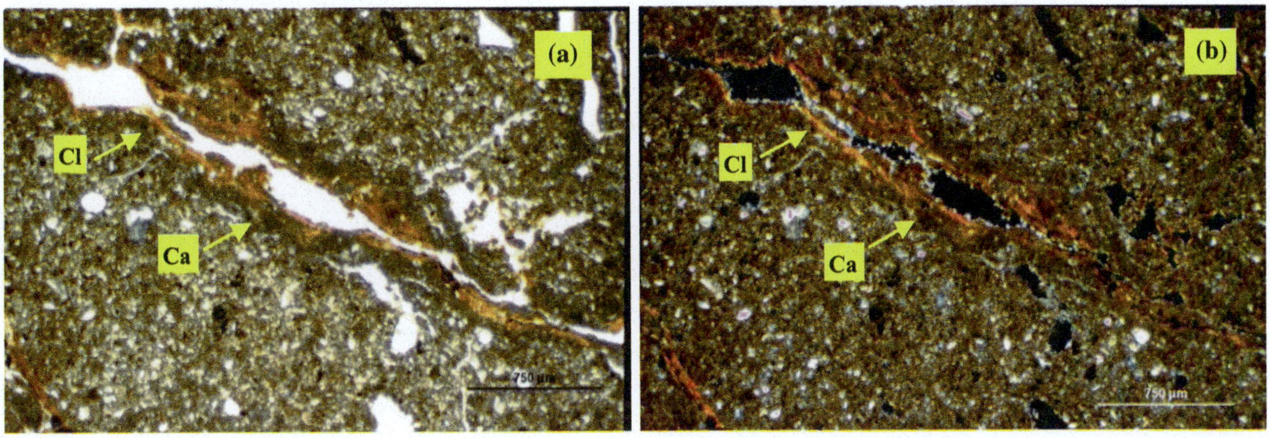

Fig. 6.2 Clay coatings (Cl) invaded by calcite (Ca) as a result of recalcification (Calcic Haploxeralfs, Btk2). **a** PPL and **b** XPL (Khormali2003)

In contrast to the high calcareous soils of major parts of Iran, argillic horizons of the northern Iran and specifically those of loess origin are less calcified. Clay coatings are best expressed in soils formed in humid regions showing the speckled b-fabric. From the semiarid zones to the humid regions, the thickness of the clay coating and the area covered by clay coatings increased. Thicker oriented clay coatings (50–300 µm) were associated with relatively higher clay content and higher relative abundance of the vermiculite. The dominance of vermiculite and illite in humid region soils (as opposed to smectite) limited the shrink/swell potential and thus helped reduce the disruption of clay coatings (Khormali et al. 2012). The type of the clay minerals present also affected the thickness and orientation of clay skins. Gunal and Ransom (2006) reported similar results in different soils of the precipitation gradient of 540–715 mm. They showed thicker clay coatings were related to the more humid regions.

Smectite constitutes the major portion of the clay minerals in well-drained Alfisols, somewhat poorly drained Mollisols, and Calcixerepts with high precipitation. It is detected in trace amounts in the soils of more arid areas. Therefore, it can be concluded that in well-drained soils, with increasing soil-available moisture, smectite increases. Increase in soil available moisture, and consequently a relatively leaching environment for the release of K^+ from micaceous minerals and mainly illite, in the calcareous environment with high Mg^{++} and high Si mobility, might provide favorable conditions for the formation of smectite through transformation (Khormali and Abtahi 2003).

The different degrees of development of the argillic horizons are greatly dependent upon the earlier degree of decalcification, and this in turn reflects the effects of soil-forming factors, especially the climate.

6.2.1.2 Clay Illuviation in Previous High Na Condition

According to Abtahi (1977), formation of argillic horizons in arid saline and alkaline environments of Lakes Shiraz and Neyriz is possibly related to the effect of past high sodium content of the lakes. These soils have undergone the following processes:

Salinization–Alkalization—Desalinization—Solonetzation—Steppification

Lakes Neyriz and Shiraz (Lake Maharlu), two of the interior Zagros mountain lakes with saline and alkaline water, are probably remnants of the post-Tethyan sea in Fars Province (Fig. 6.3). Lake Shiraz (Maharloo) contains 5200 ppm of Na and 5320 ppm of Cl (Krinsley 1970). After their maximum development about 20,000 BP, the lakes started drying up because of increasing temperature and evaporation, producing salt-affected soils. According to Abtahi (1977), the formation of argillic horizons in soils near these lakes could be related to clay dispersion resulting from the high sodium content of this saline and alkaline environment.

In such an environment, clay can migrate to lower horizons even in the presence of abundant calcium carbonate. Very few calcite depletion pedofeatures are seen as small areas with speckled b-fabric in the thin sections of Btn horizons and only fragments of clay coatings occur (Fig. 6.4). The lower limpidity of these clay coatings, which is commonly observed in natric horizons, indicates that they contain coarser clay than the coatings in other pedons. The MISECA score is 7 and 9 for these Btn horizons, suggesting weak to moderate development (Khormali et al. 2003).

6.2.1.3 Paleo-Bt Horizons in Arid Regions

According to Khormali et al. (2003), argillic horizons were also identified in the present arid climate. In Ghatrouyeh

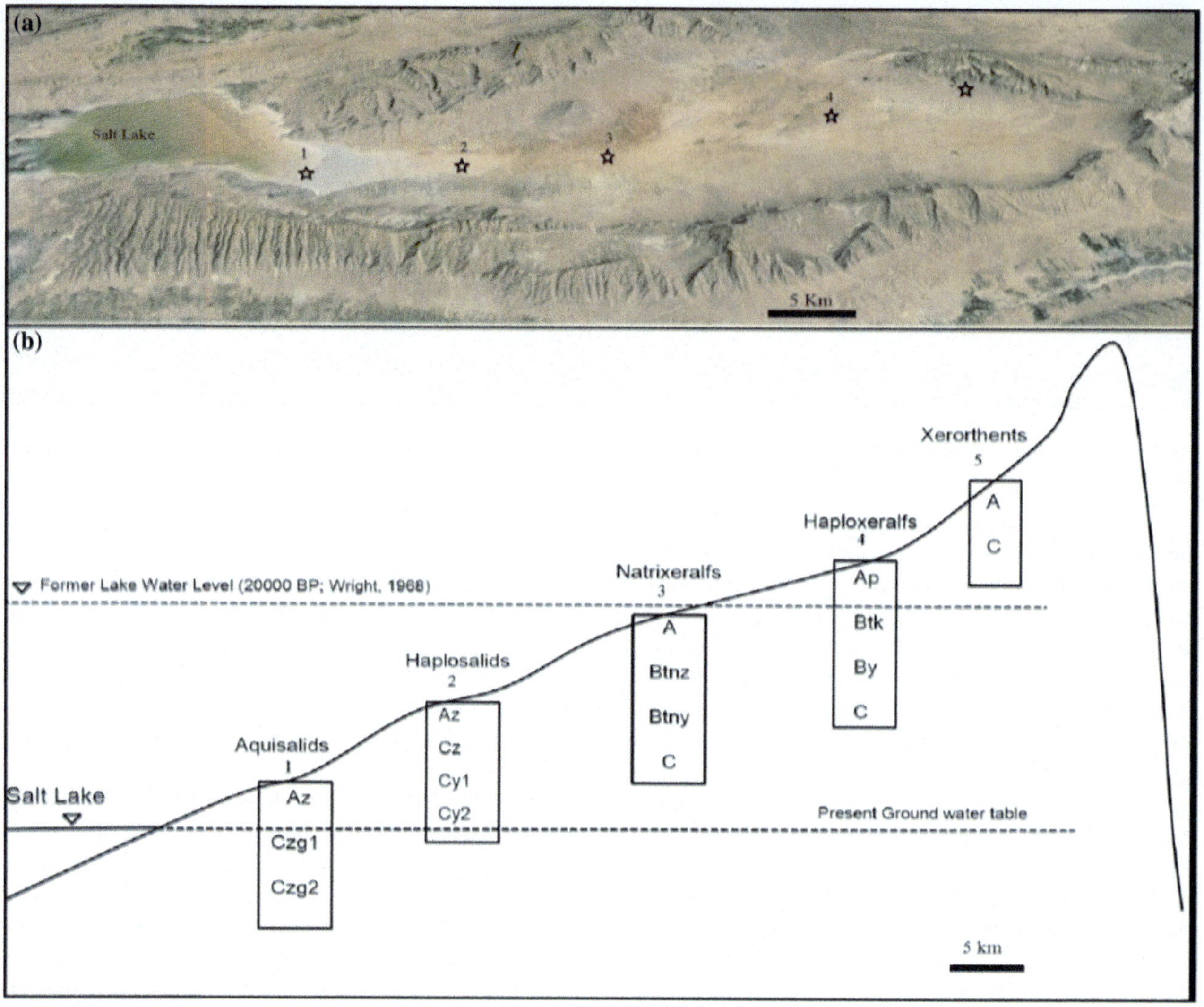

Fig. 6.3 Soil formation along a catena in Sarvestan plain of Fars Province. The sequence of the soils is representing the former saline lake retreatment (Khormali 2003)

plain of Neiriz, Fars Province, a moderately high degree of decalcification is indicated by the presence of many calcite depletion pedofeatures, but the present rainfall is about 200 mm, suggesting that in the past the climate must have been less arid (Fig. 6.5).

Preservation of the argillic horizon within these soils could have been favored by the presence of coarse limestone and other rock fragments, which would act as a stable skeleton allowing preservation of clay coatings under shrink-swell conditions. Argillic horizons seem to form much easier in a coarser than in finer textured soils. MISECA for these Bt horizons is about 7, indicating a lower degree of development than other pedons. The pebble surfaces serve as sites for silicate clay and carbonate accumulations (Gile and Grossman 1968). Several factors may contribute to the stability of pebble surfaces. In contrast to the surfaces of peds of fine earth that move inward and outward as the ped expands and contracts, the surface of pebbles remains stationary unless the pebble is displaced. Volume change of the soil on wetting should be reduced due to less fine earth per unit volume. Furthermore, the fine earth in each interstice between pebbles tends to act as a discrete unit.

Khademi and Mermut (2003) investigated the formation of argillic horizons in central Iran, Isfahan region with aridic moisture regime. They found argillic horizon as the most distinct subsurface horizon of the soils developed on colluvial fans. Evidence of illuviation is provided by the increase in the clay content and the fine to total clay ratio in the subsoil compared to the overlying horizon(s) and by the well-developed, but considerably disrupted, clay coatings observed in thin sections. Since swelling and shrinkage do not seem to be effective, the disruption of clay coatings was likely due to the gypsum and carbonate crystallization process.

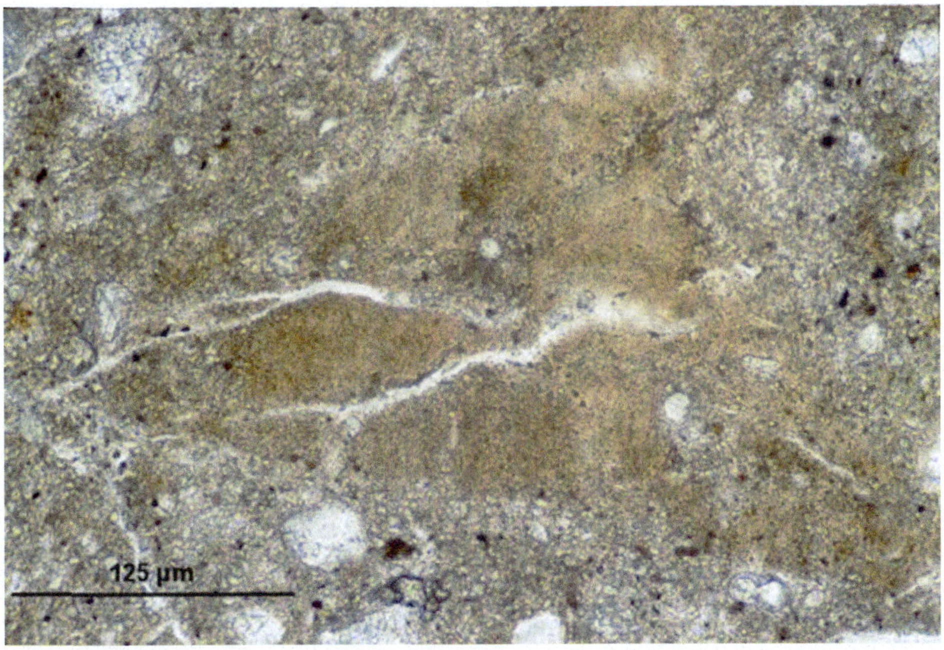

Fig. 6.4 PPL image of the Btn horizon showing speckled clay coatings typical for natric horizons (Khormali 2003)

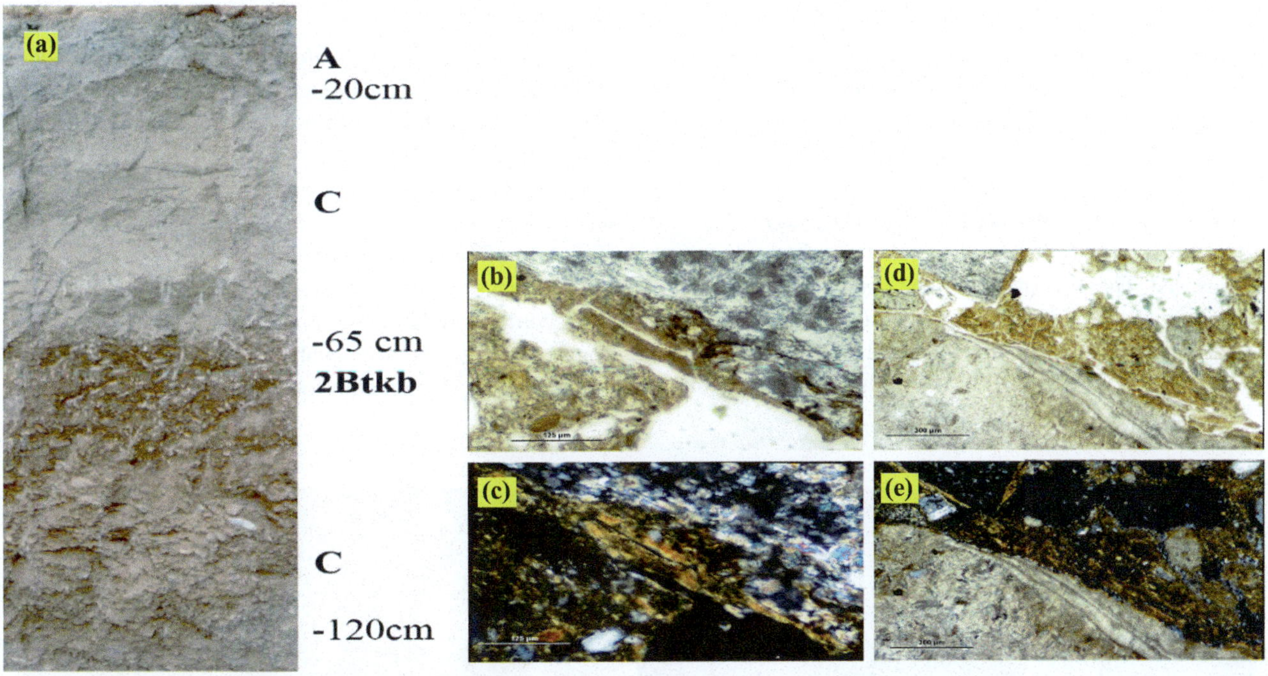

Fig. 6.5 **a** A Typic Torriorthents underlying a buried argillic horizon in the arid region of Fars Province. PPL and XPL micromorphology images show the preservation of clay coatings under (**b** and **c**) and surrounding limestone fragments (**d** and **e**) (Khormali 2003)

The coexistence of argillic, calcic, and gypsic horizons in colluvial soils is a peculiar combination, suggesting a multistage pedogenesis in this landscape. Paleo-argillic horizons were likely developed under a moister environment than today. Sufficient rainfall contributed to the removal of carbonates from the topsoil and the subsequent eluviation of clay to form the argillic horizon. Addition of more carbonates from a colluvial source or as aerosols and the gypsum deposition occurred continuously with time, and these processes slowed down considerably since the area became arid.

Based on geological and biological evidence, Moatamed (1988) has suggested a more humid climate for Iran in the early to middle Holocene. Mahmudi (1987) believes that, during the period of glaciation, central and southern parts of Iran were experiencing a climate with more rainfall than today, whereas in interglaciation periods, the climatic conditions were rather the same as today. The stable isotope geochemistry of the gypsum hydration water from the soils showed the preservation of isotopically lighter water which is an indication of an environment with more precipitation (Khademi et al. 1997). The paleo-argillic horizon was likely formed during a period of more moisture. During the transition from the moister climate to today's arid conditions, gypsum accumulation likely occurred with an isotopic signature of the paleo-environment. The present moisture regime is not sufficiently moist to move clays down in the profile.

6.2.2 Calcium Carbonate Redistribution

Soil carbonate is an important soil component because it serves as an indicator of landscape stability and paleoclimate, it affects nutrient availability, and it has implications for carbon sequestration (Gile et al. 1966; Cerling 1984; Monger et al. 2011). Without invoking biomineralization processes, many authors have traditionally, and successfully, modeled the precipitation and dissolution of calcium carbonate in soil using the classical equation below:

$$Ca^{2+} + 2HCO_3^- \rightarrow CaCO_3 + CO_2 + H_2O$$

6.2.2.1 Calcitic Features in Arid Regions

The soils in arid and semiarid regions of Iran are mainly calcareous. The presence and the mode of formation of secondary carbonates were established by the carbon and oxygen stable isotopes (Khademi and Mermut 1999). The large amount of carbonates in the surface horizons is due to addition of carbonatic debris to the surface horizon by colluviation and aeolian processes. Below the A horizon, secondary carbonates occur as soft masses and nodules. In Btky horizons, coexistence of secondary carbonates and gypsum either as pendants below the pebbles or as crystals filling the soil pores are the major macromorphological features observed within the very gravelly soil matrix.

Several lines of evidence suggest that carbonate accumulation in the Btky horizons occurred both before and after clay accumulation. These include the presence of clay skins on secondary carbonate crystals observed in micromorphological studies, the presence of both calcic and argillic horizon at depths >1 m, and the higher amount of primary calcium carbonate in the topsoil and secondary carbonates in the subsoil as proven by the stable isotope approach (Khademi and Mermut 1999). The dissolution of carbonate from the topsoil and its precipitation in the lower horizons created favorable conditions for the removal of the clay particles from the topsoil and flocculation of these particles in the zone, where carbonate accumulated. The lower content of nodules in the more arid areas can be explained by a lower precipitation slowing down the process of dissolution—recrystallization and therefore limiting the formation of nodules.

Depending upon soil texture and degree of soil development, a different development of the carbonate pendents is observed. It can only be a thin, single layer of micritic calcite underneath the gravel in the less developed soils or show an alternation of darker micrite and lighter sparite bands in the more developed and somewhat heavier textured soils (Fig. 6.6). The dark-brown bands may be clay and/or organic matter (Treadwell-Steitz and McFadden 2000).

According to Blank and Fosberg (1990), layered carbonate coatings around pebbles represent a stratigraphic sequence and are generally older than the late Pleistocene.

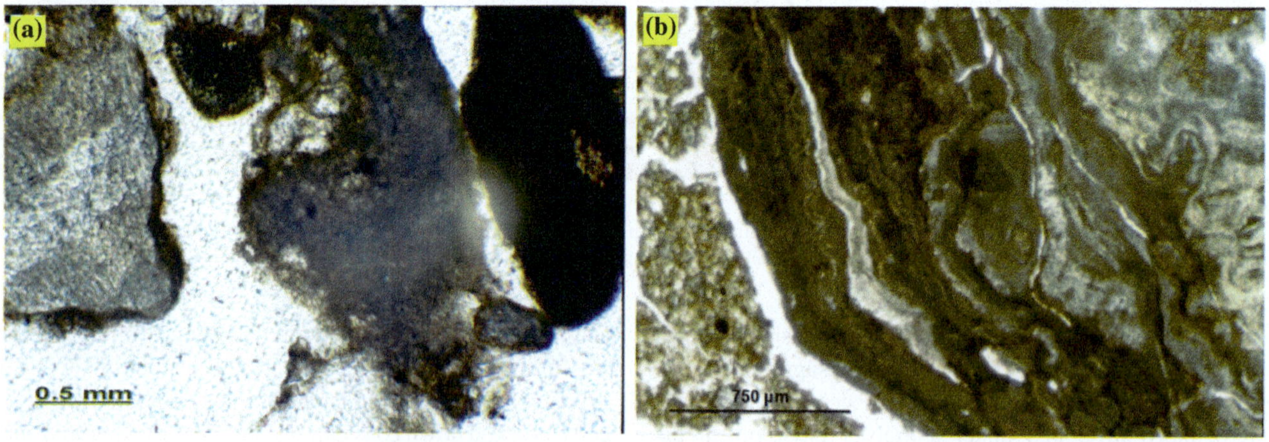

Fig. 6.6 **a** PPL images of the calcite coatings in a. less developed Torriorthents with only thin coating and **b** higher developed Haplocalcids showing many alternating bands of calcite accumulations (Khormali 2003)

Therefore, the presence of calcite pendants in these pardons can be considered as an indication of their formation during less arid climate of the past.

6.2.2.2 Calcitic features in semiarid regions

In contrast to the arid regions, the more favorable climatic conditions and denser vegetative growth in the semiarid regions could accelerate the process of dissolution–precipitation and recrystallization of calcite. The higher degree of dissolution of carbonates from the surface horizons and their precipitation in the deeper horizons are confirmed by the relatively high difference in the total $CaCO_3$ in the surface compared to the deeper horizons (Khormali et al. 2006), and the presence of only a few to common nodules in the near surface horizons compared to the large number of nodules in the deeper sola.

The calcite depletion pedofeatures, expressed partly by a speckled b-fabric, especially in the more developed Haploxeralfs are also an indication of higher dissolution and depletion of calcite in upper horizons (Khormali et al. 2003). The impregnation of some calcite nodules by Fe-oxide is the result of a higher degree of weathering, liberating iron from primary minerals. The presence of weathered limestone fragments in such soils confirms the higher degree of weathering (Fig. 6.7).

Owliaei (2012) studied the calcitic features of southwestern Iran and found that the degree of calcite impregnation with Fe/Mn oxides increases in areas with higher rainfall as a result of releasing Fe/Mn from primary minerals as well as drying and wetting cycles. The presence of pedogenic calcite coating superimposed on clay coatings suggests that decalcification of carbonates followed clay illuviation. Pendants of calcite are observed as a dominant calcitic pedofeature in the pedonsof more arid areas underneath coarser materials such as calcite nodules and small gravels. Cytomorphic and needle-like calcite were almost observed in areas with relatively higher rainfall and denser vegetative growth in the near surface horizons, confirming their biological origin.

Cytomorphic and needle-shaped calcite are the dominant features in the semiarid regions, rather than calcite nodules, pointing to different processes of accumulation, rather steered by biological factors than pure physico-chemical ones (Fig. 6.8). The occurrence of cytomorphic calcite in the studied pedons suggests the specific environmental conditions for its formation (Herrero et al. 1992): a relative high rainfall and favorable temperature resulting in denser vegetation as indicated by the formation of a mollic epipedon and a high biological activity. Needle-shaped calcite of fungal origin was found are also found beside cytomorphic calcite, in the near surface horizons with higher biologic activity (Khormali et al. 2006, 2014).

6.2.3 Gypsum Accumulation

Formation of secondary gypsum or gypsic horizon is mainly related to the process of recrystallization from previous soil gypsum and precipitation from evaporates in the presence of shallow ground water table. Two sources were responsible for the soils high in gypsum which are geologic formations, outcropped in some areas, and saline lakes high in sulfate. Morphology of gypsum was different related to the environment in which they have been formed. Pendants of gypsum were typical in the plateau soils with high stoniness (Fig. 6.9). Larger elongated crystals of gypsum were formed under the influence of saline ground water table, and small crystalline gypsum is characteristic for well-drained soils high in gypsum content in their C horizon. Gypsiferrous soils mainly occur in areas with P/ET° (ratio of precipitation to evapotranspiration) of less than 0.2, which corresponds, to more aridic climate with high evapotranspiration (Khormali 2003).

Hashemi et al. (2011) suggested that the amount of gypsum accumulation in their study area depends on soil

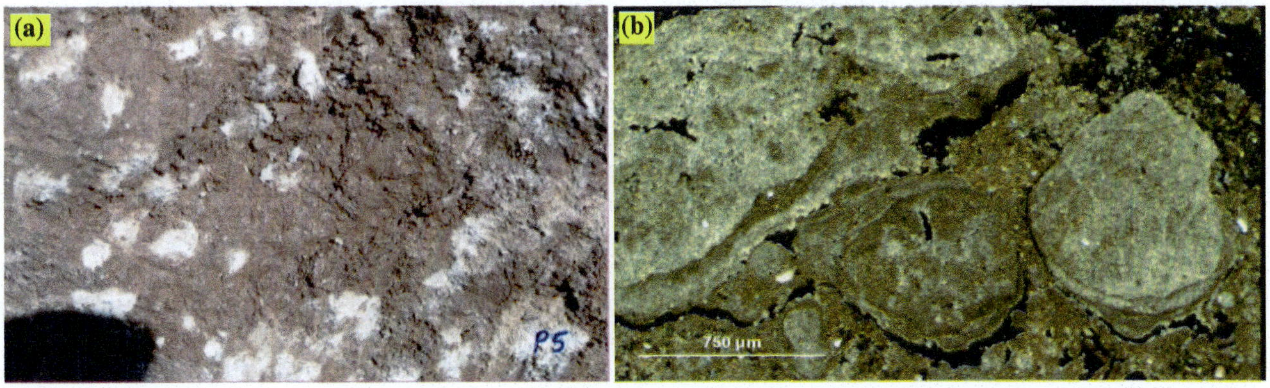

Fig. 6.7 **a** A Btk horizon of a Calcic Haploxeralf in northwestern Fars Province showing soft masses of calcite and **b** XPL images of calcite nodules in the Bk horizon of the same pedon (Khormali 2003)

Fig. 6.8 SEM image of needle-shaped calcite (left) and XPL image of cytomorphic calcite (right) (Khormali 2003)

Fig. 6.9 **a** XPL image of gypsum pendant (G) underlying the calcite pendant in a Calcic Argigypsids (**b** and **c**) Gypsum as mycelium and crystals in GypsicAquisalidsin Fars Province (Khormali, 2003)

moisture regimes rather than soil temperature regimes. They also concluded that micromorphology of gypsum crystals vary at different moisture regimes. Interlocked plates, acicular, fibrous,prismatic and blade forms in arid zones, and lenticular gypsum in semiarid zones are dominant; however, in the subsoil, clusters of lenticular, rod like, and tabular shapes are observed. Under hot and dry conditions, due to more evaporation and capillary rise, vertically arranged columnar, cubic, and needle shapes of gypsum were dominant. Toomanian et al. (2001) studied gypsum-enriched soils of central Iran and showed that the distribution, arrangement, and orientation patterns of the secondary gypsum crystals were largely depended to the physical environment. Soil fabric and microstructure, coarse to fine-related distribution pattern and voids play a considerable role in the formation of different gypsic pedofeatures.

With regard to the role of gypsum in the neoformation of clay minerals, Owliaei (2012) found close relationship between palygorskite morphology and depth of soil on gypsiferous landforms with minimum of erosion. Gypsiferous soils showed more pedogenic palygorskite compared to calcareous soils of southwestern Iran Karimi et al. (2009) provided further evidence which supports that the gypsiferous marls of hilly lands of eastern Iran contain palygorskite and the presence of this mineral was used as a proxy for origin of loess in that area. Palygorskite is considered to be inherited in plateau soils of the arid regions, whereas its occurrence in saline and alkaline soils and soils high in gypsum is mainly of authigenic origin.

Farpoor (2012) studied the soil–geomorphology relationship in the Sirjan area of southern-central Iran and mentioned that pediments with the most gypsum and palygorskite accumulation have been the peripheries of such closed water bodies of the past times. The lack of palygorskite in mantled pediments could be a clue to support the argument that pediments of the area have been located at positions that are too high to be affected by ancient water bodies. Khademi et al. (1997) and Toomanian et al. (2001) studied the genesis of gypsiferous soils on alluvial fans and fan remnants, where the gypsum source was from upper watersheds. In the study area, gypsum originates mainly from the weathering of Cretaceous limestone and Jurassic shale of the surrounding sediments. Their isotopic work on geochemistry of gypsiferous Aridisols from central Iran confirms that the most common rocks of Cretaceous and Oligo–Miocene age contain an appreciable amount of sulfate with a $\delta^{34}S$ value ranging from 11.45 to 13.96 and the sulfate in the Lower Cretaceous sediments controls the geochemistry of the younger geologic formations and consequently the sulfate in soils. From their point of view, the source of sulfates is from seawater that entered these sediments during tertiary and is a potential source for gypsum.

Considering the morphology and physical characteristics of soils and based on other observations, four genetic stages of gypsic horizons development were proposed (Toomanian et al. 2001):

- *Stage one*: Formation of gypsic lenticular crystals in voids through runoff from geologic sediments carrying gypsum in solution to the soils of the coarse textured upper fans.
- *Stage two:* Continued crystallization of gypsum crystals and formation of xenotopic or hyp-idiotopic morphology.
- *Stage three:* Formation of thin pendants and gypsic horizons with chitonic and enaulic distribution patterns.
- *Stage four:* Formation of hyper gypsic horizons.

In addition to very big-sized gypsum pendants (as large as 10 cm), gypsum occurs as microscopic forms in the gypsic horizons studied. Microscopic gypsum crystals appear as random lenticular and granular crystals, along channels and planar voids with no apparent orientation to the associated surface. Gypsum also occurs as relatively larger interlocking plates.

The sequence of horizons along with their chemical and micromorphological properties reveals that gypsum was accumulated in the deep colluvial soils through the downward water movement (per descendum mode). The formation of gypsum pendants under gravels further supports this hypothesis. Whereas gypsic and salic surface horizons in the alluvial plain are the result of an upward movement of salt-loaded brine (per ascendum mode), lagoonal conditions best describe the deposition of extremely high amount of gypsum throughout the pedons on the plateaus, as supported by the stable isotope geochemistry (Khademi et al. 1997) and clay mineralogy (Khademi and Mermut 1998).

Moghiseh and Heidari (2012) studied the co-occurrence of gypsum and halite and their morphological expressions in Bam region, Iran. Micromorphological observations demonstrate that the dominant cementing agent in the soils is halite rather than gypsum. However, due to the inexistence of petrosalic diagnostic horizon in Keys to Soil Taxonomy, these soils are to be classified as Petrogypsic Haplosalids at subgroup level in Soil Taxonomy. In WRB Taxonomy, they are classified as Petrosalic Solonchaks. Co-occurrence of gypsum and halite in the same horizon, their specific layering and vertical distribution patterns in the studied pedons might be considered as indicators for polygenetic soils in this area.

6.2.4 Salinization

The evolution of salt-affected soils has been studied by several authors (Abtahi 1977). However, the theory developed by Kovda (1973) seems to be the most logical for the calcareous soils of Iran. He has recognized four stages in the formation of soils affected by salinity, they are (i) salinization and alkalinization, (ii) desalinization, (iii) solonetzation, and (iv) dealkalinization.

In an attempt to trace the sequence of soil formation under saline conditions and clay translocation in highly calcareous soils of southern Iran, some laboratory experiments were carried out by Abtahi (1977). Based on his observations, an evolutionary sequence of soil formation under saline condition in the semiarid calcareous material is suggested. This sequence is salinization and alkalization; desalinization; solonetzation; and dealkalization (Fig. 6.10).

The evolution of soil color is important in this sequence. The transition from desalinization to solonetzation is associated with a marked change of pH. Salts in soil cause the clay to be flocculated so that water-dispersable clay content is lower in the saline soils. Dispersability of the clay probably is affected by the same factors as swelling of clay. The type of clay, the exchangeable cation on the clay, the free salts present in the soil, the concentration and composition of the electrolyte, and the presence of other materials in association with clay such as iron oxides, aluminum oxides, and organic matter are the main factors with regard to the clay dispersion. The exchangeable Na appears to be the major cause of clay dispersion and migration after most of the salt has leached. As soon as the excess salt has been leached from the saline-sodic soil, the Na could cause dispersion and migration of the clay.

Through time, Ca dissolved from the calcite in soils and parent rocks could replace the Na out of the exchange sites

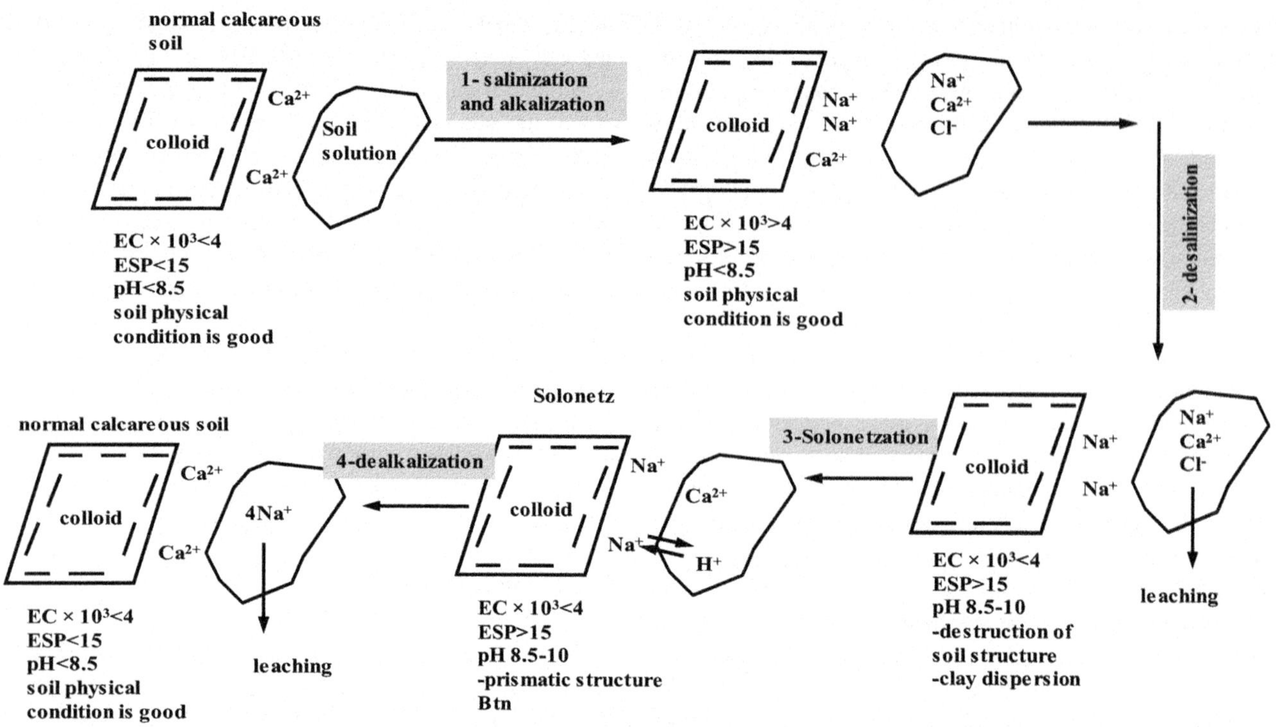

Fig. 6.10 Stages of development of saline-sodic soils under calcareous parent material

and this is the dealkalization stage where the soil physical and chemical conditions become favorable and the normal calcareous soils form.

6.2.5 Aquic Conditions in Paddy Soils

Paddy soils are a kind of artificial hydromorphism or hydragric Anthrosols; soils with prominent characteristics result from human activities, virtually any soil material modified through cultivation or by addition of materials (Fig. 6.11) (Wilding and Ahrens 2005).

Growing rice in soils on alluvial plains is very popular in northern Iran. Surface water irrigation and shallow ground water influence the soil morphology of these soils simultaneously. Hassannezhad et al. (2008) investigated the impact of anthraquic conditions on morphology, the different forms of redoximorphic features and distributions and the processes responsible for the formation or transformation of hydromorphic soils.

The most important processes have been influenced by redox condition, addition and removal of chemical components and soil particles, and changes in physical, chemical and microbiological properties through irrigation or drainage, or both. In other words, gleization and eluviations, mottle forming (oxidized illuviation and segregation of iron and manganese and separation of manganese from iron), degradation, cutan forming, redistribution of exchangeable

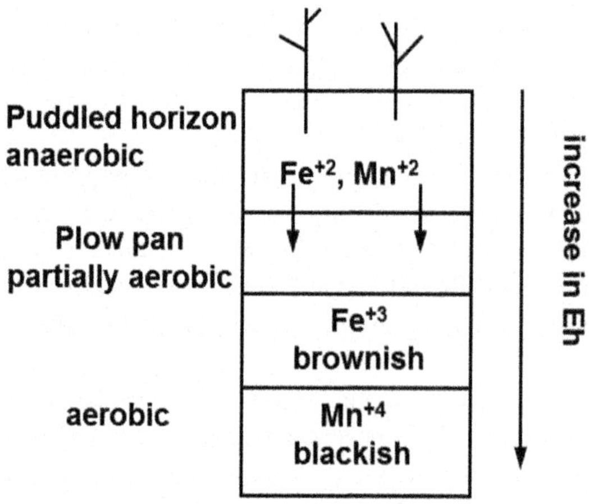

Fig. 6.11 Soil processes in a puddled pedon

bases, accumulation (or decomposition), and alteration of organic matter and other processes lead to the profile differentiation of rice soils.

Results of the study done by Hassannezhad et al. (2008) indicated that the effect of irrigation water on pedogenesis with respect to saturation, reduction, and redoximorphic features was greater than that of the shallow ground water. This study demonstrated that moisture regimes in soils with anthraquic saturation can be characterized in the same way as in soils with either episaturation or endosaturation. On alluvial plains, the soil conditions range from continuously reducing controlled by high groundwater to alternating reduction and oxidation because of artificial submergence and groundwater fluctuations. Artificially induced leaching losses, previously prevented by a high groundwater table, are partly compensated during paddy cultivation by the rejuvenation process of alluvial deposition. Soil samples, which no longer receive river-sediment, are more strongly developed.

Micromorphology provides a means to identify the presence of hydromorphic features that may otherwise be missed in the field if only the naked eye is used. The observation of hydromorphic features in thin sections suggests that careful observations in the field, using a hand lens, may increase the likelihood of identification of horizons in problem soils with aquic conditions. Fe nodules may be one of the easiest diagnostic features to find because they have sharp boundaries. However, these small features appear very similar to Fe-coated coarse fragments, and, in addition, nodules with sharp boundaries are often interpreted as relict features.

6.2.6 Organic Matter Accumulation

Accumulation of organic matter in soils and the formation of mollicepipedon is typical steppe regions with loess parent material. Mollisols are the dominant zonal soils in Golestan Province mainly formed in southern Gorgan River plains where there is a typical steppe climate with the annual precipitation of about 400–650 mm. The parent material contains high amounts of silt particles mainly derived from the loess deposits in the region. The province was formerly called white gold area due to vast cotton fields growing mainly on Mollisols or organic matter rich soils. The total area of Mollisols formed on Gorgan River plain reaches about 100,000 hectares, which reaches to more than 150,000 ha considering those of the hilly regions. Figure 6.12 shows the unique precipitation gradient (>800 mm – <200 mm) along a north-south transect extending from Alborz mountain ranges in south to the border with Turkmenistan in north. Along this transect Mollisols with different degrees of development have formed. Hapludolls/Hapludalfs (Phaeozems/Luvisols) are dominant in the forest–grassland transition zone (600–700 mm) grading to Calcixerolls (Chernozems) in the typic xeric regions (500–600 mm) and finally to Calcixerolls/Haploxerollsin (Kastanozems) the drier parts of the steppe (300–500 mm).

Figure 6.13 shows micromorphological features of a PachicArgixerolls in the Agh-Emam region northeast of the town of Azadshar on the north-facing slope of a loess covered hill rising about 270 m above the plain of Gorgan River. The profile is a typical example of Mollisols in Golestan Province. The parent material is brown, comparatively clayey loess. Soil moisture regime is xeric, and the temperature regime is thermic. The whole hill slope has been formerly under a mixture of pasture and forest. The mean slope of the study site is about 15–20%, and there is no evidence of soil erosion, runoff, and sedimentation in this area. Mollic epipedon is about 80 cm (Pachic subgroup) overlying the argillic horizon of about 30 cm over deep calcic horizon (up to 90 cm thick) overlying the loess parent material.

Very high biologic activity in the mollice pipedon is confirmed by the micromorphological studies showing the dominant excremental pedofeatures along with the well-separated granular/crumb microstructure. Root and organic residues are also dominant. In the Bt horizon, deep decalcification and subsequent clay accumulation have resulted in the development of speckled b-fabric with clay coatings along voids. Calcitic pedofeatures are observed deep in the Bk horizons mainly as nodules, needles, and coatings.

In the dry xeric regions of Golestan Province where the soil is covered by short grass vegetation especially on the north-facing slopes, thin mollic epipedons have formed. The underlying horizons are either cambic or weak calcic horizons. These soils are classified as Typic Calcixerolls or Kastanozems. These soils grade to Chernozems (or Phaeozems) in the southern typic xeric regions and to Inceptisols (Cambisols) or Aridisols (Calcisols) in its northern margin where the humidity is lower and the soil moisture regime is aridic (Fig. 6.14). In contrast to the Mollisols of the more

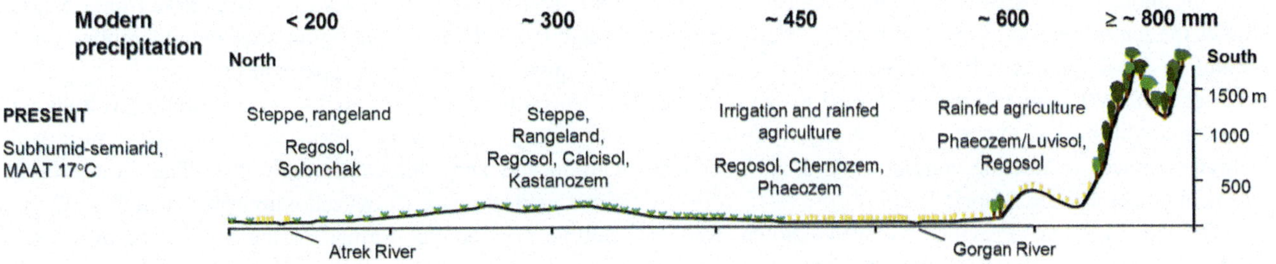

Fig. 6.12 A north-south transect extending from Alborz mountain ranges in south to the border with Turkmenistan in north showing precipitation gradient, vegetation, land use and soils (Kehl and Khormali 2014)

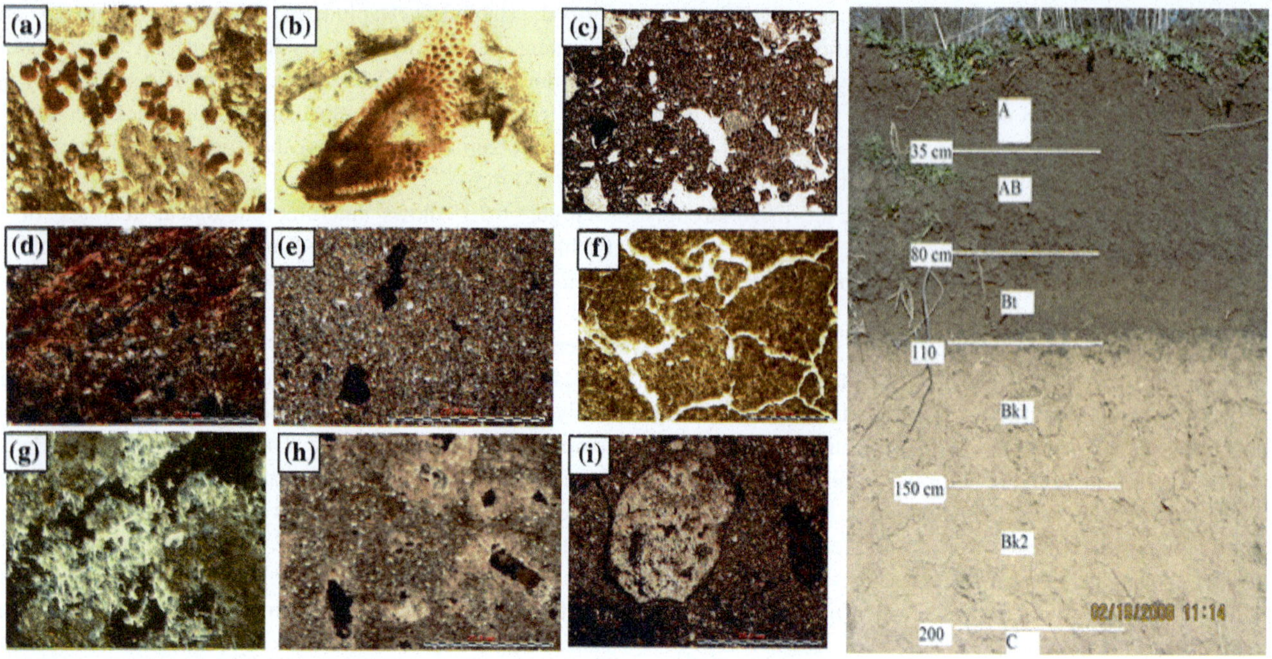

Fig. 6.13 Photograph and micromorphological features of a Pachic Argixeroll in Golestan Province; **a**, **b**, **c**: PPL image showing excremental pedofeatures, root residues, and granular/crumb microstructure in mollic epipedon, respectively; **d**, **e**, **f**: clay coatings, speckled b-fabric and well-separated angular blocky microstructure in the Bt horizon, respectively; **g**, **h**, **i**: needle-shaped, coating, and nodule of calcite in the Bk horizon, respectively under XPL

humid regions, there is no sign of decalcification or clay illuviation in this drier part of the steppe.

Khormali and Kehl (2011) studied the clay mineralogical evolution along a north-south transect in northern Iran with the precipitation gradient of >800 mm in south to <200 mm in north with the border with Turkmenistan. In the aridic regions of northern parts, the dominant clay minerals are illite and chlorite mainly of inherited from loess parent materials. In the typic xeric regions where there is higher soil available moisture (P/ET_ > 0.4), smectite content increases especially in the Bt horizons and is believed to be mainly of transformed origin. Increasing soil available moisture, and consequently a relative leaching environment for the release of K from micaceous mineral sand mainly illite, in the calcareous environment with high Mg^{+2} and high Si^{+4} mobility may provide favorable conditions for the formation of smectite through transformation.

In the more humid areas, vermiculite appears and constitutes the major portion of the clay fraction. Vermiculite is present in small amounts in the loess material, and the occurrence of vermiculite in the forest land is due to higher leaching conditions and the removal of K mainly from mica. In high chemical weathering conditions, hydroxy-interlayer vermiculite can also be formed.

Although Golestan Province is well known for its Mollisols, these soils are also found in other regions where the conditions are suitable. A Mollisol catena in highly calcareous parent material under semiarid conditions of southern Iran was studied to determine the effects of water table depth and its fluctuations on the organic carbon content of mollic epipedons, genesis of subsurface horizons, and mineralogical variations in these horizons. The soils formed on depressions (microlows) have the shallowest water table, the longest time of saturation, and greatest organic carbon content and have 50 cm thick mollic epipedons. Subsurface horizons have characteristics of cambic horizon. No calcic horizon has formed in these soils, mainly due to the lack of wetting and drying cycles due to the permanent saturation.

The soils with a water table depth of deeper than 1 m have cyclic saturated conditions. Organic matter content and thickness of mollic epipedons of these soils are less than that of the soils on microlows. They show developed calcic horizons. Secondary carbonates present in B horizons of these soils are related mainly to the discharge from a shallow water table. Evapotranspiration was mainly reponsible for precipitation of seconday carbonates.

The soils formed on the higher landscape positions with very deep water tables show lower amounts of organic carbon and very thin mollic epipedon. They are not saturated and do not show redoximorphic features. Only a cambic horizon has formed in these soils, because of organic matter addition and transformation, and translocation of $CaCO_3$.

Fig. 6.14 A Kastanozem (Typic Calcixerolls) showing a Mollic Epipedon (Location 37° 35′ 49.7″ N, 55° 25′ 36.3″E)

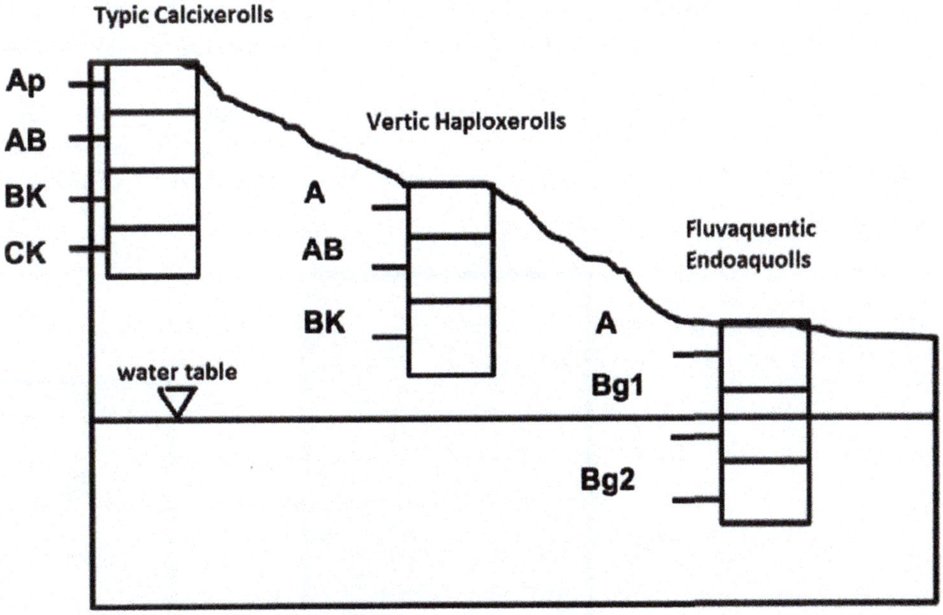

Fig. 6.15 A Mollisols catena in Kurdestan, western Iran (Nabiollahy et al. 2006)

A Mollisols catena was studied in semiarid region of western Iran as illustrated by Fig. 6.15 (Khormali and Nabiollahi 2009; Nabiollahi et al. 2006). Soils formed on the upper section of the alluvial valley bottom i.e., Typic Calcixerolls with the deepest water table, had lower OC and thinner mollic epipedon comparing to other soils. The soils were not saturated and showed no evidence of redoximorphic featuers. Presence of secondary carbonate and calcic horizons were mainly related to the dissolution of the carbonate in the upper soil horizons and its downward leaching and precipitation in the lower horizons. Speckled and striated b-fabric of the near surface soil horizons are the

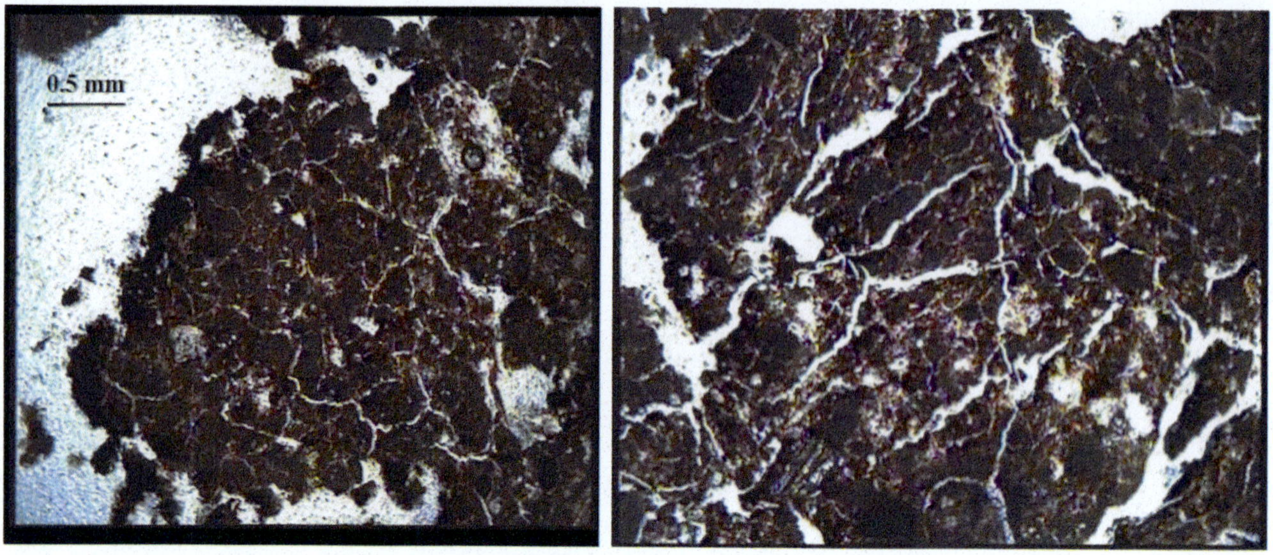

Fig. 6.16 PPL images of the mollic epipedon in Typic Calcixerolls (left) and vertic Haploxerolls (right) showing granular and angular blocky microstructure separated by planar voids, respectively (Khormali and Nabiallahi 2009)

evidences of carbonate depletion from surface layers. Soils of the mid-valley section i.e., Vertic Haploxerolls, with water table depth (1–2 m) have periodic saturation (Fig. 6.16). Thickness of the mollic epipedons and OC were lower than the soils of the lower valley bottom. Soils of the lower valley bottom or low lands classified as Fluvaquentic Endoaquolls had thick dark mollic epipedons and the highest OC comparing to other soils. In the well-drained soils, illite was dominant. In contrast, in poorly drained soils of the lower valley bottom, smectite was the dominant clay mineral. Abtahi and Khormali (2001) found smectite as a major clay mineral in poorly drained Calciaquolls of southern Iran.

6.2.7 Vertic Features

Vertisols in Iran were developed in regions with different climatic conditions and bedrock compositions. Khormali (2003) reported the formation of slickensides in Vertisols and Vertic Haploxeralfs in Fars Province containing high smectite (Fig. 6.17). Heidari et al. (2005b) carried out a mineralogical and micromorphological study to characterize Vertisols from separate regions, revealing an aberrant soil composition for one study area, where the clay fraction is not dominated by smectite. Vertisols are mainly characterized by the occurrence of planar voids, a porphyricc/f (coarse/fine)

Fig. 6.17 High shrink-swell properties resulted in the formation of slickensides in a Vertic Haploxeralfs in Fars Province (Khormali 2003)

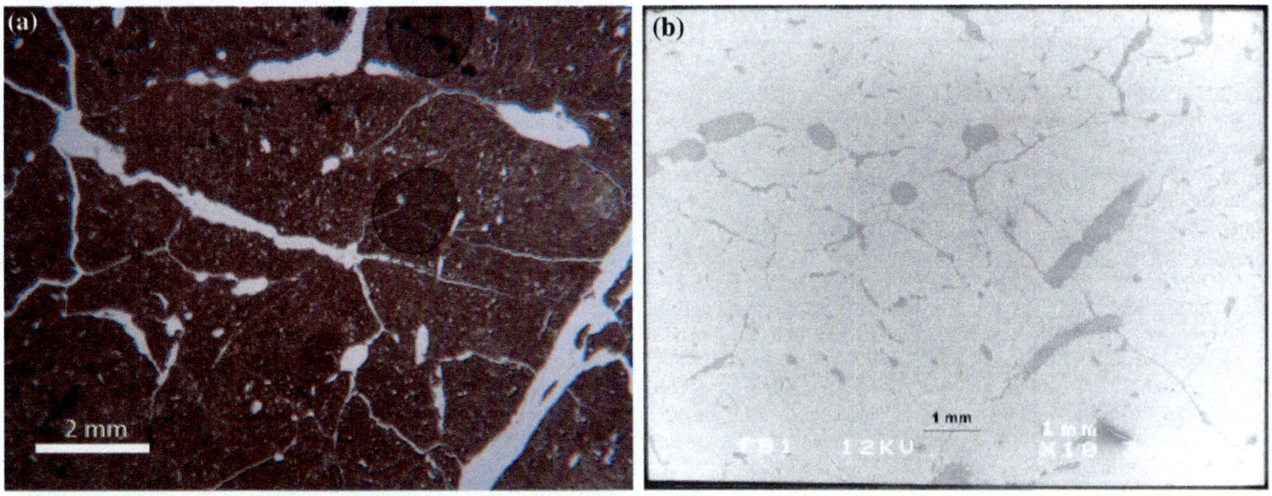

Fig. 6.18 Planar voids associated with moderately separated angular blocky microstructure as wedge-shaped peds in a Haploxerert (**a**. PPL, and **b**. Backscattered SEM, Khormali, 2003)

related distribution pattern and striated b-fabrics, as far as this type of b-fabric is not masked by microcrystalline calcite (Fig. 6.18). Common pedofeatures are various types of carbonate and Fe/Mn hydroxide accumulations.

The composition of the clay fraction of the studied Vertisols shows that development of vertic properties is not always determined by the presence of smectite, but that it can also be related to high fine clay content. The presence of high amounts of fine clay, which is not necessarily smectitic, can be sufficient to cause vertic behavior (Heidari et al. 2005b).

6.3 Conclusion

Major parts of Iran are of arid and semiarid climate, and there is a high amount of calcium carbonate in soils. Climatic zones therefore seems best describe the formation and distribution of soils, and the major soil-forming processes are those associated with soil carbonate redistribution and soluble salts. Local conditions determine which process is dominant. In arid regions therefore, calcitic features, salinization, and gypsification are dominant. In the semiarid zones, clay illuviation, organic matter accumulation, and vertic processes are observed. In the subhumid and humid regions, the illuviation of clay is more intense and gleization is the other major process. To better understand how the soil-forming processes act, the geomorphic background and paleo-environmental reconstruction seems also of great importance.

References

Abbaslou H, Abtahi A, Peinado FJM, Owliaei HR, Khormali F (2013) Mineralogy and characteristics of soils developed on Persian Gulf and Oman Sea Basin, Southern Iran: implications for soil evolution in relation to sedimentary parent material. Soil Sci 178:568–584

Abtahi A (1977) Effect of a saline and alkaline ground water on soil genesis in semiarid southern Iran. Soil Sci Soc Am J 41:583–588

Abtahi A (1985) Synthesis of sepiolite at room temperature from SiO_2 and $MgCl_2$ solution. Clay Miner 20:521–523

Abtahi A, Khormali F (2001) Genesis and morphological characteristics of Mollisols formed in a catena under water Table influence in southern Iran. Commun Soil Sci Plant Anal 32:1643–1658

Blank RR, Fosberg MA (1990) Micromorphology and classification of pedogenic calcium carbonate accumulations that surround or occur on the undersides of coarse fragments in Idaho (USA). In: Douglas LA (ed) Soil micromorpholoy: a basic and applied science, Development in Soil Science, vol 19. Elsevier, Amsterdam, pp 341–346

Cerling TE (1984) The stable isotopic composition of modern soil carbonate and its relationship to climate. Earth Planet Sci Lett 71:229–240

Emadi M, Baghernejad M, Bahmanyar MA, Morovvat A (2012) Changes in soil inorganic phosphorous pools along a precipitation gradient in northern Iran. International Journal of Forest, Soil and Erosion 2(3):143–147

Farpoor MH, Neyestani N, Eghbal MK, Borujen IE (2012) Soil–geomorphology relationships in Sirjan playa, south central Iran. Geomorphology 138:223–230

Frechen M, Kehl M, Rolf C, Sarvati R, Skowronek A (2009) Loess chronology of the Caspian Lowland in Northern Iran. Quatern Int 128:220–233

Gharaee HA, Mahjoory RA (1984) Characteristics and geomorphic relationships of some representative Aridisols in southern Iran. Soil Sci Soc Am J 48:115–119

Gile LH, Peterson FF, Grossman RB (1966) Morphological and genetic sequences of carbonate accumulation in desert soils. Soil Sci 101:347–360

Gile LH, Grossman RB (1968) Morphology of the argillic horizon in desert soils of southern New Mexico. Soil Sci 106:6–15

Givi J, Abtahi A (1985) Soil genesis as affected by topography and depth of saline and alkaline groundwater under semiarid conditions in southern Iran. Iran Agric Res 4:11–27

Gunal H, Ransom MD (2006) Clay illuviation and calcium carbonat accumulation along a precipitation gradient in Kansas. Catena 68:59–69

Hashemi SS, Baghernejad M, Khademi H (2011) Micromorphology of gypsum crystals in Southern Iranian soils under different moisture regimes. J Agr Sci Tech 13:273–288

Hassannezhad H, Pashaee A, Khormali F, Mohammadian M (2008) Morphology and micromorphology of paddy soils under different soil moisture regime and ground water table in Mazandaran Province, Northern Iran, Amol. Int J Soil Sci 3:149–156

Heidari A, Mahmoodi Sh, Stoops G, Mees F (2005a) Micromorphological characteristics of Vertisols of Iran, including nonsmectitic-soils. Arid Land Res Manag 19: 29–46

Heidari A, Mahmoodi Sh, Roozitalab MH, Mermut AR (2005b) Diversity of clay minerals in the Vertisols of three different climatic regions in Western Iran. J Agri Sci Tech 10: 269–284

Herrero J, Porta J, Fedoroff N (1992) Hypergypsic soils. micromorphology and landscape relationship in northeastern Spain. Soil Sci Soc Am J 56:1188–1194

Jafarzadeh AA, Burnham CP (1992) Gypsum crystals in soils. J Soil Sci 43:409–420

Karimi AR, Khademi H, Kehl M, Jalalian A (2009) Distribution, lithology and provenance of peridesert loess deposits in northeastern Iran. Geoderma 148:241–250

Kehl M, Khormali F (2014) Excursion Guide Book. International Symposium on Loess, Soils and Climate Change in southern Eurasia Gorgan, Iran. 15–19 October 2014

Kemp RA, Zarate MA (2000) Pliocene pedosedimentary cycles in the southern Pampas, Argentina. Sedimentology 47:3–14

Khademi H, Mermut AR (1998) Source of palygorskite in gypsiferous Aridisols and associated sediments from central Iran. Clay Miner 33:561–575

Khademi H, Mermut AR (2003) Micromorphology and classification of Argidsand associatedgypsiferousAridisols from central Iran. Catena 54:439–455

Khademi H, Mermut AR (1999) Submicroscopy and stable isotope geochemistry of pedogenic carbonates and associated palygorskite in selected Aridisols from Iran. Eur J Soil Sci 50:207–216

Khademi H, Mermut AR, Krouse RH (1997) Isotopic composition of gypsum hydration water in selected landforms from central Iran. Chem Geol 138:245–255

Khormali F, Abtahi A (2001) Soil genesis and mineralogy of three selected regions of Fars, Bushehr and Khuzestan Provinces of Iran, formed under highly calcareous conditions. Iran Agri Res 20:67–82

Khormali F (2003) Clay mineralogy, micromorphlogy and development of soils in arid and semiarid regions of Fars Province, Southern Iran. PhD. Thesis. Shiraz University, Iran

Khormali F, Kehl M (2011) Micromorphology and development of loess-derived surface and buried soils along a precipitation gradient in Northern Iran. Quatern Int 234:109–123

Khormali F, Ghergherech Sh, Kehl M, Ayoubi Sh (2012) Soil formation in loess-derived soils along a sub-humid to humid climate gradient, Northeastern Iran. Geoderma 179–180:113–129

Khormali F, Abtahi A, Stoops G (2006) Micromorphology of calcitic features in highly calcareous soils of Fars Province, Southern Iran. Geoderma 132:31–46

Khormali F, Abtahi A, Mahmoodi Sh, Stoops G (2003) Argillic horizon development in calcareous soils of arid and semi-arid regions of Southern Iran. Catena 53:273–301

Khormali F, Abtahi A (2003) Origin and distribution of clay minerals in calcareous arid and semiarid soils of Fars Province, Southern Iran. Clay Miner 38:511–527

Khormali F, Nabiallahi K (2009) Degradation of Mollisols as affected by land use change. J Agri Sci Tech 11:363–374

Khormali F, Monger C, Feng Y (2014) Experimental micropedology- a technique for investigating soil carbonate biogenesis along a desert-grassland-forest transect I New Mexico, USA. Spanish J Soil Sci, 130–147

Kovda VA, Van Den Berg C, Hagan RM (eds) (1973) International Source Book on Irrigation, Drainage and Salinity. FAO-UNESCO, Paris

Krinsley DB (1970) A geomorphological and paleoclimatological study of the playas of Iran. Geological Survey, U.S. Department of Interior, Washington, DC

Mahjoory RA (1975) Clay mineralogy, physical and chemical properties of some soils in arid regions of Iran. Soil Science Society of America Proceedings 39:1157–1164

Mahmoodi SH (1994) Properties and management of gypsiferous soils. 4th Soil Science Congress of Iran, Isfahan Univ. of Technology, 29–31 Aug. (in Persian)

Mahmoudi F (1987) Changes in topography of Iran in Quaternary. J Geogr Res 22:5–43 (in Persian)

Moatamed A (1988) Climatic conditions of central Iran in Quaternary. J Sci, Univ. Tehran 17: 115–129 (in Persian)

Moghiseh E, Heidari A (2012) Polygenetic saline gypsiferous soils of the Bam region, Southeast Iran. J Soil Sci Plant Nutr 12:729–746

Monger HC, Southard RJ, Boettinger JL (2011) Classification of soils. Aridisols. In: Huang PM, Li Y, Sumner ME (eds) Handbook of Soil Science, 2nd Edn. Boca Raton, CRC Press. pp 33–1 to 34–1

Nabiollahy K, Khormali F, Ayoubi Sh (2006) Formation of Mollisols as affected by landscape position and depth of groundwater in Kharkeh Research Station, Kurdestan Province. J Agri Sci Natural Resources of Gorgan University 13(4):20–30

Owliaie HR (2012) Micromorphology of pedogeniccarbonate features in soils of Kohgilouye, Southwestern Iran. J Agri Sci Tech 14: 225–239

Roozitalab MH (1994) Aridisols in Iran and their sustainable utilization. 4th Soil Science Congress of Iran, Isfahan Univ. of Technol, pp 29–31. Aug. (in Persian)

Siadat H, Bybordi M, Malakouti MJ (1997) Salt affected soils of Iran: A country report. International symposium on "Sustainable management of salt affected soils in the arid ecosystem". Cairo. Egypt, pp 9–14

Soil Survey Staff (2014) Keys to Soil Taxonomy. USDA, NRCS, Washington, DC

Toomanian N, Jalalian A, Eghbal MK (2001) Genesis of gypsum enriched soils in north-west Isfahan, Iran. Geoderma 99: 199–224

Treadwell-Steitz C, McFadden LD (2000) Influence of parent material and grain size on carbonate coatings in gravely soils, Palo Duro Wash, New Mexico. Geoderma 94:1–22

Wilding LP, Ahrens R (2005) Provision for anthropogenically impacted soil. European Bureau Res p 7

Wilson MJ (1999) The origin and formation of clay minerals in soils: past, present and future perspectives. Clay Miner 34:7–24

World Reference Base for Soil Resources, WRB (2014). Food and Agriculture Organization of the United Nations, Rome

Author Biographies

Dr. Farhad Khormali is a Professor of pedology at Gorgan University of Agricultural Sciences and Natural Resources, Iran. He is also a committee member of the Soil Science Society of Iran as well as editor in chief of the *Journal of Soil Management and Sustainable Agriculture*. His main research interests are soil genesis, loess–paleosols studies, micromorphology, clay mineral weathering, and soil quality and soil carbon storage.

Dr. Norair Toomanian is an Associate Professor of Soil Science in Soil and Water Research Department, Isfahan Agricultural and Natural Resources Research and Education Center, AREEO, Isfahan, Iran. His current research interests are digital soil mapping, suitability mapping, and data basing of Iranian soils. He has been teaching graduate students at various universities in Iran and has supervised many theses and dissertations.

Major Soils, Properties, and Classification

7

Mohammad Hassan Roozitalab, Norair Toomanian,
Vahid Reza Ghasemi Dehkordi, and Farhad Khormali

Abstract

Iran is a large territory with diverse climatic condition, geology and vegetation cover which have led to formation of different soils that cover about 58% of the total land surface. The remaining areas are composed of bare mountains, sand dunes, playas/salinas, etc. Four major agro-climatic zones are distinguished: (i) arid region, (ii) semiarid region, (iii) humid and sub-humid region of the southern coasts of the Caspian Sea, and (iv) tropical semiarid region of the South. The Aridisols (*Calcids, Gypsids, Salids, and Cambids*), Entisols (*Orthents and Fluvents*), and Inceptisols (*Xerepts, Ustepts, Udepts, and Aquepts*) constitute about 97% of the total soil resources of the country, i.e., in the arid and semiarid zones. Other soil orders including Mollisols, Alfisols, and Ultisols, which have mainly developed in the Caspian Sea region, constitute less than 3% of the total soil cover. In this chapter, we define the extent of the soil orders, suborders, and dominant great groups in each region and describe main morphological and physicochemical properties of soils representing various soil great groups and their potentials for agricultural use.

Keyword

Soil properties • Soil classification • Soils of the arid to semi-arid regions
Soils of Caspian region

7.1 Introduction

The first general schematic and small-scale soil map of Iran was prepared in 1942 by the Russian soil scientists VA Kovda and YP Lebedev at a scale of 1:6,000,000 without any field studies. The first initiative to study the soil resources of the country based on the field observations and laboratory analyses of the soils was conducted during 1955–1961 as a part of collaboration between the Soil Department of the Bongah-e-Abyari (Irrigation Company) of the Ministry of Agriculture and FAO. As the result, the book titled "The Soils of Iran" with accompanying maps at the scale of 1: 2,500,000 was published in 1964 (Dewan and Famouri 1964).

The soil survey study in Iran began in early 1950s, and official projects for preparing soil maps and land classification for irrigation were carried out in 1953. Until 2016, more than 20 million hectares (mha) of the land, mostly in the inter-mountain plains/valleys, have been studied at reconnaissance scale at 1:1,000,000 to semi-detailed scale at 1: 50,000. Detailed soil maps are only available for about 750,000 ha of the land. In 1995, after several years of fieldworks, a project to assess the land potentials of various regions for major kinds of land use such as irrigated, dry farming, range and forestry is completed and subsequently the land potentiality maps for all the provinces were published at the scale of 1: 250,000.

M.H. Roozitalab (✉) · V.R. Ghasemi Dehkordi
Soil and Water Research Institute, Tehran, Iran
e-mail: mroozitalab@gmail.com

N. Toomanian
Soil and Water Research Department, Isfahan Agricultural and Natural Resources Research and Education Center, AREEO, Isfahan, Iran

F. Khormali
Gorgan University of Agricultural Sciences and Natural Resources, Gorgan, Iran

M.H. Roozitalab et al. (eds.), *The Soils of Iran*, World Soils Book Series,
https://doi.org/10.1007/978-3-319-69048-3_7

These initiatives were generally based on the two main publications titled "Manual of land classification system for irrigation" (Soil Institute of Iran 1979) and "Manual of multipurpose land classification" (Soil Institute of Iran 1970). Both manuals were developed and prepared through a joint project with FAO during 1970s (see Chap. 2), by PA Mahler, a senior FAO's consultant, and his Iranian colleagues at the Soil Institute of Iran.

In 2000, Soil and Water Research Institute published another soil map of Iran at a scale of 1:1,000, 000 after several years of field observations and soil survey (SWRI 2000; Banaei 2005). This map was prepared using all relevant soil survey reports of different regions as well as information on geology, geomorphology, topography, vegetation, agro-climates. Satellite imageries of the whole country were analyzed and interpreted. Initially, soils developed in various landforms were delineated at a scale of 1: 500,000, and later compiled to produce the soil map which was published at the scale of 1:1,000, 000 for the country. Late Mohammad Hassan Banaei (Fig. 7.1), a prominent soil scientist, coordinated and supervised the project for preparing the soil map. This soil map shows distribution of the soils at the subgroup level, as defined by the USDA Soil Taxonomy (Soil Survey Staff 2014). Each map unit is composed of association of the subgroups of the soils. Miscellaneous land types (rocks, dunes, etc.) are included in the map unit's symbol as associated or inclusion, if dominantly present (Fig. 7.5).

Despite the valuable efforts and planning made, this soil map is not supported and linked to a dataset on properties of the soils representing the various map units. Therefore, quantitative analysis, interpretation, or preparation of thematic maps using properties of representative soils are not feasible. However, in this chapter, attempts were made to select and show the properties of representative soils developed in various climatic zones of the country.

Fig. 7.1 Mohammad H. Banaei 1938–2005

7.2 Soils and Crop Diversity

As indicated in the previous chapters, Iran benefits from very diverse physiographic units and climatic conditions, which have contributed to development of various vegetation cover and agricultural production systems (see Chaps. 3, 4, and 5). The country is very rich in biodiversity and is the origin or main center of several major crop species such as wheat, barley, and many medicinal and herbal plants as well as fruit trees such as almond, pistachio, pomegranate.

Agricultural production systems since antiquity have largely been depended on irrigation to produce food, feed, and fiber. Soils are generally calcareous and mostly developed in the arid and semiarid regions of Iran. Most of the soils are usually N deficient with low organic carbon content of less than 0.6%. Thus, they need application of N-fertilizers for viable and economic production (see Chap. 10). Calcium carbonate content of the soils is usually more than 25%. The available potassium content is medium to high, and the amount of available P in cultivated soils is variable depending on the rate and frequency of application of P-fertilizers in the soils. Deficiency of micronutrients such as Fe, Zn, and Cu has been reported in many soils, and application of the micronutrients fertilizers is customary practice in many orchards such as citrus, apples, and pistachios.

Sheet and rill erosions are very common on steep slopes in cultivated land on which appropriate tillage practices, management of crop residue, and proper crop rotation are not practiced (see Chaps. 12 and 13). Dry farming of wheat, barley, and food legumes is a widespread practice on the soils in western and central highlands of Zagros Mountain (Roozitalab et al. 2013).

7.3 Generalities on Soil Moisture and Temperature Regimes

Soil moisture and temperature regimes have been used in the Soil Taxonomy to define various categories. Soil moisture regime is generally used at the higher category of the soil classification to define suborders of the soils.

The soil moisture regime is defined in terms of the level of groundwater or the seasonal changes of water held at a tension of less than 1500 kPa in the soil moisture control section (Soil Survey Staff 2014). The upper boundary of the control section is the depth to which a dry soil will be moistened by 2.5 cm of water within 24 h, and the lower boundary is the depth to which a dry soil will be moistened by 7.5 cm of water within 48 h (Soil Survey Staff 2014). The soil temperature regime is used in defining the family or in some cases in other categorical level in Soil Taxonomy. The soil temperature regimes are defined in terms of the mean annual soil temperature at a depth of 50 cm from the surface or at the upper boundary of a root-limiting layer. In practice, a computer program is used to determine the soil moisture and temperature regimes.

In 1976, M.H. Banaei prepared a generalized map of Iran at 1:4,000,000 scale, showing the generalized soil moisture and temperature regimes of different regions based on climatic data of more than 100 metrological stations using the computerized Newhall method. He published a revised map at the scale of 1:2,500,000 in 1997 (Fig. 7.2) using a new method and larger data obtained from 300 metrological stations (Banaei 1998). A summary on the prevalent soil moisture and temperature regimes in different regions of Iran is presented in the following section.

7.3.1 Soil Moisture Regimes

About 97% of the country are affected by arid to semiarid climatic conditions from which about 30% of the land are located in the semiarid and 67% are situated in extreme arid to arid climatic conditions (De Pauw et al. 2004). Consequently, aridic and xeric soil moisture regimes are predominant in the country (Fig. 7.2).

The **aridic** moisture regime is mostly prevalent in the soils of the central, eastern, and southern parts of the country, along the Persian Gulf, and covers about 63% of the soil resources.

The **xeric** moisture regime covers about 25% of the area and is common in the soils developed in the semiarid highland regions of the west, northwest, and northeast of the country where dry-farming agriculture is a customary practice.

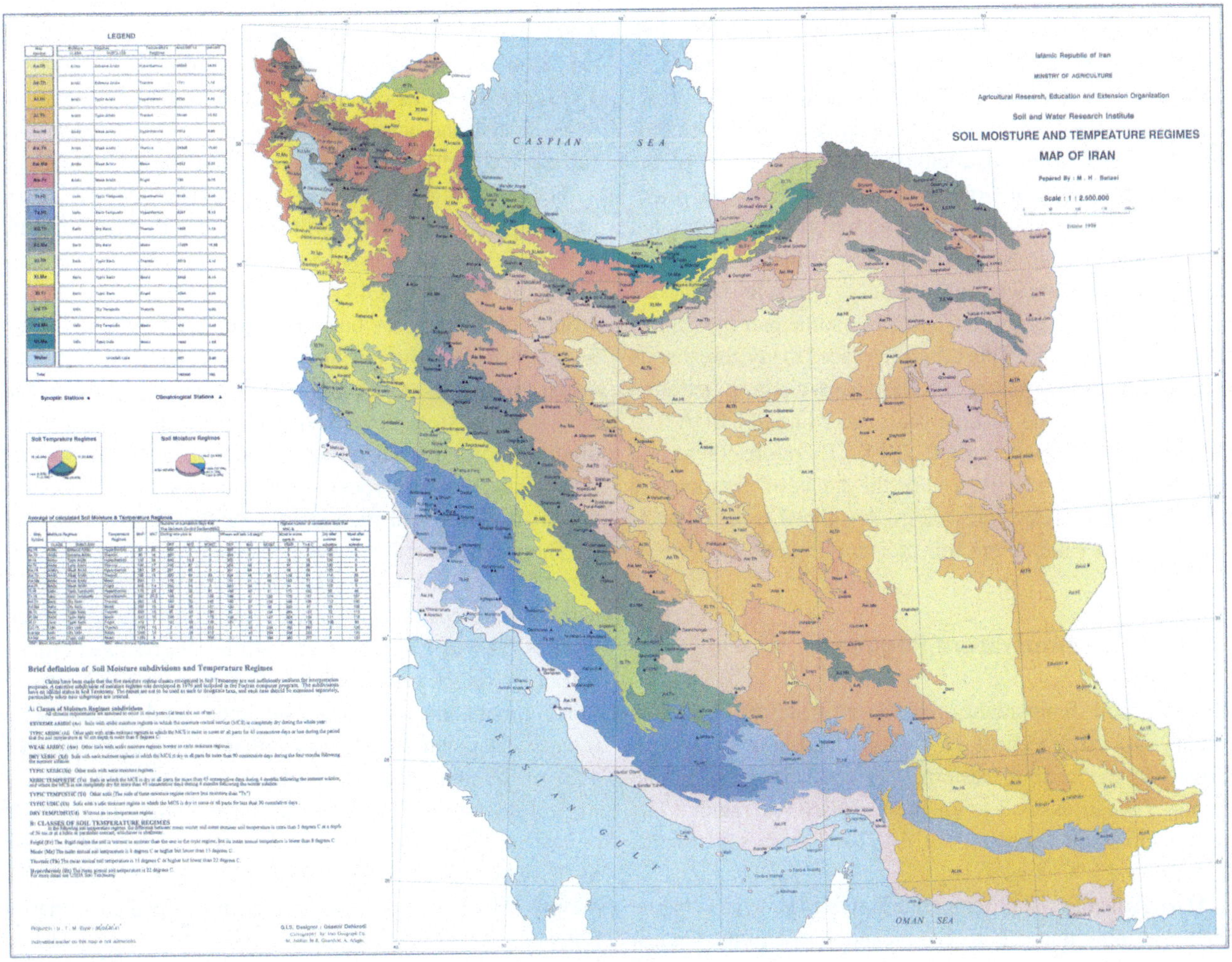

Fig. 7.2 Map of soil moisture and temperature regimes in Iran (Banaei 1998)

The **ustic** moisture regime generally occurs in the soils developed in the narrow strip region in the south of the country close to the warm tropical coastal regions of the Persian Gulf. It covers about 10% of the soils mainly in parts of Khuzestan, Hormozgan, Fars, Ilam, and "Kohgiluyeh and Boyer-Ahmad" provinces, where thermic and hyperthermic temperature regimes are dominant. Rainfall occurs mainly during warm winter and spring seasons, and dry-farming of wheat and barley is a widespread practice in some areas (Fig. 7.2).

The **udic** moisture regime is mostly dominant in the soils developed in narrow strip of the coast of the Caspian Sea region in Mazandaran and Gilan provinces. It covers only 1.7% of the soils of the country (Fig. 7.2). However, in the Caspian Sea region, soils formed on the floodplains and lowlands are usually under rice cultivation and have an aquic soil moisture regime.

Soil moisture regime of the transitional zone: In Golestan Province, the average annual precipitation is decreased from about 1000 mm in the mountainous area with elevation of more than 1800 m asl to about 200 mm in lowland of Turkmen Sahara with an elevation of 25 m below the sea level. The soil moisture regime ranges from udic in the high elevation of the mountainous areas and upper terraces to aridic in the lowland of the Turkmen Sahara.

A few studies on the climatic data of several meteorological stations in the mountainous areas of the east of the Caspian Sea region in Golestan and Mazandaran provinces indicated that the soil moisture regime in this transitional zone is mainly ustic rather than udic (Sarmadian 1997). It means that moisture control section of the soils is dry for more than 90 cumulative days in all or some part in 6 out of 10 years in this zone. This criterion places many forest soils in central to eastern part of the Caspian Sea region under an ustic moisture regime in some years according to USDA Soil Taxonomy (Soil Survey Staff 2014). Therefore, there is a need for further study to monitor the status of the moisture control section of the forest soils developed in this transitional zone to validate the soil moisture regime.

7.3.2 Soil Temperature Regimes

Thermic soil temperature regime is widespread in the country and covers about 37% of the land, mainly in soils formed in the central, eastern, and southern parts of the Iranian Plateau as well as in the Caspian Sea region. The mean annual soil temperature (MAST) of these soils ranges from 16 to 22 °C (Fig. 7.2).

Hyperthermic soil temperature regime, which has a MAST of more than 22 °C, covers about 36% of the land. It is mainly prevalent in the east and southeast of the country in Kerman and Sistan Baluchestan provinces and in the coastal regions of the Persian Gulf in Bandar Abbas, Bushehr, and Khuzestan provinces where hot tropical climatic condition is prevailing.

Mesic soil temperature regime covers about 24% of the land. It is mainly developed in the mountainous highland regions in the west, northwest, and northeast areas of the country where MAST ranges from 8 to 15 °C. **Frigid** soil temperature regime only occurs in about 3% of the soils, mainly in high altitude region of Zagros and Alborz Mountain Ranges where MAST is less than 8 °C (Fig. 7.2).

7.4 Soil Mapping and Classification

Soil mapping and land classification have been carried out in the country since 1952 (see Chap. 2). These studies indicate that many soils have an ochric epipedon with low organic matter content of less than 0.6. Calcic, gypsic, and cambic horizons are the dominant subsurface horizons in many soils except for the soils developed in the southern coasts of the Caspian Sea region. Gypsic horizon usually occurs in the soils developed on old terraces and upper plateaus in central and eastern regions, which generally receive less than 250 mm mean annual precipitation and have a mean annual evaporation of more than 1500 mm. The soils developed in the Caspian Sea region are more diverse and have either ochric or mollic epipedon in the surface and argillic or cambic horizons in the subsurface.

For preparing this chapter, we digitized the soil map of Iran prepared in 2000 by Soil and Water Research Institute (SWRI 2000) and arranged a dataset using geographic information system (GIS). In addition, we revised the subgroups of Inceptisols based on the latest version of Soil Taxonomy (Soil Survey Staff 2014). The extent of the soil orders, suborders, great groups, subgroups, and other mapping units was calculated from 4116 mapping units of this digitized soil map. Therefore, all data on soils and non-soils (miscellaneous land types) reported in this chapter are based on this digitized soil map and the related dataset.

The digitized soil map and the relevant data show that about 58% of the total areas of the country (about 96 million hectares) are covered by different soil orders and suborders (Tables 7.1 and 7.2). The extent of the subgroups of different soil orders is presented in Table 7.3. The non-soils, such as rocky mountains, outcrops, badlands, sand dunes, salinas (*Dagh*), lake, and others, cover about 69 million hectares or 42% of the total land area of the country (Table 7.4).

Soil resources of Iran are mostly classified in seven soil orders of Aridisols, Entisols, Inceptisols, Vertisols, Mollisols, Alfisols, and Ultisols, according to USDA Soil Taxonomy. Most of the soils are classified in three orders of Entisols, Aridisols, and Inceptisols, which respectively, constitutes 42%, 41%, and 14% of the total soil areas. In total, these soil orders make up about 97% of the whole soil resources (Table 7.1, Figs. 7.3 and 7.4).

Table 7.1 Extent of soil orders and suborders in Iran

No.	Order	Area 1000 ha	% of soils	Suborder	Area hectare	% of soils
1	Alfisols	510	0.53	Aqualfs	76,000	0.08
2				Udalfs	274,200	0.28
3				Ustalfs	9000	0.01
4				Xeralfs	151,100	0.16
5	Aridisols	39,516	41.14	Argids	52,400	0.05
6				Calcids	7,583,600	7.89
7				Cambids	2,387,600	2.49
8				Gypsids	17,311,800	18.02
9				Salids	12,181,100	12.68
10	Entisols	39,700	41.33	Aquents	163,200	0.17
11				Fluvents	6,682,600	6.96
12				Orthents	31,915,100	33.22
13				Psamments	939,100	0.98
14	Inceptisols	13,999	14.57	Aquepts	361,000	0.38
15				Udepts	410,600	0.43
16				Usteps	2,117,000	2.20
17				Xerepts	11,110,500	11.57
18	Mollisols	2314	2.41	Aquolls	163,700	0.17
19				Rendolls	506,300	0.53
20				Udolls	654,500	0.68
21				Ustolls	35,900	0.04
22				Xerolls	953,800	0.99
23	Ultisols	4.5	0.005	Humults	4,500	0.005
24	Vertisols	13.2	0.014	Xererts	13,200	0.014
Sum soils		96,057.9	100		96,057,900	100

Table 7.2 Soil orders and major soil great groups

Orders	Area million ha	% Total land area	Major soil great groups
Entisols	39.7	24.1	Torriorthents, Torrifluvents, Torripsamments, Xerorthents, Xerofluvents, Ustorthents, Ustifluvents, Ustipsamments, Udorthents, Udifluvent, Fluvaquents, Endoaquents, and Psammaquents
Aridisols	39.5	23.9	Haplocalcids, Haplogypsids, Calcigypsids, Petrogypsids, Haplocambids, Haplosalids, Aquisalids, Haplargids, and Natrargids
Inceptisols	14.0	8.5	Calcixerepts, Haploxerepts, Calciustepts, Haplustepts, Dystrudepts, Eutrudepts, Endoaquepts, Halaquepts, and Humaquepts
Mollisols	2.3	1.4	Calcixerolls, Argixerolls, Haploxerolls, Hapludolls, Argiudolls, Calciudolls, Calciustolls, Endoaquolls, Haprendolls, Endoaquolls, and Calciaquolls
Alfisols	0.5	0.3	Hapludalfs, Endoaqualfs, Haploxeralfs, and Haplustalfs
Vertisols	<0.01	<0.01	Calcixererts and Haploxererts
Ultisols	<0.01	< 0.01	Haplohumults and Hapludults,
Total soil area	96.0	58.2	
Miscellaneous land types	68.7	41.8	Bare mountains, outcrops, badlands, sand dunes, salt flats, and others

Table 7.3 Kind and extent the soil orders and their subgroups in Iran

No	Order	Suborder	Subgroups	Area in ha.	% of land	% of soils
1	Alfisols	1-Aqualfs	Mollic Endoaqualfs	15,000	0.009	0.016
			Typic Endoaqualfs	49,700	0.030	0.052
			Typic Natraqualfs	11,300	0.007	0.012
		2-Udalfs	Lithic Hapludalfs	26,100	0.016	0.027
			Typic Hapludalfs	246,700	0.150	0.257
			Ultic Hapludalfs	1400	0.001	0.001
		3-Ustalfs	Aridic Haplustalfs	9000	0.005	0.009
		4-Xeralfs	Calcic Haploxeralfs	131,600	0.080	0.137
			Aquic Haploxeralfs	11,200	0.007	0.012
			Typic Haploxeralfs	8,200	0.005	0.009
2	Aridisols	1-Argids	Ustic Haplargids	30,700	0.019	0.032
			Xeric Natrargids	21,700	0.013	0.023
		2-Calcids	Lithic Xeric Haplocalcids	397,100	0.241	0.413
			Aquic Haplocalcids	3200	0.002	0.003
			Typic Haplocalcids	1,697,700	1.030	1.767
			Ustic Haplocalcids	53,900	0.033	0.056
			Xeric Haplocalcids	5,431,700	3.296	5.655
		3-Cambids	Fluventic Haplocambids	370,300	0.225	0.385
			Ustertic Haplocambids	66,900	0.041	0.070
			Vitrandic Haplocambids	9300	0.006	0.010
			Aquic Haplocambids	52,300	0.032	0.054
			Sodic Haplocambids	71,800	0.044	0.075
			Sodic Xeric Haplocambids	51,200	0.031	0.053
			Typic Haplocambids	49,300	0.030	0.051
			Ustic Haplocambids	18,200	0.011	0.019
			Ustifluventic Haplocambids	100,200	0.061	0.104
			Xeric Haplocambids	1,591,400	0.966	1.657
			Xeric Aquicambids	6800	0.004	0.007
		4-Gypsids	Vitrandic Haplogypsids	287,600	0.174	0.299
			Lithic Haplogypsids	464,700	0.282	0.484
			Leptic Haplogypsids	981,900	0.596	1.022
			Aquic Haplogypsids	5200	0.003	0.005
			Sodic Haplogypsids	38,600	0.023	0.040
			Typic Haplogypsids	8,391,700	5.091	8.736
			Ustic Haplogypsids	301,400	0.183	0.314
			Xeric Haplogypsids	2,773,900	1.683	2.888
			Typic Calcigypsids	1,682,200	1.021	1.751
			Ustic Calcigypsids	37,800	0.023	0.039
			Xeric Calcigypsids	1,914,200	1.161	1.993
			Typic Petrogypsids	384,000	0.233	0.400
			Xeric Petrogypsids	48,400	0.029	0.050
		5-Salids	Gypsic Haplosalids	2,665,300	1.617	2.775
			Leptic Haplosalids	1,831,600	1.111	1.907
			Aquic Haplosalids	2700	0.002	0.003
			Typic Haplosalids	4,127,500	2.504	4.297

(continued)

Table 7.3 (continued)

No	Order	Suborder	Subgroups	Area in ha.	% of land	% of soils
			Gypsic Aquisalids	1,922,300	1.166	2.001
			Typic Aquisalids	1,631,700	0.990	1.699
3	Entisols	1-Aquents	Mollic Endoaquents	6500	0.004	0.007
			Typic Endoaquents	42,100	0.026	0.044
			Typic Psammaquents	22,500	0.014	0.023
			Aquandic Fluvaquents	4700	0.003	0.005
			Mollic Fluvaquents	9600	0.006	0.010
			Typic Fluvaquents	77,700	0.047	0.081
		2-Fluvents	Mollic Udifluvents	25,800	0.016	0.027
			Typic Udifluvents	69,700	0.042	0.073
			Aquic Torrifluvents	218,700	0.133	0.228
			Typic Torrifluvents	3,156,300	1.915	3.286
			Ustic Torrifluvents	993,200	0.603	1.034
			Xeric Torrifluvents	1,590,600	0.965	1.656
			Typic Ustifluvents	91,700	0.056	0.095
			Mollic Xerofluvents	30,100	0.018	0.031
			Typic Xerofluvents	506,700	0.307	0.527
		3-Orthents	Lithic Udorthents	141,400	0.086	0.147
			Typic Udorthents	351,300	0.213	0.366
			Vitrandic Torriorthents	24,200	0.015	0.025
			Lithic Torriorthents	4,284,700	2.600	4.461
			Lithic Ustic Torriorthents	103,500	0.063	0.108
			Lithic Xeric Torriorthents	2,840,300	1.723	2.957
			Typic Torriorthents	8,880,600	5.388	9.245
			Ustic Torriorthents	983,700	0.597	1.024
			Xeric Torriorthents	4,894,200	2.969	5.095
			Xerertic Torriorthents	93,200	0.057	0.097
			Aridic Ustorthents	979,900	0.595	1.020
			Lithic Ustorthents	675,500	0.410	0.703
			Aquic Ustorthents	33,900	0.021	0.035
			Typic Ustorthents	1,072,500	0.651	1.117
			Lithic Xerorthents	3,579,200	2.172	3.726
			Aquic Xerorthents	4200	0.003	0.004
			Typic Xerorthents	2,152,200	1.306	2.241
			Aridic Ustorthents	820,700	0.498	0.854
		4- Psamment	Typic Torripsamments	566,300	0.344	0.590
			Ustic Torripsamments	124,000	0.075	0.129
			Xeric Torripsamments	89,900	0.055	0.094
			Aridic Ustipsamments	113,200	0.069	0.118

(continued)

Table 7.3 (continued)

No	Order	Suborder	Subgroups	Area in ha.	% of land	% of soils
4	Inceptisols	1-Aquepts	Typic Endoaquepts	224,100	0.136	0.233
			Typic Halaquepts	102,500	0.062	0.107
			Typic Humaquepts	34,500	0.021	0.036
		2-Udepts	Fluventic Humudepts	2800	0.002	0.003
			Typic Humudepts	9900	0.006	0.010
			Humic Dystrudepts	68,800	0.042	0.072
			Lithic Dysrtudepts	3900	0.002	0.004
			Aquic Dystrudepts	32,000	0.019	0.033
			Typic Dystrudepts	144,800	0.088	0.151
			Dystric Fluventic Eutrudepts	16,200	0.010	0.017
			Fluventic Eutrudepts	600	0.000	0.001
			Aquic Eutrudepts	1300	0.001	0.001
			Typic Eutrudepts	130,300	0.079	0.136
		3-Ustepts	Aridic Haplustepts	272,600	0.165	0.284
			Aridic Lithic Haplustepts	217,200	0.132	0.226
			Fluventic Haplustepts	91,800	0.056	0.096
			Gypsic Haplustepts	23,500	0.014	0.024
			Typic Calciustepts	947,400	0.575	0.986
			Lithic Haplustepts	442,700	0.269	0.461
			Typic Haplustepts	119,800	0.073	0.125
			Vertic Haplustepts	1900	0.001	0.002
		4-Xerepts	Typic Calcixerepts	7,869,700	4.775	8.193
			Fluventic Haploxerepts	538,500	0.327	0.561
			Gypsic Haploxerepts	403,200	0.245	0.420
			Lithic Haploxerepts	386,500	0.234	0.402
			Aquic Haploxerepts	59,900	0.036	0.062
			Typic Haploxerepts	1,708,600	1.037	1.779
			Vertic Haploxerepts	143,900	0.087	0.150
5	Mollisols	1-Aquolls	Fluventic Endoaquolls	15,400	0.009	0.016
			Fluvaquentic Endoaquolls	41,100	0.025	0.043
			Typic Endoaquolls	65,500	0.040	0.068
			Typic Calciaquolls	41,600	0.025	0.043
		2-Rendolls	Lithic Haprendolls	171,200	0.104	0.178
			Typic Haprendolls	335,000	0.203	0.349
		3-Udolls	Calcic Argiudolls	6700	0.004	0.007
			Aquic Argiudolls	8600	0.005	0.009
			Typic Argiudolls	223,600	0.136	0.233
			Fluventic Hapludolls	5100	0.003	0.005
			Aquic Hapludolls	3700	0.002	0.004
			Typic Hapludolls	53,600	0.033	0.056
			Typic Calciudolls	353,300	0.214	0.368

(continued)

Table 7.3 (continued)

No	Order	Suborder	Subgroups	Area in ha.	% of land	% of soils
		4-Ustolls	Typic Calciustolls	35,900	0.022	0.037
		5-Xerolls	Calcic Argixerolls	10,700	0.006	0.011
			Typic Argixerolls	57,500	0.035	0.060
			Fluventic Haploxerolls	4100	0.002	0.004
			Calcic Haploxerolls	9100	0.006	0.009
			Lithic Haploxerolls	317,700	0.193	0.331
			Aquic Haploxerolls	6700	0.004	0.007
			Typic Haploxerolls	48,400	0.029	0.050
			Aquic Calcixerolls	8000	0.005	0.008
			Typic Calcixerolls	479,500	0.291	0.499
			Vertic Calcixerolls	12,200	0.007	0.013
6	Ultisols	Humults	Typic Haplohumults	4500	0.003	0.005
7	Vertisols	Xererts	Chromic Haploxererts	8600	0.005	0.009
			Aridic Calcixererts	4600	0.003	0.005
Total soil area				96,057,500	58.2	100.00
Total area of other miscellaneous land types				68,762,000	41.8	
Total land area				164,819,500		

Table 7.4 Types and generalized extent of non-soils or miscellaneous land types in Iran

No	Miscellaneous	Area in ha.	% total land
1	Beaches	60,700	0.037
2	Badlands	7,468,800	4.532
3	Dune lands	3,549,700	2.154
4	Kalut	699,300	0.424
5	Lava flow	144,900	0.088
6	Marsh	237,200	0.144
7	Playa	3,812,300	2.313
8	Rocky mountains	45,751,900	27.759
9	Rock outcrop	2,942,800	1.785
10	River wash	129,200	0.078
11	Sand dune	53,300	0.032
12	Salt plug	106,500	0.065
13	Urban	254,400	0.154
14	Water	3,551,000	2.154
	Total	68,762,000	%41.2

Less than 3% of the soil resources are classified in two soil orders of Mollisols and Alfisols (Table 7.1). They are mostly developed under sub-humid to humid climates of the Caspian Sea. Ultisols are only formed in a very limited and scattered area under forest cover in the mountainous parts of the Caspian Sea region. Vertisols, which make up less than 0.1% of the total soils, are mostly developed on calcareous sediments and materials under xeric and aridic soil moisture regimes in central plateau and plains of Zagros Mountain. Distribution of dominant soil orders and major land types is shown in Fig. 7.3

Orthents (*Regosols*), which are young soils and developed in different climate zones, are the most prevalent soil suborders in the country and cover about 33% of the total soils. Gypsids (*Gypsisols*), which are the most widespread Aridisols, make up about 18% of the total soils (Table 7.1). In total, these two soils cover about 51% of the total soil

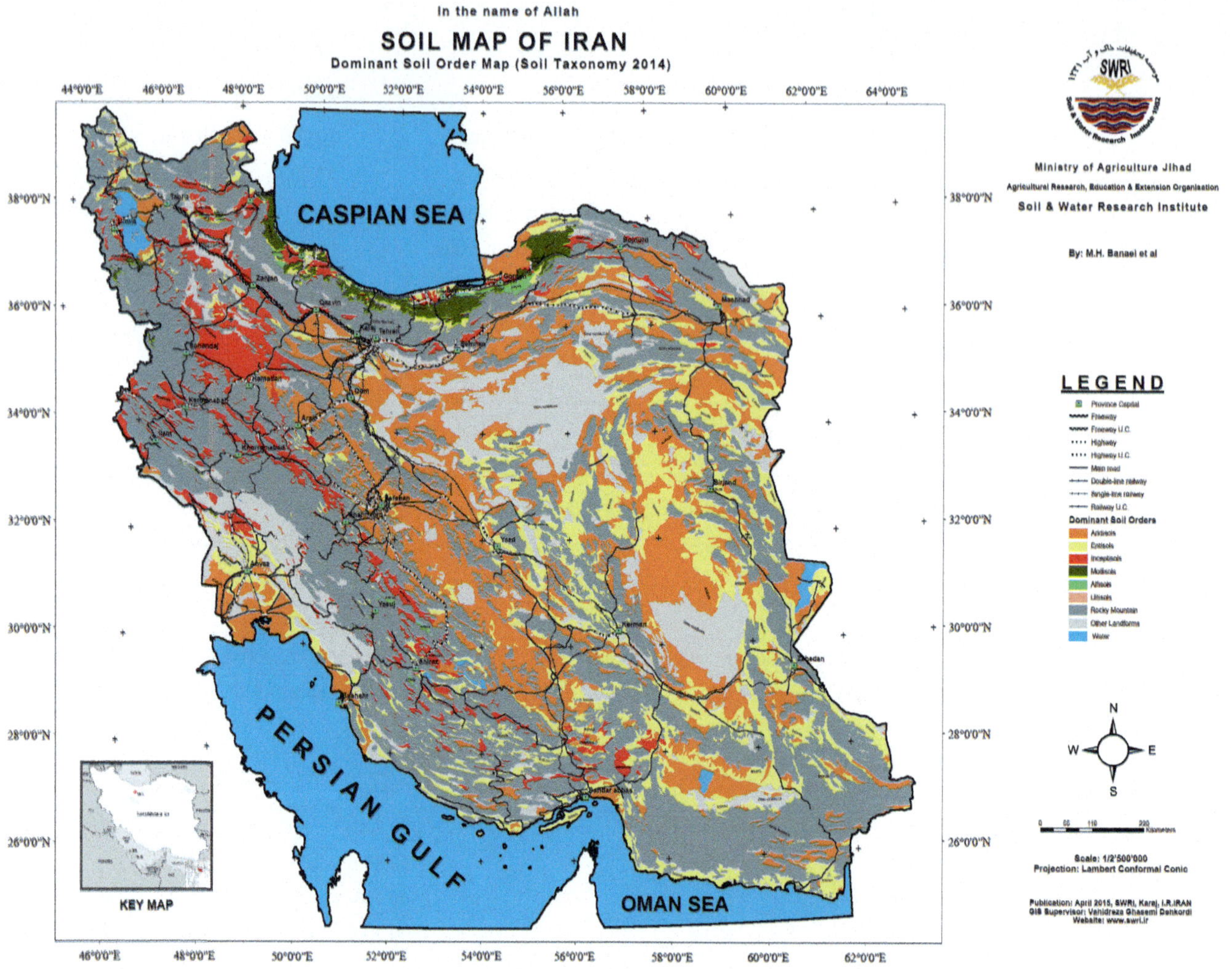

Fig. 7.3 Generalized distribution of dominant soil orders and miscellaneous land types

areas. Table 7.3 shows the total areas of each subgroup of the soil orders in the country (Fig. 7.4).

Figure 7.5 shows the association of the subgroups of the soils as defined by USDA Soil Taxonomy 2014. The taxonomic units of the soil map are either subgroups of the soil orders or miscellaneous land types as delineated at that scale. Each map unit generally includes a dominant component estimated to constitute more than %50 of the unit, a minor component estimated to constitute less than %35, and an inclusion which was generally estimated to cover less than %15 of the map unit. For each map unit, the texture of topsoil and overall slope of the dominant soil component are reported for defining the general land use.

Table 7.3, which was prepared using the digitized soil map according to 2014 version of the Keys to Soil Taxonomy (Soil Survey Staff 2014), shows the extent of each subgroup and its proportion to the total land and the total soil areas of the country. The most prevalent subgroup delineated on the map is "Typic Torriorthents" (*Regosols*), which covers a total area of 8.88 million hectares (m ha) or about 9% of the total soil areas. The second major subgroup is "Typic Haplogypsids" (*Gypsisols*) which covers an area of 8.39 mha or 8% of the total soil areas. The other major subgroup is "Typic Calcixerept" (Calcisols) which covers a total area of 7.87 m ha or 8% of the total soil area. These three subgroups which belong to the three soil orders of Entisols, Aridisols, and Inceptisols constitute about 25% of the total soil areas of the country.

The total non-soil or other miscellaneous land types of the country are about 68.7 million hectares or 41.8% of the total surface area of the country. The largest miscellaneous land type is "bare or rocky mountains" that covers about 27.8% of the total land area or 45.7 m ha. Table 7.4 shows the different kinds and extent of other miscellaneous land types in the country.

7.5 Soils of the Arid Region

As indicated before, about 67% of the land resources of Iran, mainly in central, eastern, and southern parts are affected by hyperarid to arid climatic conditions. **Hyperarid** conditions

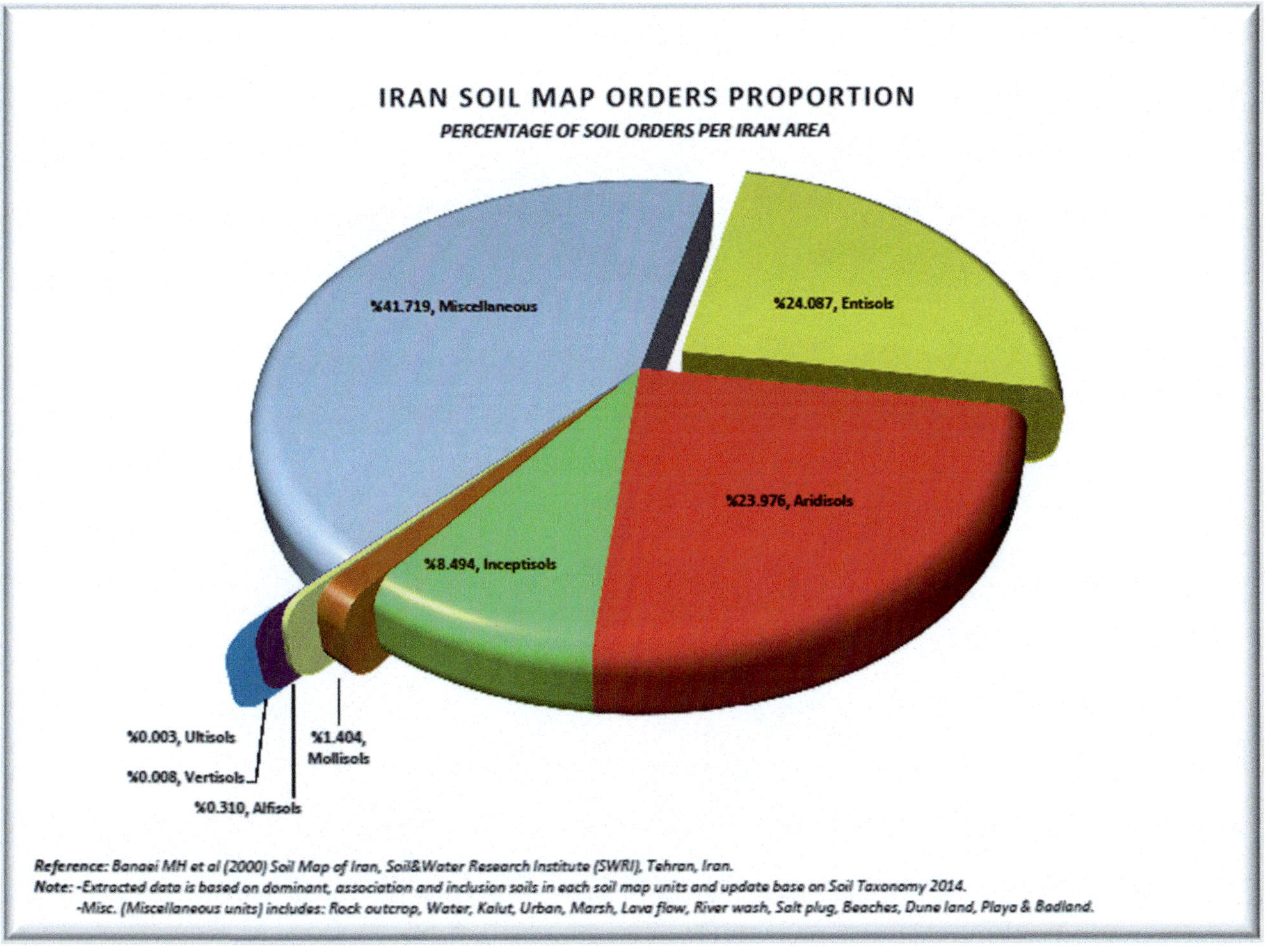

Fig. 7.4 Proportion of soil orders in relation to the total land surface area (SWRI 2015)

with mean annual rainfall of less than 50 mm constitute only 2.7% of the land area of the country where aridity index (AI) is less than 0.03. **Arid** conditions with mean annual rainfall of between 80 and 150 mm up to between 200 and 350 mm constitute about 65% of the land area of the country with an aridity index between 0.03 and 0.2 (De Pauw et al 2004). The aridic soil moisture regime is dominant and covers about 63% of the total area or about 103 m ha of the land area of the country (Banaei 1998).

Most of the soils developed in the arid region are classified in Aridisols and Entisols, constituting respectively about 41 and 30% of the soil resources of the country (Table 7.5). In addition, large areas of rocky mountains, rock outcrops, playas, and sand dunes formed in the central and southeastern part of the country are located in the arid regions of the country (Table 7.3).

7.5.1 Aridisols

These soils constitute about 39.5 m ha or 41% of the total soil resources of the country (Table 7.1). They are generally developed on the stable landforms where prevailing soil-forming factors and processes have led to the formation of different soils (see Chap. 6). The main subsurface diagnostic horizons developed in these soils are cambic, gypsic, calcic, argillic, salic, petrocalcic, or petrogypsic (Mahjoory 1975; Abtahi 1977; Gharaee and Mahjoory 1984; khademi and Mermut 2003; Khormali and Abtahi 2003; Khormali et al. 2003; Owliaie et al. 2006; Mahmoodi et al.2007; Toomanian et al. 2001; Farpoor et al. 2012 and Abbaslou et al. 2013).

Gypsids are the most widespread soils of Aridisol which constitute about 18% of the total soil cover (17 m ha). The other major suborders, i.e.,Salids, Calcids, and Cambids

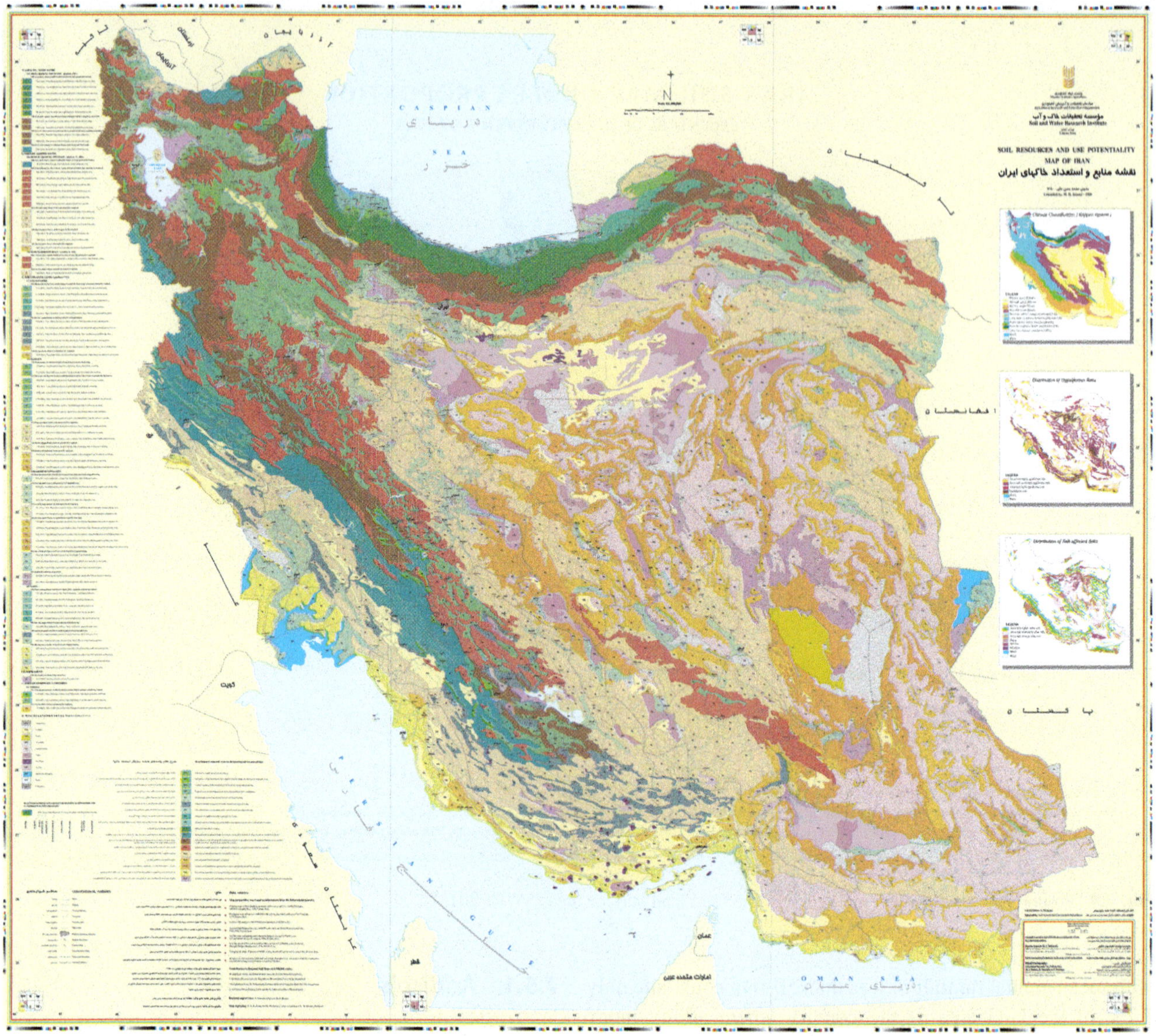

Fig. 7.5 Soil map of Iran at scale of 1:1,000,000 (SWRI 2000)

cover about 12.5% (12 m ha), 8% (7.5 m ha), and 2.5% (2.4 m ha) of the total soil areas, respectively (Tables 7.1 and 7.5).

Many Gypsids, Calcids, and Salids are not suitable for cultivation of crops and orchards due to varieties of soil and land limitations. They are usually used as low productive rangelands or for non-agricultural uses.

Gypsids (*Gypsisols*) which have a gypsic horizon within 100 cm of the soil surface are the most widespread Aridisols and cover about 17.3 m ha or 18% of the soil resources of the country (Tables 7.1 and 7.5). Haplogypsid is the dominant great group and covers about 13 m ha of the land or about 14% the total soil area. Gypsids are usually developed on upper plateaus and old terraces (Khademi and Mermut 1998; Farpoor et al. 2012; Toomanian et al. 2001) and are not generally suitable for crop and fruit production due to major limitations such as high salinity, high gypsum content, land subsidence, coarse texture, low water holding capacity, shallow depth, and steep slope and/or undulating landforms.

One of the most widespread great groups of Gypsids is Haplogypsids. Accumulation of gypsum is one of the prominent pedogenic processes occurring in these soils. Gypsum can be readily found in all geomorphic surfaces (Figs. 7.6 and 7.7) including fans, dissected old terraces, piedmonts, and playas (Toomanian et al. 2003; Toomanian 2011). It also occurs in upper terraces of some river alluvial plains or in back swamps.

Morphology: Morphological properties and genesis of Haplogypsids are fully described by Toomanian et al. (2001, 2003) in a sequence of landforms from a Cretaceous

Table 7.5 Suborders and great groups of Aridisols and Entisols developed in the arid region of Iran

Order	Suborder	Great group	Area 1000 ha	% Total soils
Aridisols	Gypsids	Haplogypsids	13,245	13.79
		Calcigypsids,	3634	3.78
		Petrogypsids	432	0.45
	Salids	Haplosalids	8627	8.98
		Aquisalids	3554	3.70
	Calcids	Haplocalcids	7583	7.89
	Cambids	Haplocambids	2380	2.48
		Aquicambids	7	0.01
	Argids	Haplargids	31	0.03
		Natrargids	22	0.02
		Subtotal	39,515	41.13
Entisols	Orthents	Torriorthents	22,051	22.97
	Fluvents	Torrifluvents	5952	6.20
	Psamments	Torripsamments	778	0.81
		Subtotal	28,781	29.98
		Total Soils	68,296	71.11

Fig. 7.6 Typical landscapes on which Haplogypsids are mostly developed (Toomanian et al. 2003)

mountain front to adjacent piedmont plain. This can be considered as a representative sequence for most Iranian sloping landscapes in arid regions. Gypsum crystals (whitish glassy, shining) start to be easily visible (with un-aided eyes) in soils when its amount roughly exceeds 10%. Lenticular mineral grows in big interpedal voids with gypsum enrichment. However, in gravelly or non-gravely sandy old alluviums, it occupies the whole soil groundmass. In some cases of this stage, other soil materials are not clearly visible with un-aided eyes. Table 7.6 presents an example of the key morphological features of such soils. One of the specific characteristics of Haplogypsids developed on old alluviums is the formation of a vesicular crust layer (Av) under A horizon. This crust breaks easily down when someone walks on such soils.

Physicochemical properties: The sequence of soils in studied transects reveals that the physical characteristics of soils are influenced by clay, gravel, and gypsum contents. In this case, crystal sizes have good co-relationship with its content. Equilibrium of gypsum and calcite minerals controls the chemical properties of such soils. Therefore, soil extract is super-saturated with Ca^{2+} and SO_4^{2-} ions in such soils. The physicochemical properties of these soils are presented in Tables 7.7 and 7.8.

Fig. 7.7 A typical profile of a Gypsid (Haplogypsid) with pendants of gypsum developed on a plateau near Isfahan (Toomanian et al. 2003)

Micromorphological features: The form and size of gypsic pedofeatures vary with changes in soil environment (Stoops 2003; Toomanian et al. 2003). The tabular pseudo-rhombohedral, tabular pseudo-hexagonal, prismatic pseudo-hexagonal, prismatic pinacoidal and twinned crystals can form columnar, prismatic, lenticular, rose-like, needle-like, fibrous or tubular gypsic pedofeatures in different soils.

The gypsum crystals are pockets of crystals and pendants in soil profile (Fig. 7.8). In most cases of upland soils, at starting stages of gypsification process, microscopic gypsum crystals tend to form random lenticular and granular crystals, along the channels and planar voids with no apparent orientation to the associated surfaces. The euhedral crystals with xenotopic and hyp-idiotopic fabrics are concentrated in soil pores forming channel or chamber internal coatings, infillings, and grain external coatings. In such cases, the gypsum crystals are pockets of crystals and pendants in soil profile.

In more concentration of gypsum, on old dissected surfaces, gypsum crystals are oriented in vertical threads or bundles. Vertical gypsum threads are somewhat connected laterally to form a firm network (Fig. 7.8). Gypsum crystals and aggregates constitute the groundmass of these soils, and other soil materials remain as isles between. Internal fabric of crystals in the threads and the knots are hyp-idiotopic and xenotopic. The formation, growth, and connection of pendants in such gravelly soils establish a firm continuous crystalline network (porous media) that constructs the whole profile. Formation of lenticular gypsum crystallitic and isles fabrics is the results of these processes.

Table 7.6 Morphologic characteristics of a Haplogypsid in Isfahan area (Toomanian et al. 2003)

Geomorphic position	Hor.	Depth (cm)	Layer boundary[a]	Color (dry)	Structure[b]	Consistency[c] (dry)	Secondary gypsum[d]	Reaction[e]
Loamy skeletal, gypsic, thermic, Leptic Haplogypsid								
Old gravelly alluvium (plateau)	A	0–8	aw	10YR5/6	g	so	1f sm	ev
	Av	8–15	aw	10YR5/6	v	sh	1f sm	Ev
	By1	15–73	gr ir	10YR7/3	m	sh	3 m pnd	Ev
	By2	73–115	gr ir	10YR7/3	m	sh	3 m pnd	Ev
	2By3	115–150	–	10YR7/3	m	sh	1 m hm	Ev

[a]Boundary: *a* abrupt, *g* gradual, *w* wavy, *ir* irregular; [b]Structure: *g* granular, *m* massive, *v* vesicular; [c]Consistency: *so* soft, *sh* slightly hard; [d]Secondary gypsum: 1 = small, 3 = big, *f* few, *m* many, *sm* soft masses, *pnd* pendant, *hm* hard masses; [e]Reaction: *ev* very effervescent

Table 7.7 Physical properties of a representative Haplogypsid soil in Isfahan area (Toomanian et al. 2001)

Horizon	Depth (cm)	Clay	Silt	Sand	VFS[a]	Texture[b]	SP[c] (%)	Gravel (%)
		%						
Loamy skeletal, gypsic, thermic, Leptic Haplogypsid								
A	0–8	12	33	55	11.2	g sl	19	30
Av	8–15	15	35	50	15.6	Sl	20	15
By1	15–73	6	22	72	11.5	g sl	16	45
By2	73–115	8	27	65	8.1	g sl	21	45
2By3	115–150	6	24	70	7.3	Sl	25	5

[a]*VFS* very fine sand; [b]*g* gravelly, *sl* sandy loam; [c]*SP* saturation percentage

Table 7.8 Chemical properties of a representative Haplogypsid soil in Isfahan area (Toomanian et al. 2001)

Hor.	OM (%)	Gy (%)	CCE[a] (%)	CEC cmol/kg	pH	EC (dS/m)	Soluble anions and cations (meq/l) ρb (gr/cm^3)							
							Ca^{2+}	Mg^{2+}	Na^+	K^+	$Co3^{2-}$	$HCO3^-$	$SO4^{2-}$	Cl^-
Loamy skeletal, gypsic, thermic, Leptic Haplogypsid														
A	0.253	2.6	44.5	6	7.7	3.2	33	1	3.3	0.3	0	1	33.1	3.5
Av	0.253	2.6	44.5	6.1	7.7	3.2	33	1	3.3	0.3	0	1	33.1	3.5
By1	0.137	77.92	10.2	5.1	7.9	4.3	35	7	7.6	0.8	0	0.8	38.7	10.9
By2	0.163	68	18.1	4.9	7.8	4.4	33	10	9.8	0.45	0	0.6	41.6	11
2By3	0.22	65	24.2	3.1	7.7	3.7	33	7	3.4	0.3	0	2.1	35.2	6.1

[a]*CCE* Calcium Carbonate Equivalent, measured with ammonium acetate

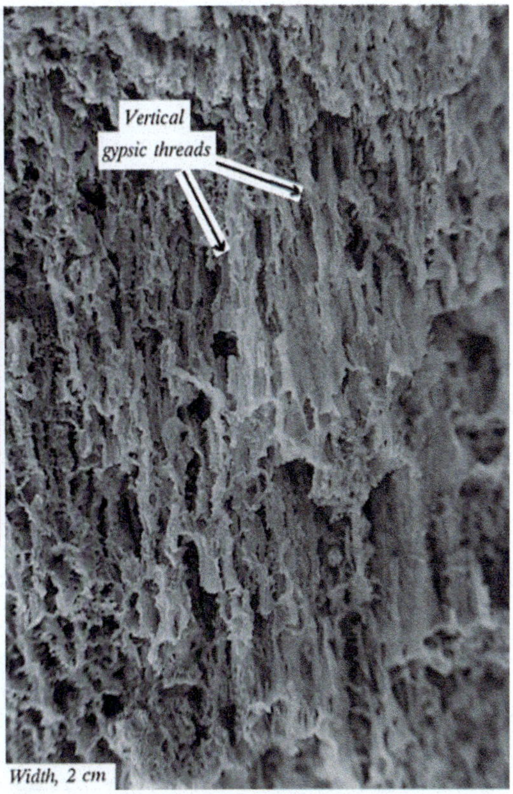

Fig. 7.8 Macromorphology of gypsum crystals in soils developed (left) on hyper-enriched gravelly or (right) on non-gravely old alluviums (Toomanian et al. 2003)

Salids (*Solonchaks*) have a salic horizon within 100 cm of the soil surface and are highly affected by salt accumulation throughout the soil profile. They cover a large area of the region, about 12 m ha (Table 7.5) and are not suitable for crop production due to a very high salt content.

The dominant great group is Haplosalids, which cover about 8.6 m ha of the land or about 9% of the total soil resources (Table 7.5). They are mostly formed in lowland, playas, and piedmonts of the arid region (Fig. 7.11). They are generally used as low productive rangelands. In fine-textured soils, the soluble mineral accumulations which are mostly originated from water table and salic horizons tend to form in the soil surfaces via capillary rise (Fig. 7.9). However, in gravelly and coarse-textured soils (piedmonts), salinity is mostly caused by percolating water, and the depth of salic horizon is governed by the depth of rainwater percolation (Moghiseh and Heidari 2012). Salic horizon formation coincides mostly with calcic and/or gypsic horizons in extreme aridic and ustic moisture regimes in the country. The formation of these

Fig. 7.9 A typical profile of a Haplosalid in cultivated area of Segzy lowland, Isfahan Province

Fig. 7.10 Polygons in salt crust geomorphic position left (Farpoor et al. 2012); desert pavement on saline soils near Bam, Kerman Province

Fig. 7.11 Salt dome and its nearby salt-affected soils (Typic Haplosalid) near Lar in Fars Province

three horizons at once is not considered in Soil Taxonomy. Moghiseh and Heidari (2012) reported high extent of soils with gypsic and salic horizons in Bam region, southeast of Iran. These horizons are cemented, occurring as petrogypsic and petrosalic horizons in the soils (Fig. 7.10 right).

Fig. 7.12 Gypsum (G) and halite (H) crystals in petrogypsic/salic horizon as coatings and infillings **a** PPL, **b** XPL (Moghiseh and Heidari 2012)

Morphology: Salids have evolved with two major morphologic characteristics. First, in sloping piedmonts around the Great Kavir in east of Bam, the soil surface is covered by igneous and sandstone gravels. A thin discontinuous surface horizon overlays abruptly on a powdery hypergypsic horizon. Underneath is a very hard cemented petrosalic-gypsic horizon. This gravelly horizon gradually merges into a massive and clayey horizon with very few gypsum crystals, which extends down to any depth. Second, in lowland or saline areas (*Dagh*) contained whitish powdery mixture of soluble minerals with gypsum crystals, the fine-textured salic-gypsic horizon extends to depth. The salt-crusted soil surface has a unique polygonal shape, which is seen in Fig. 7.10, left (Farpoor et al. 2012). These polygonal shapes are formed by growing of salt and gypsum crystals in crusted layer.

Physicochemical properties: Table 7.9 shows the results of physicochemical analysis of a representative profile of a Haplosalid developed in the upland near Bam.

Micromorphological features: The bimodal distribution of gypsum and halite with depth most probably point to a polygenic origin of the profiles. At first, during previous less arid phase, gypsum and halite must have been leached down to form the present subsurface horizons. Increasing drought reduces the activity of gypsum mineral compared with minerals that are more soluble. Therefore, the gypsum occurs mainly as coatings on the sand grains, and interstitial voids are filled by halite (Fig. 7.13). It is clear that halite has crystallized later, probably when the climate became more arid (Moghiseh and Heidari 2012). In most cases, salt accumulations are zonal assemblages of single cubic grains or its unshaped compacted crystals, which engulfs the other soil materials (Fig. 7.12).

Calcids (*Calcisols*) which have a calcic horizon within 100 cm from the soil surface are the third most widespread soils of the Aridisols and cover about 7.6 m ha or about 8% of the soils of the country. Haplocalcid is the only great group of the Calcids delineated on the map (Table 7.5). Calcids are not generally suitable for crop and fruit tree cultivation due to high calcium content, shallow depth, coarse texture, steep slope, and/or undulating surface (Gharaee and Mahjoory 1984). However, limited areas of these soils, which have suitable texture, slope, and depth, could be irrigated and used for crop production.

One of the most widespread subgroups of Calcids developed in the arid region of Iran is Haplocalcids (Fig. 7.14). These Calcids are one of the most widespread

Table 7.9 Physicochemical characteristics of a Haplosalid soil in Bam area (Moghiseh and Heidari 2012)

Hor.	Depth (cm)	Gravel	Clay	Silt	Sand	CCE[a]	Gypsum	pH	EC (dSm^{-1})
			%						
Fine-loamy, gypsic, hyperthermic, Petrogypsic Haplosalid									
A	0–4	5	37.5	18.3	44.2	9.68	0.77	8.19	90.2
Byz1	4–15	10	35.4	25.6	39.0	3.0	69.45	8.01	129.2
Byzm	15–37	20	12.4	18.2	69.4	1.06	12.71	7.92	187.7
Byz2	37–56	20	13.6	15.4	71.0	1.95	6.6	8.77	40.8
Byz3	56–88	40	13.4	14.3	72.3	1.9	12.6	8.65	40.6
Byz4	88–108	10	30.2	30.2	39.6	3.4	3.33	8.41	41.3
Byz5	108–130	0	34.4	38.4	27.2	16.25	4.5	8.52	56.4

[a]*CCE* Calcium Carbonate Equivalent, measured with ammonium acetate

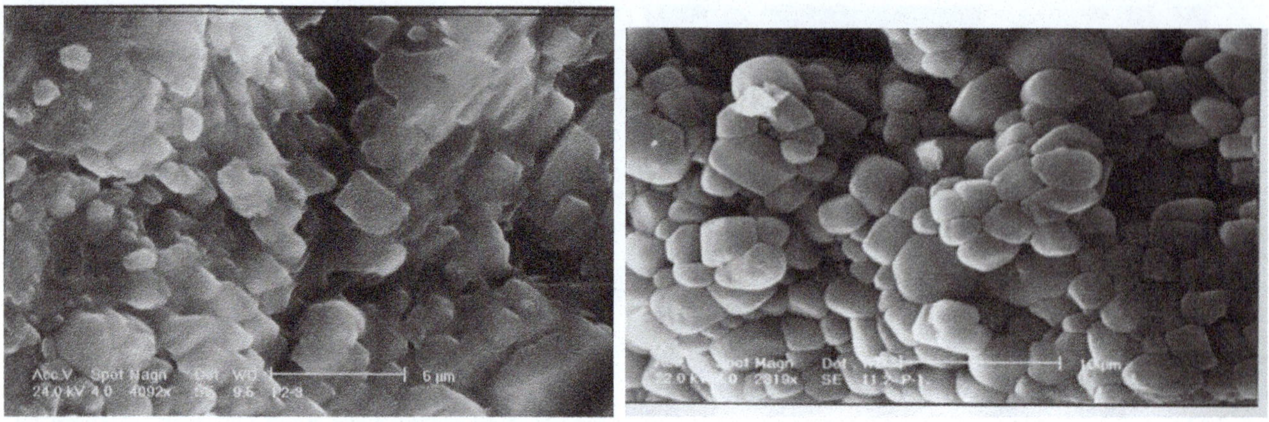

Fig. 7.13 Scanning electron micrograph of cubic salt crystals formed in a Haplosalid soil in Bam region (Moghiseh 2005)

Fig. 7.14 A typical landscape of Calcids (Typic Haplocalcid) developed on a plateau near Isfahan) left) and (Lithic Petrocalcid) formed in Fars Province close to Kazeroon (right)

soils in the arid region, where accumulation of secondary calcium carbonate occurs in subsurface horizons of the soils.

Morphology: In such soils, if not truncated, secondary carbonate accumulated horizons (Bkx,) are underlain an A horizon and underneath is a C or R. The secondary carbonates (calcites) are mostly concentrated in mycelium, infilling, powdery pocket, concretion, and pendant forms (Fig. 7.15). Zamanian (2005) describes a typical morphology of such soil in Qazvin Province (Table 7.10).

Physicochemical properties: The physicochemical properties of these soils are resulted from the activities of calcium and carbonate ions in the soils (Table 7.11). As it is clear from the Table 7.11, the clay content in this profile decreases and the gravel increases with depth.

Micromorphological features: Micromorphology properties of these soils showed sequential accumulations of carbonates. Large voids were the main sites for calcite precipitation, because such voids dry more rapidly than smaller ones. Large voids are usually in direct contact with lower atmospheric concentration of CO_2. Moreover, calcite has preferences for self-nucleation, therefore plugs progressively large voids by preferential precipitation on previously deposited calcite crystals.

The most characteristic pedofeatures are calcite depletion zones. They can be identified as areas of speckled or striated b-fabric in a micromass dominated by a crystallitic b-fabric. The dominant b-fabric thus varies from crystallitic to striated. As the decalcification features are localized phenomena in a groundmass with a crystallitic b-fabric, they are compatible with the presence of the relatively large amounts of carbonates indicated by chemical analyses of bulk samples. Various forms of calcitic pedofeatures are present in most Haplocalcids. Most notable are microcrystalline impregnative and pure calcite nodules (Fig. 7.16), which occur in matrix of the soils. Nodules of needle shaped calcite are occasionally observed. Cytomorphic calcite is mostly associated with the decalcified zones (Manafi et al. 2008).

Fig. 7.15 Infillings of secondary carbonates in soil vogue and channels (Toomanian 2007, right; Razavipour 2009 left)

Table 7.10 Morphologic characteristics of representative soil in Qazvin Province (Zamanian 2005)

Geomorphic position	Hor.	Depth (cm)	Layer boundary[a]	Color	Structure[b]	Consistency[c] (dry)	SCC[d]
				dry			
Clayey skeletal, mixed, thermic, Typic Haplocalcid							
Old fan	A1	0–14	cs	7.5YR4/4	1 gr	sh	es
	A2	14–34	gs	7.5YR5/4	1 sbk	h	es
	ABk	34–51	as	7.5YR5/4	1 sbk	h	es 1f sm
	Bk1	51–69	gw	7.5YR7/4	1 sbk	vh	es 1f sm
	2Bk2	69–82	cw	7.5YR6/4	m	eh	es 1f sm
	2Bk3	82–92	as	7.5YR7/4	m	eh	es 2 m
	3Bk4	92–122	cs	7.5YR7/3	m	eh	es 2 m
	4Bk5	122–152	aw	7.5YR7/3	m	eh	es 2 m
	5Bk6	>152	–	7.5YR7/3	m	eh	es 2 m

[a]Boundary: *c* clear, *g* gradual, *a* abrupt, *w* wavy, *s* smooth; [b]Structure: *gr* granular, *sbk* subangular blocky, *m* massive; [c]Consistency: *sh* slightly hard, *h* hard, *vh* very hard, *eh* extremely hard; [d]Secondary Carbonate Calcium: *es* slightly effervescent, *f* fine, *m* medium, *sm* soft mass, *sc* concretion, 1 = few, 2 = common

Table 7.11 Physicochemical properties of soils in Qazvin Province (Zamanian 2005)

Horizon	Depth (cm)	ECse (dS/m)	Clay	Silt	Sand	Gravely	CCE[a] (%)	pH
			%					
Clayey skeletal, mixed, thermic, Typic Haplocalcid								
A1	0–14	0.58	45.96	32.04	22.00	27.59	12.27	7.9
A2	14–34	0.52	62.00	20.00	18.00	40.55	19.94	7.8
ABk	34–51	0.39	50.00	20.08	29.92	49.04	29.14	7.8
Bk1	51–69	0.48	43.96	16.04	40.00	55.72	30.67	7.8
2Bk2	69–82	0.67	34.00	14.00	52.00	64.54	33.74	8.0
2Bk3	82–92	0.78	32.04	16.04	51.92	57.35	34.36	7.9
3Bk4	92–11122	0.71	31.96	18.04	50.00	67.76	38.34	8.0
4Bk5	122–152	0.63	26.00	14.04	59.96	62.04	23.62	8.1
5Bk6	>152	2.48	25.96	26.04	48.00	51.59	34.05	7.8

[a]*CCE* Calcium Carbonate Equivalent

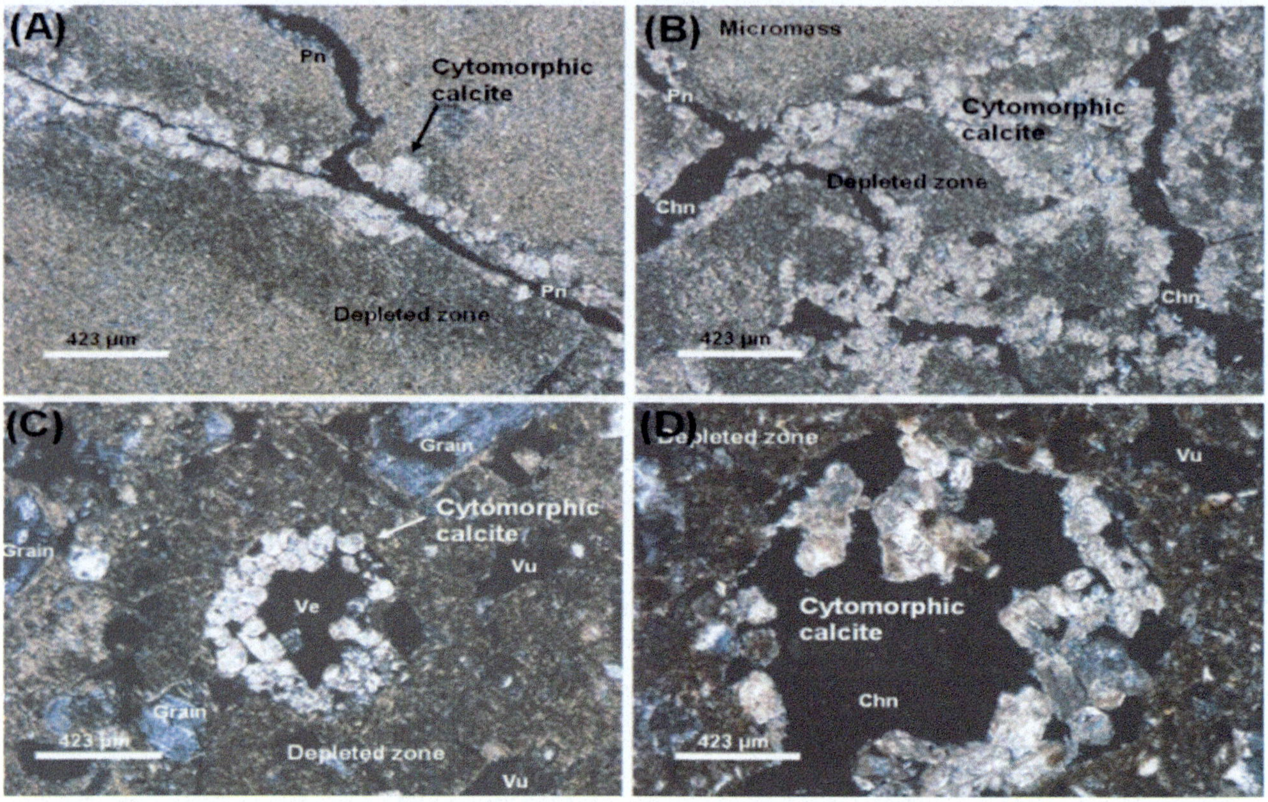

Chn: Channel, Vu: Vugh, Ve: Vesicle, Pn: Plain, OM: Organic matter, NFC: Needle fiber calcite

Fig. 7.16 Infillings of calcite crystals in planar, vugh, and channels, XPL (Manafi et al. 2008)

Cambids (*Cambisols*) are less extensive Aridisols and cover only about 2.4 m ha or 2.5% of the soils of the country. These soils have a cambic horizon within 100 cm of the soil surface. Haplocambid is the dominant great group. They are suitable for agricultural production where good quality water resources are available for irrigation, and the soil and land qualities such as salinity, water holding capacity, nutrient availability and slope are appropriate (Fig. 7.17).

Argids are not extensive and cover only a limited area of about 52,000 ha or 0.05% of the soil resources (Table 7.5). They are not usually under crop production due to their undulating landscape or other limitations. They are generally developed on the stable landscape of the upper terraces or plateaus and have an argillic (Haplargids) or a natric horizon (Natrargids) within 100 cm of the soil surface (Khademi and Mermut 2003; Khormali et al 2003). They are generally used as a low productive rangeland.

7.5.2 Entisols

These soils have been extensively developed in the arid region on young and unstable landforms. These soils cover about 28.8 m ha of the arid region or 30% of the total land areas (Table 7.5). Entisols, which are developed under other climatic conditions in xeric, ustic, and udic soil moisture regimes, are described in the relevant sections. However, it should be emphasized that about 73% of the total Entisols in the country are developed in the aridic soil moisture regime (Table 7. 1).

Entisols lack any distinct subsurface horizons except for an ochric epipedon in the surface. The main suborders, great groups, and subgroups of Entisols are presented in Tables 7.3 and 7.5 and are briefly described as followings:

Torriorthents (*Regosols*) are predominant soils and over about 22 m ha or about 23% of the total soil area of the country. This great group constitutes about 55% of the total Entisols and 76% of the Entisols developed in the arid region. They are usually formed on piedmont plains, undulating upper plateaus, or uplands with steep slope. They are mostly shallow, gravelly, or coarse-textured soils (Figs. 7.18 and 7.19). Therefore, most of them are not suitable for crop production and are generally used as low productive rangelands. Limited areas of Torriorthents are under tree or crop production, where water is available and land qualities are suitable.

These soils have only an ochric epipedon that is developed directly on hard rocks, regolith, and any kind of young

Fig. 7.17 A typical profile and land use of a Haplocambid under irrigation developed on the terraces of Zayandeh Rud near Isfahan

unconsolidated deposits in the vast areas of the arid region. They may be very shallow or very deep, gravelly or not, coarse or fine textured, with low to high potentiality for different uses.

Morphology: The absence of well-developed horizons in these soils is the result of the limited influence of climate, lithology, or chronological events. However, the nature of the bedrock appeared to determine the additional presence of certain characteristics. Figure 7.18 shows a typical position and characteristics of a Torriorthents developed in central Iran in Isfahan region. In most regions, the Torriorthents contain only an ochric epipedon over C-horizons, except for the cases where well-developed buried soils are covered by recent deposits (Esfandiarpour et al. 2013). The morphologic characteristics of a Torriorthent soil in Isfahan are presented in Table 7.12.

Physicochemical properties: Selected physicochemical and morphological characteristics of a typical Torriorthents of Isfahan region are shown in Table 7.13. In this soil, the light-textured young materials of the upper fan consist of coarse angular fragments. Periodic sedimentation inhibits the development of the recent soil, which buries a paleosol that has a gypsic horizon.

Stereoscopic interpretation of landforms in the watershed indicated that materials with gypsum have been eroded from the mountain and moved as colluvial and alluvial fans to plateaus and piedmont plains.

Micromorphological features: Thin sections of two Torriorthents developed in different geomorphic positions show very little evidence of pedogenesis in the soils. However, few euhedral or sub-hedral gypsum crystals are present in the pores or in the groundmass of soils. In the buried gypsic horizon, gypsic pedofeatures occur as little spots and pendants.

Torrifluvents (*Fluvisols*) *are* the second most extensive Entisols in the arid region and cover about 6 m ha of the land or 6.2% of the total soils. They are usually the most productive soils in the region where adequate water resources are available for irrigating crops and trees. They are developed on alluvial or piedmont plains and generally have suitable texture and depth for plant growth.

Torripsamments (*Arenosols*) are developed in small areas of about 780,000 ha in the arid region on very sandy alluvial plains closed to the rivers or on sand blown deposits (Tables 7.3 and 7.5). These soils are generally poor in soil nutrients and have a very low water holding capacity. They are not generally used for agricultural production due to unavailability of water resources and poor soil qualities. If water resources are available, limited areas of these soils may have potential for cultivation of vegetables and fruit trees using high efficient and pressurized irrigation system and/or fertigation.

In summary, soils of the arid region are productive where water resources are available for irrigation and the soil qualities are suitable for crop production. Presently, in the arid region of the country, there are areas of Aridisols, which are under cultivation of crops, such as wheat, barley, sorghum, saffron, different vegetables, and fruit trees such as pistachio, pomegranate, almond, date palm, citrus, and figs. Almost all Iranian pistachio, saffron, pomegranate, and date palm production, which are largely exported to other countries, are produced under irrigation on these soils (see

Fig. 7.18 Geomorphic position and profile of a Torriorthent soil in Isfahan region (Toomanian 1995)

Chap. 5). Where water is not available, these soils have very limited capacity for agricultural production and are mostly used as low potential pastoral rangelands. Some areas of these soils have been degraded due to overgrazing and mismanagement and are now facing severe wind and water erosion, salinity, and desertification processes.

Fig. 7.19 A typical profile of a Torrifluvent developed on the lower terraces of Zayandeh Rud in Isfahan

7.6 Soils of the Semiarid Region of Alborz and Zagros Mountains

According to the agro-climatic map of the country (see Chap. 9), semiarid regions mostly in the west, northwest, and northeast of the country constitute about 30% of the total land area. Soils developed under this climatic condition mostly have a xeric moisture regime. The total soil resources developed in the semiarid region cover an area of around 16 m ha or about 17% of the total soil areas of the country (Table 7.14). Most of the dryland agriculture is located in the highlands of the semiarid regions of Zagros and Alborz Mountain Ranges.

Soils developed in the xeric moisture regime are generally classified into five orders of Entisols, Inceptisols, Mollisols, Vertisols, and Alfisols. Table 7.14 shows the estimated total areas of the suborders of the soils with xeric moisture regime in the semiarid region.

About 97% of the soils developed in the xeric moisture are classified in two orders of Entisols and Inceptisols (Table 7.14).

Table 7.12 Morphological properties of a Torriorthent in Isfahan region (Toomanian 1995)

Hor.	Depth (cm)	Color (moist)	Texture	Structure	Consistence (dry)	Boundary	Gy (%)	Gravel (%)
Loamy skeletal over sandy, mixed, thermic, Typic Torriorthent								
A	0–15	7.5YR4/4	SL	sg	So	d ir	–	60
C1	15–70	7.5YR5/4	SL	sg	Sh	d w	–	60
2C2	70–100	10YR5/4	LS	sg	Sh	a b	–	30
2Byb	100–140	10YR4/4	L	sg	Sh	–	2 mcs	20

Symbols are applied according to soil survey manual

Table 7.13 Selected physicochemical properties of a Typic Torriorthent (Toomanian et al. 2001)

Horizon	Depth (cm)	Sand	Silt	Clay	OM	Gypsum[a]	CaCO3	pH	CEC[b] (cmol/kg)	EC (dS/m)
		%								
Loamy skeletal over sandy, mixed, thermic, Typic Torriorthent										
A	0–15	53	30	17	0.66	1.2	55.45	8	5.69	1.1
C1	15–70	68	21	11	0.54	1.1	59.65	8	2.79	1.06
2C2	70–100	81	13	6	0.3	1	59.27	8.15	2.1	0.84
2Byb	100–140	32	44	24	0.4	13.2	33.8	7.8	4.75	2.9

[a]corrected on dry weight and crystal water of gypsum; [b]Calcium Carbonate Equivalent, measured with ammonium acetate

7.6.1 Inceptisols

Inceptisols are the most widespread soils in the semiarid region and have either a cambic or calcic subsurface horizon. They are generally formed on relatively stable landforms, on upper river terraces or the piedmont plains with cambic or calcic subsurface horizons (Moazallahi and Farpoor 2012; Owliaie 2012). Xerepts (*Cambisols and Calcisols*) are the only suborder and the most dominant soils which cover an area of about 10.4 m ha in the region, that is about 11% of the soils of the country or 65% of the soils of the region. They have been formed on relatively stable landforms. These soils are usually under production for different crops and fruits where annual precipitation is adequate, or water resources are available for irrigated agriculture (Roozitalab et al. 2013). Calcixereptis, the most dominant great group followed by Haploxerept, is generally the most productive soils (Table 7.3) (Figs. 7.20 and 7.21).

Haploxerepts (*Cambisols*)

These soils occur in the xeric moisture regime with both thermic and mesic temperature regimes in the semiarid region of Iran. The soils have a cambic horizon below an ochric epipedon. Table 7.15 shows selected chemical and physical properties of a Typic Haploxerept developed under range on calcareous materials in Kermanshah Province, the Merek Watershed. The region is a part of the Central Zagros Mountain Chain, stretching in a northwest–southeast direction. Dry farming is a customary practice, especially on the hills. The main dry-farming crops are wheat, barley, and chickpeas. The irrigated crops include wheat, barley, rapeseeds, chickpeas, clover, alfalfa, and sugar beet. A small area is covered by orchards such as walnut, apples, cherries, and other fruit trees (Milani et al. 2010).

Calcixerepts (*Calcisols*) which have a calcic subsurface horizon is the major great group of Inceptisols in the region.

Table 7.14 Soils developed in the semiarid region with xeric moisture regime

Order	Suborder	Area/1000 ha	Soils of the region (%)	Total soils (%)
Entisols	Orthents Fluvents	4966 220	30.7 1.4	5.2 0. 22
Inceptisols	Xerepts	10,466	64.9	10.9
Mollisols	Xerolls	324	2.0	0.34
Alfisols	Xeralfs	131	0.8	0.13
Vertisols	Xererts	13	< 0.1	<0.01
	Total Soils	16,120	100	16.8

Fig. 7.20 A typical landscape of Inceptisols (Calcixerepts) under dryland wheat cultivation in Maragheh Research Station with xeric moisture regime (Photo by M. H. Roozitalab)

Most of Calcixerepts have developed on piedmonts, plateaus, and river alluvial plains in the mountainous regions with xeric moisture regime. This great group is the most prominent soil in many provinces of the country (Table 7.3).

Morphology: In sloping landforms with calcareous parent materials, soils are highly calcareous throughout, with an average $CaCO_3$ content of more than 40%, increasing with depth. Pendants, powdery pockets, concretions, and calcic nodules are the main morphological features of these soils. As Yusefifard (2012) has reported, the Calcixerepts that occur on gravelly materials are derived from extrusive igneous rocks (andesite and dacite) in Ardabil (Fig. 7.22). Soils of this kind generally have weak-to-medium sub-angular blocky or massive structure. Morphologic characteristics of a Clacixerept developed on back slope of a catena in Ardekan area of Fars province are shown in Table 7.16.

Physicochemical properties: In contrast to the arid regions, the more favorable climatic condition and denser vegetative which occur in the mesic–xeric regions facilitated the accelerated process of dissolution–precipitation and recrystallization of calcite. Higher degree of dissolution of carbonates takes place in surface horizons. This is confirmed by relatively high difference in the total $CaCO_3$ in the surface compared to the deeper horizons and the presence of only a few to common nodules in the soil surface compared to the considerable number of nodules in the deeper layers.

In gravelly soils, formation of calcitic pendants is the dominant calcitic pedofeature. A considerable amount of calcite occurs also as coatings along channels. Soil texture in Calcixerepts varies from sandy loam to clay, depending on the parent material (lithology) and position in the landscape. Some of the key physicochemical properties of a representative Calcixerept developed in Ardekan area of Fars province are presented in Table 7.17.

Micromorphological features: The microstructure varies from weakly to well separate sub-angular blocky, and the c/f-related distribution is porphyric. The course materials consist mainly of quartz and feldspar grains and fragments of limestone. Since soils are highly calcareous, the b-fabric is mostly calcitic crystallitic (Khormali 2003). Microcrystalline calcite nodules are widespread especially in Bk horizons. Micritic calcites appear as nodules or as coatings in channels, and laminar calcite coatings around coarser limestone fragments are the other types of calcitic features in some soil. The same micromorphological pedofeatures were observed (Fig. 7.23) in Calcixerepts as described on external

Fig. 7.21 Dryland agriculture on Calcixerepts and Haploxerepts at the two main research stations representing highlands of Zagros Mountains in (left) Maragheh and (right) Kermanshah (Photo by M.H.Roozitalab)

Table 7.15 Selected chemical and physical properties of a Typic Haploxerept soil developed in Kermanshah Province, Merek Watershed (Milani et al. 2010)

Hor.	Depth (cm)	Color (moist)	Sand	Silt	Clay	OC	pH	ECe	$CaCO_3$	>2 mm (%)	CEC
			%					(dS/m)	($\%^-$)		($cmol^+kg^-$)
Fine, mixed, superactive, thermic Typic Haploxerept											
Ap	0–20	7.5YR4/4	19.2	38	46.8	2.67	7.84	0.4	22	10	27.4
Bw1	20–40	7.5YR4/4	15.2	38	44.8	1.69	7.73	0.38	21	20	28.8
Bw2	40–70	7.5YR4/4	17.2	34	42.8	1.28	7.95	0.34	25	20	29.8
Bw3	70–103	7.5YR4/4	23.2	34	36.8	0.82	7.96	0.33	31.8	30	25.2
Bw4	103–125	7.5YR4/4	29.2	38	42.8	0.65	8	0.28	44.5	50	–

Fig. 7.22 A typical gravelly Calcixerept soil developed on andesite in Ardabil province (Yusefifard 2012)

igneous-derived parent materials (andesite and dacite) in Ardabil area (Yusefifard 2012).

Table 7.18 shows selected chemical and physical properties of a Typic Calcixerept developed under range on limestone in a mesic soil temperature regime in the Honam Watershed, located in Zagros Mountain Chain, south of Alashtar city of Lorestan province (Milani et al. 2010). The area has cold winters and relatively modest summers. The mean annual air temperature is 8.8 °C, the maximum mean monthly temperature is 20.3 °C, and the minimum mean monthly temperature is −4 °C. The average monthly potential evapotranspiration is 35 mm in winter and 176 mm in summer.

7.6.2 Entisols

Here too, Entisols are mostly developed on unstable landforms with steep slope or on the recent river deposits or alluvial plains. They make up about 32% of the soils in the region (Table 7.14) and are classified as Xerofluvents and Xerorthents.

Xerofluvents (*Fluvisols*) constitute only about 1.4% of the total soils of the region that is about 220,000 ha. They are generally formed on river terraces and are the most productive soils for crop production. They are either under irrigated or dry-farming.

Xerorthents (*Regosols*), which are mostly used as low productive rangelands, occur on unstable and slopping landscapes. They are one of the most prevalent soils in the xeric moisture regime. They cover an area of about 5 million ha or around 31% of the soils of the region or about 5.2% of the total soil resources of the country (Table 7.14). These soils are not well developed, because the parent materials are composed of recent deposits or made up of hard materials that resist further weathering. The relief and steepness of the slopes may not allow development of the soils.

Physicochemical properties: A Typic Xerorthent that occurs on a colluvial–alluvial fan in Kamfirouz in Fars Province has a gravelly clay loam texture and a massive structure throughout the soil profile (Razavipour 2009). Selected properties of these soils are shown in Table 7.19.

7.6.3 Mollisols (*Kastanozems*)

These soils are generally developed on stable landforms in association with or as inclusion in Inceptisols, in small areas of the semiarid zones of central and southern Zagros Mountains. They occur mostly on the highlands of Chaharmahal and Bakhtiari, Kermanshah, Lorestan, and Fars provinces where annual precipitation is rather high, native vegetations mainly *Quercus brantii*, and short grasses are dense enough to promote formation of a mollic epipedon. They make up about 2% of the soil resources in the region (Table 7.14).

Mollisols occur usually on undulating landforms on limestone or other calcareous parent materials in erosional terraces where cultivation of agricultural crops is not usually

Table 7.16 Selected morphologic characteristics of a Calcixerept in the Fars Province (Khormali 2003)

Hor.	Depth (cm)	Color (moist)	Structure[a]	Consistence[a] (moist)	Boundary[a]	Other components[+]
Fine, smectitic, calcareous, mesic, Typic Calcixerept						
Ap	0–25	10–7.5YR4/4	m2 gr	Fr	gs	Few to common fine roots
Bw	25–70	7.5YR4/4	c 1 abk ± (m1abk)	fi	cl	Few irregular calcic powdery pockets, few to common fine roots
Bk1	70–100	7.5YR4/4	c 1 abk ± (m1abk)	fi	ab	Common to many irregular calcic powdery pockets and concretions
Bk2	100–130	7.5YR4/4	m	fi	–	Common to many irregular calcic powdery pockets and concretions, 40% gravel

[a]Symbols used according to abbreviations given in Soils Survey Manual, + all soils are calcareous throughout and ± Indicates primary and secondary structure

Table 7.17 Selected physicochemical properties of a Typic Calcixerept formed in Ardekan, Fars Province (Khormali 2003)

Hor.	Depth (cm)	Sand	Silt	Clay	pH	Textural class	EC (dSm^{-1})	SP	CEC ($cmol^{+}$ kg^{-1})	Gravel	CCE	OC
		%								%		
Fine, smectitic, calcareous, mesic, Typic Calcixerept												
Ap	0–25	17.2	43.4.4	39.4	7.4	sicl	0.3	50	18	–	31.5	0.6
Bw	25–70	17.2	39.4	43.4	7.4	sic	0.3	56	20	–	32.5	0.5
Bk1	70–100	15.2	47.4	37.4	7.5	sicl	0.3	58	16	–	36.5	0.2
Bk2	100–130	21.2	37.4	41.4	7.5	c	0.3	59	20	40	48	0.1

CCE Calcium Carbonate Equivalent, measured with ammonium acetate

Fig. 7.23 Calcite infilling and coating in a Calcixerept developed on andesite-derived parent mater in Ardabil Province (Yusefifard 2012)

common. Degradation of the mollic epipedon due to improper land use changes is remarkable (Khormali and Nabiollahi 2009).

Calcixerolls, Haploxerolls, and Argixerolls are the three great groups of Xerolls, with total surface area of about 320,000 ha that are reported (Tables 7.3 and 7.14).

Calcixerolls have a calcic horizon below the mollic epipedon and are mostly found in limited areas in association with or as inclusion of Calcixerepts.

Haploxerolls have developed in lesser extent in the region and have a cambic horizon. These soils are often in association with Haploxerepts. The xerolls of the semiarid region

Table 7.18 Selected properties of a Typic Calcixerept developed in Lorestan Province in Honam Watershed (Milani et al. 2010)

Hor.	Depth (cm)	Color (moist)	Sand	Silt	Clay	OC	pH	ECe	$CaCO_3$	>2 mm (%)	Av. K^+	Av. P
			%					(dS/m)	(%¯)		$mgkg^{-1}$	
Fine, mixed, mesic, Typic Calcixerepts												
Ap	0–30	10YR6/3	5	43	52	1.31	7.55	0.54	26.3	–	10.5	10.2
Bw	30–55	10YR6/3	4	41	55	0.51	7.67	0.43	26.8	–	1.9	2.8
B	55–80	10YR5/4	5	46	49	0.29	7.63	0.41	28	–	1.6	2.8
Bk	80–130	10YR5/4	6	47	47	0.16	7.73	0.31	29.5		2.3	2

Table 7.19 Selected physicochemical properties of a Typic Xerorthent (Razavipour 2009)

Horizon	Depth (cm) 0–15	Sand	Silt	Clay	TNV	OC	Gravel	EC	pH	CEC
		%						dS/m		cmol/kg
Loamy skeletal, mixed, sub-active, thermic, Typic Xerorthents										
A	0–15	30.8	41	28.2	15.2	0.45	50	0.63	7.8	18.2
C1	15–60	43.8	28	28.2	30.1	0.17	40	0.53	7.9	15.4
C2	60–120	38.8	32	29.2	33.6	0.15	50–60	0.63	7.9	19.3

CCE Calcium Carbonate Equivalent, measured with ammonium acetate

of Zagros and Alborz Mountains occur in a climatic zone bordering the soils developed under ustic or aridic moisture regime.

Argixerolls (*Luvic Kastanozem*) have an argillic horizon below the mollic epipedon, developed in a tiny area in the semiarid region. They are generally under native plant cover.

The selected Argixeroll developed in the Ardekan region in Fars Province is described by Khormali (2003). The soil has a medium-textured, high-base saturation and is slightly calcareous in the surface, but calcareous in the subsurface horizons. The key morphological characteristics of the soil are shown in Table 7.20. Soil structure is very well-developed, strong granular crumb with excremental pedofeatures in surface. The soil has a well-developed angular blocky structure in subsurface horizons.

Physicochemical properties: *Selected* characteristics of a soil profile from Ardekan region in Fars Province are shown in Table 7.21.

In Xerolls of Fars province, the most common pedofeatures are calcite depletion, which increases in importance with development of the soils and can be identified as areas of speckled or striated b-fabric (Khormali 2003). This pattern varies from much extended depletion zones in Argixerolls to very few depletion zones in Haploxerolls. Therefore, the b-fabric varies from crystallitic to striate. Microlaminated limpid clay coatings (striated b-fabric) are observed in Bt horizons of the Argixerolls in Fars Province (Fig. 7.24).

7.6.4 Alfisols (*Luvisols*)

They occur in small and scatted areas of about 130,000 hectares in the region (Table 7.14). Haploxeralf is the most important great group of the Alfisols, mainly formed on the stable surfaces under natural vegetation, mixture of grasses, forbs, and woody shrubs in the semiarid region and under warm deciduous forests of more humid zones in the country. The landforms around Kor River in Fars Province (Fig. 7.25) and in Gorganrud River in Golestan Province are typical landforms for development of dry and moist Haploxeralfs, respectively.

Morphology: The selected morphological features of a Calcic Haploxeralf formed in the Gazvin Plain are shown in Table 7.22. Most of the soils are heavy-textured, ranging from loam and clay loam to silty clay and clay soils.

Calcium carbonate accumulation as powdery pockets and nodules are the main morphological features of somewhat developed soils with calcic horizons. High amount of gypsum is formed within 100 cm with or without a calcic horizon in some soils of Fars Province (Khormali 2003)

Physicochemical properties: Electrical conductivity varies in these soils depending on the location and geomorphic position of the soils. CEC in such soils differs mostly depending upon the amount and type of clay minerals. Some physicochemical properties of the same profile in Gazvin plain are shown in Table 7.23.

Table 7.20 Selected morphologic properties of a Typic Argixeroll developed in Ardekan, Fars Province (Khormali 2003)

Horizon	Depth cm	Color (moist)	Structure[a]	Consistence[a] (moist)	Boundary[a]	Other components[+]
Fine, smectitic, superactive, calcareous, mesic, Typic Argixerolls						
A	0–25	7.5YR3/3	gr	fr	cs	Common fine roots
Bt	25–75	7.5YR4/4	c1abk ± m2abk	fi	cs	Common clay skins
C	75–120	7.5YR4/4	m ± (m1abk)	fi	–	–

[a]Symbols used according to abbreviation given in Soils Survey Manual, ± Indicates primary and a secondary structure

Table 7.21 Physicochemical characteristics of a Typic Argixeroll developed in Ardekan, Fars Province (Khormali 2003)

Hor.	Depth (cm)	Particle size distribution			Textural class	pH	OC	[a]CCE	SP	Gypsum	CEC cmol/kg^{-1}	EC dSm^{-1}
		Sand	Silt	Clay								
		%					%					
Fine, smectitic, superactive, calcareous, mesic, Typic Argixerolls												
A	0–25	30	38	32	cl	7.1	2.1	9	38	tr	22	0.4
Bt	25–75	20	34	46	c	7.3	1.4	10	43.1	tr	32.1	0.3
C	75–120	42	30	28	cl	7.4	0.6	74	30.6	tr	18.8	0.2

[a]*CCE* Calcium Carbonate Equivalent (%)

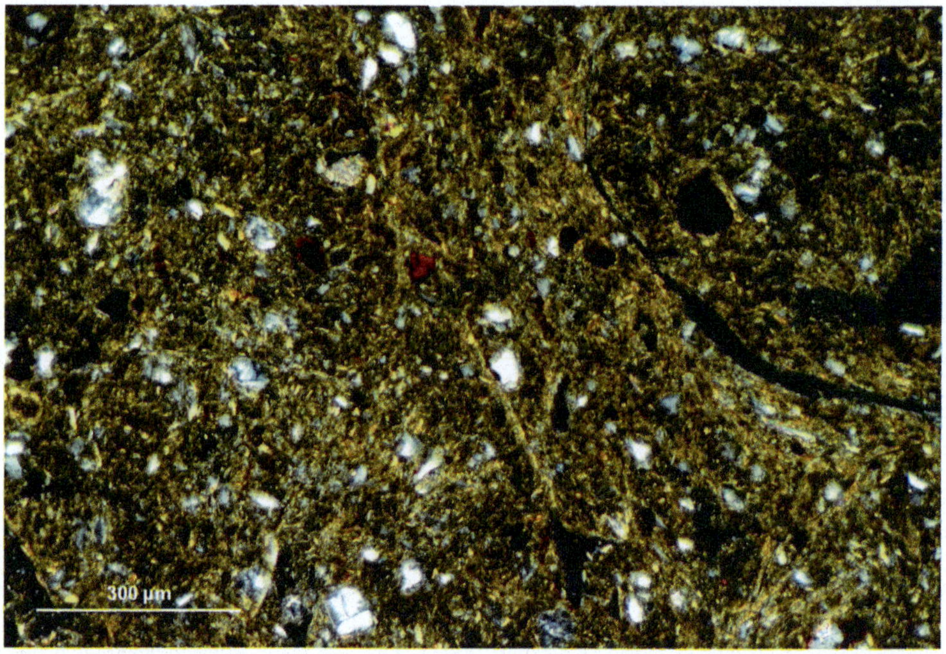

Fig. 7.24 Striated b-fabric in a Typic Argixeroll formed in Fars Province, circular polarized light (Khormali 2003)

Micromorphological features: Most micromorphological studies of Argillic horizons in arid and semiarid to Mediterranean regions of Fars Province have shown few to no clay coatings (Khormali et al. 2006). Although shiny surfaces considered as clay skins were developed in some soils, but in thin sections, microlaminated clay coatings occur mostly in Haploxeralfs (Fig. 7.26), and few or none were present in other soils.

The soils show evidence for illuviation in the fine to total clay ratio in Bt horizons compared to overlying horizons. In Haploxeralfs, occurrence of calcite depletions pedofeatures is common in various parts of Iran. These pedofeatures increase in importance with development of the soils and can be identified as areas of speckled or striated b-fabric. The calcite depletion pedofeatures, expressed partly by a speckled b-fabric, especially in the more developed Haploxeralfs,

Fig. 7.25 A typical landscape of Haploxeralf developed in Fars Province (Razavipour 2009)

Table 7.22 Selected morphologic characteristics of a Calcic Haploxeralf, Gazvin Plain (Manafi 2009)

Horizon	Depth (cm)	Color	Structure	Consistency			Gravel (%)
				Dry	Moist	Wet	
Fine-loamy, mixed, active, thermic, Calcic Haploxeralfs							
Ap	0–17	10YR5/4	f1gr, c1sbk	so	lo	ss-ps	5
Bt	17–36	7.5YR5/6	c1sbk, c2abk	sh	vfr	ss-ps	20
2Bk	36–89	10YR6/4	m2abk	sh	fi	ss-p	35
2Btk1	89–120	7.5YR6/4	m2abk	h	fi	ss-ps	15
2Btk2	126–160	7.5YR5/4	m2abk	vh	vfi	ss-ps	47
2Btk3	160–200	10YR7/4	m2abk	vh	vfi	ss-p	25

Table 7.23 Selected physicochemical characteristics of a Calcic Haploxeralf, Gazvin Plain (Manafi 2009)

Horizon	Particle size distribution (%)				pH	EC (dSm^{-1})	CEC (cmol/kg)	OC (%)	CCE (%)	SP (%)	ρb ($grcm^{-3}$)
	Sand	Silt	Clay	Text.							
Fine-loamy, mixed, active, thermic, Calcic Haploxeralf											
Ap	34.8	29	36.2	C L	7.58	0.51	20.14	0.92	1.34	41.39	1.67
Bt	41.8	13	45.2	C	7.5	0.51	17.1	0.85	4.78	39.46	1.78
2Bk	50.8	12	37.2	S C	7.66	0.39	18.18	0.56	17.43	38.97	1.78
2Btk1	31.8	15	53.2	C	7.6	0.4	17.45	0.49	34.74	39.02	1.55
2Btk2	42.8	18	39.2	C L	7.69	0.45	14.28	0.37	32.65	49.26	1.72
2Btk3	26.8	26	47.2	C	7.71	0.64	14.93	0.27	41.84	46.62	1.58

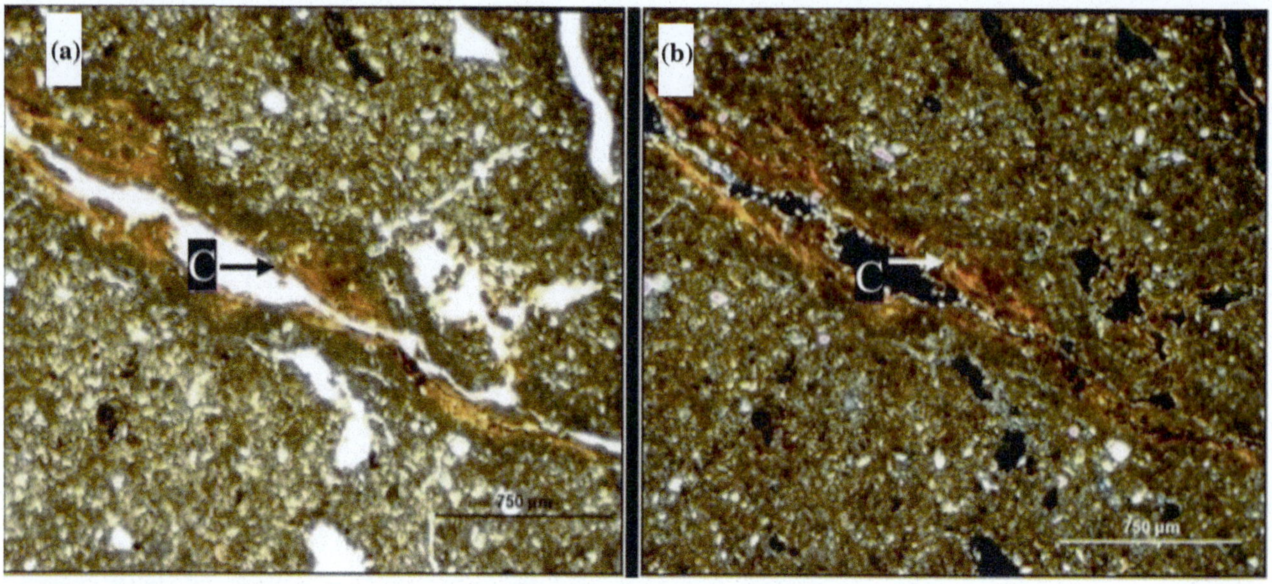

Fig. 7.26 Highly oriented clay coatings (C) of a channel in Btk3 horizon (**a** ppl and **b** xpl) (Khormali 2003)

are also an indication of higher dissolution and depletion of calcite in upper horizons (Khormali 2003). This pattern varies from much extended depletion zones to very few depletion zones. Therefore, the b-fabric varies from crystallitic to striated. Other soils mostly displayed a crystallitic b-fabric combined with a speckled b-fabric depending on the degrees of calcification or decalcification. The impregnation of some calcite nodules by Fe-oxide is the result of a higher degree of weathering, liberating iron from primary minerals (Khormali 2003).

7.6.5 Vertisols (Xererts)

These soils occur in a very small area, less than 0.1% of the total soils in the semiarid region. Calcixererts and Haploxererts are the major great groups of the Xererts (Table 7.14). These soils usually have an ochric epipedon over a cambic or calcic subsurface horizon. The mineralogy class is usually mixed or montmorillonic (Heidari et al. 2008). They are generally developed on calcareous deposits or limestone, mainly in the provinces of Chaharmahal and Bakhtiari, Kermanshah, and Lorestan (Fig. 7.27).

Calcixererts: It is the main great group of *Vertisols* of the region (Table 7.14). Calcixererts are very deep soils with high shrink–swell potential. Composition of parent materials and their geomorphic positions has immense role in formation of these soils. The Calcixererts of Chaharmahal and Bakhtiari provinces are developed in a limited area on toe-slops of a catena spread in more humid parts (Farpoor 1995).

Morphology: These soils exhibit high clay content, slickensides, or a wedge-shaped structure at specific depths and cracks that periodically open and close. Some soils exhibit a distinct gilgai microtopography and granular microaggregates at the surface. The Calcixererts of Lorestan and Kermanshah provinces are very deep soils with an ochric or sometimes mollic epipedon in the surface horizons (Heidari et al. 2008). Subsurface horizons show a moderately to strongly developed, medium to coarse prismatic structure with slickensides and wedge-shaped aggregates. Most of them have shown moderate to strongly developed calcic horizons, which are usually overlain by a cambic horizon. In Ardabil province, these soils are characterized with a mollic epipedon with fine-to-medium granular structure and moderately to strongly developed fine to medium wedge-shaped peds and angular blocky structure with slickenside in the subsoils. Irregular medium carbonate nodules have been observed in the soils of all locations (Heidari et al. 2008).

Physicochemical properties: The main physicochemical characteristics of a Chromic Calcixerert soil in Kermanshah province are presented in Table 7.24. The clay content is high throughout the profiles, with higher values in the subsurface horizons. The soils have slickenside and cracks with large amounts of illites in the clay fraction. Smectite clay minerals constitute only a small portion of the clay fraction. The typical morphology of these soils is probably due to the fineness of the clay fraction coupled with the influence of climatic regime.

Fig. 7.27 A Vertisol (Chromic Calcixerert) with a calcic horizon (left) developed in association with a Inceptisol, Typic Calcixerept (right) in the xeric moisture regime in Chaharmahal and Bakhtiari provinces (Courtesy by Dr. A Mehnatkesh)

Table 7.24 Selected physicochemical properties of a Chromic Calcixerert in Kermanshah area (Heidari et al. 2008)

Horizon	Depth (cm)	Structure	Clay	Silt	Sand	pH	EC (dS/m)	$CaCO_3$	OC	CECs	FC/TC	COLE
			%					(%)				
Ap	0–20	m2gr	58.4	32.4	9.2	7.9	0.4	27.9	1.39	22.8	0.53	0.12
Bss	20–44	c3pr	56.8	34.0	9.2	8.0	0.3	28.3	1.01	24.57	0.47	0.11
Bkss1	44–65	c3pr	67.2	24.8	8.0	7.7	0.3	32.4	1.02	21.96	nd	nd
Bkss2	65–100	c2pr	73.279.2	24.8	2.0	7.6	0.3	35.7	1.23	21.96	0.73	0.14
Bkss3	100–155	c2abk	79.2	18.8	2.0	7.7	0.5	39.9	1.13	21.96	0.67	0.14
Bkss4	155–175	c2sbk	72.6	24.4	3.0	7.6	0.2	43.0	0.7	19.50	0.77	0.13

FC/TC fine clay over total clay. *CCE* is measured with ammonium acetate

7.7 Soils of the Caspian Sea Region

Caspian Sea is the biggest lake in the world and has an area of about 371,000 square kilometers. The entire southern coast is located in Iran and is mainly influenced by humid to sub-humid climatic condition (see Chap. 9). The region that comprises three provinces of Gilan, Mazandaran, and Golestan constitutes an area of more than 58,200 sq km or 3.5% of the total land area of the country. Because of its favorable climatic condition and fertile soils, the region is generally suitable for cultivation of various crops and fruit trees such as rice, wheat, oilseeds, tea, maize, cotton, citrus, kiwi, hazelnuts, and olive.

The soils are generally fertile and play a key role in agricultural production and food security of Iran. In 2015, about 1.4 million hectares, which is about 10% of the total cropland and orchards of the country, were under irrigated and rainfed farming systems in the three above-mentioned provinces (SCI 2015).

Soil development in the Caspian Sea region is highly influenced by the high amount of annual precipitation, which decreases from West to East. The annual precipitation changes from more than 1500 mm in extreme west in Gilan Province to less than 250 mm in the northeast of the region near the Turkmenistan border at the northern part of the Golestan Province (see Chaps. 3 and 9).

The major soil moisture regimes are udic and xeric, and the prevalent soil temperature regime is thermic. The udic moisture regime is common in Gilan and western part of Mazandaran provinces. The xeric moisture regime is dominant in the central to eastern part of the region in Mazandaran and Golestan provinces. The mountainous areas to the south of the plains, which are under forest, have predominately udic (or ustic) soil moisture regime. The aquic soil moisture regime is prevalent in the lowland of the region, which is mainly under paddy cultivation. Aridic soil moisture regime is dominant in the northern part of the Golestan Province.

Soil resources are very diverse and are classified into six soil orders. Soils cover about 4.84 million hectares or 83% of the total land area. Table 7.25 shows the major soil orders and their great groups in the region. The five orders of Entisols, Inceptisols, Mollisols, Alfisols, and Aridisols are the most widespread soils (Table 7.26). Ultisols have only developed in a limited area under natural forests on the mountains in the south of the region, which receive higher precipitation and have a udic soil moisture regime.

Many studies on the effect of the various soil-forming factors on the development of the soils in the Caspian Sea region have been carried out (Khormali et al 2012 and Rezaei et al. 2012). These studies indicate that the development of the soils is highly influenced by topography, parent materials, vegetation, and rainfall.

7.7.1 Mollisols

These soils are the most prevalent soils in the region. They occur on various landforms and constitute about 1.9 m ha or 39% of the total soil resources of the region. The soils have a mollic epipedon and a cambic, calcic, and/or argillic horizon. The base saturation is 50% or more in all horizons. The main suborders of the Mollisols in the region are aquolls, udolls, xerolls, and rendolls (Table 7.26)

Aquolls (*Gleysols*) are mostly under paddy cultivation and are formed in Gilan and Mazandaran provinces. They have an aquic soil moisture regime and are classified in the two great groups of Endoaquolls and Calciaquolls (Table 7.3). The mollic epipedon generally overlies the cambic or calcic horizon. These soils cover an area of about 122,000 ha and are generally formed in association with Aquepts (Endoaquepts) or Aqualfs (Endoaqualfs).

Udolls are the most widespread Mollisols and cover a total area of about 645.000 ha in the region. They are generally classified in three great groups of Calciudolls (*Chernozems*), Hapludolls, and Argiudolls (*Phaeozems*). In Calciudolls, a calcic horizon has been developed below the mollic epipedon and in Hapludolls; a cambic horizon is formed below the mollic epipedon (Fig. 7.28). Argiudolls, which are more prevalent in the west of the Caspian Sea region in the Gilan Province, have the argillic horizon below the mollic epipedon.

Udolls are developed under mountainous beech, oak, and hornbeam forests in Lahijan (Bijarbagh and Amircola), Gilan Province on phyllite and granite (Bakhshipour 1993), and on mountains and hills of Narmab in Gorganrud

Table 7.25 Suborders and major great groups of the soils developed in Caspian Sea region

Orders	Suborders	Great Groups
Entisols	Aquents Fluvents Orthents	Endoaquents, Fluvaquents, Udifluvents, Xerofluvents Udorthents, Xerorthents
Inceptisols	Udepts Xerepts Aquepts	Dystrudepts, Eutrudepts, Humudept Halpoxerepts, Calcixerept Humaquepts, Endoaquepts
Mollisols	Udolls Xerolls Aquolls Rendolls	Hapludolls, Argiudolls, Calciudolls Haploxerolls, Calcixerolls, Argixerolls Endoaquolls Haprendolls
Alfisols	Udalfs Xeralfs Aqualf	Hapludalfs, Paleudalfs Haploxeralfs, Endoaqualfs
Ultisols	Udults Humults	Hapludults, Haplohumults
Aridisols	Calcids Salids Cambids	Haplocalcids, Aqusalids, Haplosalids Haplocambids

Table 7.26 Extent of soil orders and suborders in the Caspian Sea region

Order	Area 1000 ha	Soils of region %	Suborder	Area in 1000 ha
Alfisols	367	7.5	Xeralfs Udalfs Aqualfs	20 272 75
Aridisols	532	11.0	Calcids Salids Cambids	171 199 162
Entisols	1140	23.6	Orthents Fluvents Aquents	769 317 54
Inceptisols	895	18.5	Xerepts Udepts Aquepts	322 410 163
Mollisols	1903	39.3	Xerolls Udolls Rendolls Aquolls	630 645 506 122
Ultisols	5	<.01	Humults	5
Total soils	4842	100		4842

Fig. 7.28 A Mollisol with a cambic horizon (Typic Hapludoll) developed on calcareous parent material in Gilan under udic soil moisture regime (left, courtesy of Dr. H. Ramezanpour) and (right) a Mollisol (Typic Calcixeroll) with a calcic horizon formed on calcareous material in Gorgan area under xeric moisture regimes (By F. Khormali)

Watershed from loess deposit in Golestan province (Khormali et al. 2005).

Morphology: The Hapludolls in Narmab area are formed on a hill slope with parent material consisted mainly of loess deposits. Formation and development of mollic epipedon and leaching of calcium carbonate were the major processes. The mollic epipedon is thicker in soils under natural forest. Intensive tillage and water erosion are the main factors responsible for reduction in thickness of the mollic epipedons in farmland areas. While calcium carbonate content of the subsurface horizon of the forested soils is significantly higher than surface horizon due to the leaching process, it is considerably high in the surface horizon of the deforested soils and the amount does not change significantly with soil depth. This indicates the effect of deforestation and cultivation practices which has led to the soil erosion and the subsequent exposure of the highly calcareous subsurface horizon. Although the Argiudolls of Bijarbagh and Amircola regions are developed in mountainous area with slopes of more than 70%, the soils are thick and contain well-developed argillic horizons. This indicates the influential role of the forest and humid climate on the soil development and physicochemical properties.

Selected morphological and physicochemical properties of a Typic Argiudoll developed in Amircola and Lahijan are shown in Tables 7.27 and 7.28.

Xerolls (*Kastanozem*). These soils are also prominent Mollisols mostly developed in Mazandaran and Golestan provinces. They cover a total area of about 630,000 ha. The main great groups of Xerolls are Haploxerolls, Argixerolls, and Calcixerolls. The soils classified as Haploxerolls have generally cambic horizons below the mollic epipedon. The Argixerolls have argillic horizons below the mollic epipedon. Large areas of Argixerolls (Fig. 7.29) with thick mollic epipedons have been developed on the loess deposit in Mazandaran and Golestan provinces. Calcixerolls are the most dominant Xerolls and have calcic horizon within 150 cm of the mineral soil surface.

Table 7.29 shows selected properties of a Calcic Argixeroll developed in the Gonbad area, Golestan Province, under natural vegetation mainly trees consisting of Fagus orientalis and Quercus castaneifolia.

Selected properties of a Typic Calcixerolls developed in the Gorganrud Watershed, Golestan Province, are presented in Table 7.30. This soil has a calcic horizon below the argillic horizon. Gorganrud Watershed in Golestan Province has an average annual soil temperature and precipitation of 16 °C and 600 mm respectively. The major parts of the forest are grown on mountains and hills with loess parent material. The main plant species of the forest are Alnus subcordata, Parrotia persica, Carpinus betulus, and Crataegus sp. The farmlands are mainly under wheat cultivation.

Rendolls (*Mollic Leptosols*) have been extensively formed in the Caspian Sea region under the udic soil moisture regime. The mollic epipedon is less than 50 cm thick that overlies calcareous materials with more than 40% Calcium Carbonate Equivalent. These soils, which cover a total area of about 500,000 hectares (Table 7.26), do not have argillic or calcic horizons. The main great group of Rendolls is Haprendolls (Fig. 7.30).

In Golestan and Eastern Mazandaran, loess deposits cover a large area of about 300,000 hectares extending from

Table 7.27 Selected morphologic characteristics of a Typic Argiudoll developed in Lahijan area (Bakhshipour 1993)

Horizon	Depth (cm)	Color (moist)	Texture	Structure	Consistence	
					Dry	Wet
Loamy skeletal, mixed, thermic, Typic Argiudoll						
A1	0–13	10YR3/2	SiCL	1 m gr	sh	ss/sp
A2	13–31	10YR3/3	SiCL	1 m gr	h	s/p
Bt1	31–62	7.5YR4/6	SiC	2 m abk	vh	vs/vp
Bt2	62–86	5YR4/6	SiC	3 m abk	vh	vs/vp
Bt3	86–122	5YR4/4	SiC	2 m abk	vh	s/p
C	122–175	5YR4/4	CL	1 m abk	vh	s/p

Table 7.28 Selected physicochemical properties of a Typic Argiudoll developed in Lahijan area (Bakhshipour 1993)

Hor.	Depth (cm)	Particle size distribution			SP	Gr.	OC	pH	CEC	EC
		Clay	Silt	Sand						
		(%)							Cmol/kg	dS/m
Loamy skeletal, mixed, thermic, Typic Argiudoll										
A1	0–13	31.3	55.1	13.6	72	43	3	5.8	24.4	0.3
A2	13–31	36	50.7	13.3	61	40	2.2	5.9	16.9	0.3
Bt1	31–62	46.6	44.8	8.6	62	25	0.77	5.9	16.9	0.2
Bt2	62–86	50	42.9	6.9	63	46	0.57	5.9	18.6	0.1
Bt3	86–122	49.3	43.7	7	62	51	0.35	5.9	17.2	0.1
C	122–175	39.1	36.7	24.3	50	78	0.35	5.7	17.4	0.1

Hor. Horizon; *SP* Saturation percentage; *Gr* Gravel; *OC* Organic carbon; *CCE* Calcium Carbonate Equivalent, measured by ammonium acetate the upland to the river terraces and piedmont plains. The depth of the loess deposits is more than 35 m in some places such as in Nekah, Nodeh, and Tooshan (see Chap. 8). Paleosols which reflect climatic changes in the region during the last 130,000 years have been widely formed during the wet periods (see Chap. 8). Recent soils (Fig. 7.31) have been developed on the surface of the loess deposit on various physiographic units and climatic conditions (Khormali et al. 2012).

7.7.2 Alfisols

These soils are one of the important soil orders in the Caspian Sea region. They occur in the udic, xeric, and aquic moisture regimes and cover a total area of about 370,000 hectares or 7.5% of the total soils (Table 7.26). The main suborders of the Alfisols in the region are **Xeralfs** (*CalcicLuvisols*), **Udalfs** (*Haplic Luvisols*), and **Aqualfs** (*Gleysols*).

The Aqualfs (Endoaqualfs) are found in lands under rice cultivation and cover an area of about 75,000 ha. The Udalfs (Hapludalfs) are the most widespread Alfisols in the region and cover an area of about 270,000 ha (Fig. 7.33). The Xeralfs (Haploxeralfs) occur in smaller areas of about 20,000 ha in the eastern Mazandarn and in Golestan provinces, mainly on the loess deposit (Fig. 7.32).

Haploxeralfs (*Haplic Luvisols*) : This is a major great group of Alfisols in the xeric moisture regime developed mostly in the Mazandaran and Golestan provinces.

Fig. 7.29 Mollisols with thick mollic epipedon and argillic horizon (Pachic Argixeroll) developed in the upper terraces and plateaus on loess deposits near city of Gorgan in the east of the Caspian Sea region (By F Khormali)

Table 7.29 Selected morphological and physicochemical properties of a Calcic Argixeroll developed in Gonbad area, Golestan Province (Khormali et al. 2012)

Hor.	Morphologic properties				pH	OC (%)	SP	CCE (%)	CEC (cmol$^+$kg$^-$)	*Particle size distribution*		
	Depth (cm)	Color (moist)	Structure	Calcite nodules						*Clay*	*Silt*	*Sand*
Calcic Argixeroll												
A	0–30	10YR2/2	3c sbk	–	7.0	2.1	57	8	35	*33*	*49*	*18*
Bt	30–60	7.5YR4/4	2 m abk	–	7.5	1.1	60	10	32	*42*	*50*	*8*
Bk1	60–105	10YR6/4	2 m abk	c2sc	7.9	0.2	44	30	21	*26*	*61*	*13*
Bk2	105–145	10YR6/4	1f abk	m2sc	7.9	0.2	46	33	20	*26*	*61*	*13*
C	145–190	10YR6/5	m	–	7.8	0.1	46	28	20	*26*	*63*	*11*

CCE Calcium Carbonate Equivalent

Table 7.30 Selected morphological and physicochemical properties of a Typic Calcixeroll developed in the Gorganrud Watershed, Golestan Province (Khormali and Ajami 2011)

Hor.	Morphologic properties			pH	OC (%)	CCE (%)	CEC ($cmol^{+}kg^{-}$)	Particle size distribution		
	Depth (cm)	Color (moist)	Structure					Clay	Silt	Sand
Fine-silty, mixed, calcareous, thermic, Typic Calcixeroll										
Ap	0–20	10YR/3	m2 abk	7.27	2.47	4	23.2	19.2	67.5	13.3
Bt1	20–60	10YR4/4	m2 abk	7.17	0.69	2.5	24.1	32.5	61.7	5.8
Bt2	60–100	10YR5/6	f1 abk	7.51	0.62	13.5	29.2	15.8	75	9.2

Fig. 7.30 Typic Haprendoll developed on calcareous parent material under udic soil moisture regime in Gilan (right, courtesy of Dr. H Ramezanpour) and Mazandaran Provinces (left, courtesy of Dr. A Heidari)

Fig. 7.31 Recent soils (Argixerolls) have been developed on the loess deposit (right) and paleosols developed at the upper and lower parts of the loess deposit in Gorgan area (left), by M. H. Roozitalab

Fig. 7.32 Alfisol (Typic Haploxeralf) developed under the forest cover on a loess deposit in the Gorgan region (right) and paleosols were formed in the lower part of the underlying loess deposit (left), photo by M H Roozitalab

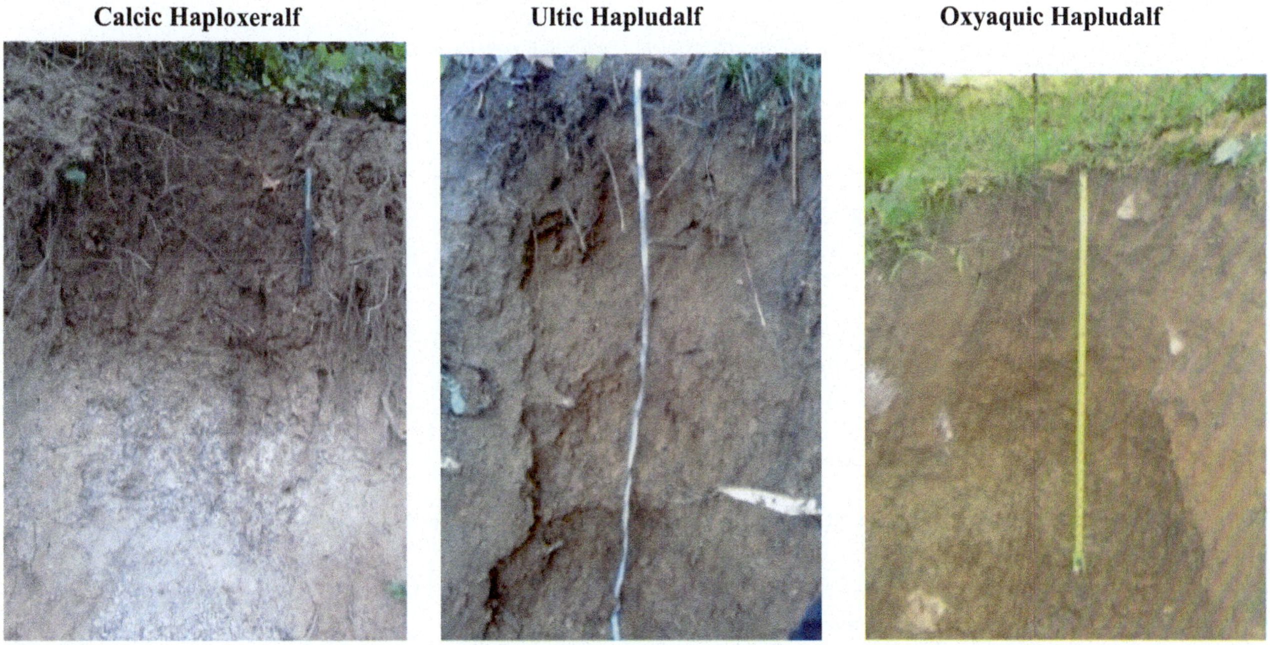

Fig. 7.33 Alfisols developed under xeric and udic soil moisture regimes in Golestan (by Dr. F. Khormali) and Gilan provinces (by H Ramezanpour)

Tables 7.31 and 7.32 show selected properties of two Haploxeralfs developed in Golestan Province developed under natural forest and croplands.

Udalfs are developed in the udic moisture regime in the Caspian Sea region in areas of about 270,000 (Tables 7.3 and 7.26). ***Hapludalfs*** (*Luvisols*) is the only prevalent great group of Udalfs. These soils are developed on phyllites, granites, and other igneous and metamorphic rocks in the Lahijan region (Bakhshipour 1993) and on thick sedimentary and metamorphic rocks of Tertiary and Quaternary periods in Gilan Province (Rezaei et al. 2012). These soils are also developed in Gilan and Mazandaran provinces on limestone and calcareous sedimentary deposits (Emadi et al. 2012) and in Golestan Province on loess deposits (Khormali and Kehl 2011, Khormali et al. 2012). Some of these soils are reported to be under paddy cultivation in Nowshahr, Rasht, and Ramsar (Raheb and Heidari 2011).

Physicochemical properties: The studied soil profiles were slightly acidic, and their pH increased with depth due to extensive leaching and removal of basic cations from the upper horizons. Organic carbon is the highest in the surface layers and decreased regularly with depth. The soil texture ranged from sandy loam, sandy clay loam to clay. The soils were well drained in all physiographic positions except in the foot slopes. Despite a considerable variation within soil profiles, the pH, EC, and CCE gradually decreased, and clay content, OC, and CEC increased in the soils across precipitation gradient. Some physicochemical properties of selected profiles of Hapludalfs developed in three provinces of Gilan, Mazandaran, and Golestan are shown in Tables 7.33, 7.34, and 7.35.

Micromorphological features: Most of the Udalfs identified in Northern Iran have a well-developed sub-angular blocky microstructure. The argillic horizon has moderately to well-separated sub-angular blocky microstructure associated with planes, channels, and vughs. The occurrence of well-developed sub-angular blocky microstructure is related to the clay content, the type of dominant clay mineral, and the shrink–swell properties. The c/f-related distribution is open-porphyric in the argillic horizon, with quartz (10–60 μm) as the major coarse component. Open-porphyric distribution provides the free translocation and orientation of clay particles from upper horizons around voids.

In general, major pedogenic processes including the clay illuviation and calcite accumulation affect the voids shape and volume. The planar voids are the only types, which increase in abundance with depth. This phenomenon can be attributed to the shrink–swell activity during the wetting and drying periods (Khormali et al. 2012). The dominant b-fabric of the Bt horizons is speckled and striated in the Udalfs.

Table 7.31 Morphological and physicochemical properties of a Calcic Haploxeralf developed in Golestan Province (Khormali et al. 2012)

Hor.	Morphologic properties				pH	OC (%)	SP	CCE (%)	CEC ($cmol^{+}kg^{-}$)	Particle size distribution		
	Depth (cm)	Color (moist)	Structure	Calcite nodules						Clay	Silt	Sand
Fine-silty, mixed, Thermic, Calcic Haploxeralfs												
A	0–10	10YR3/3	1f sbk	–	6.9	2.6	59	7	42	30	57	13
Bt	10–50	7.5YR3/4	2 m abk	–	7.0	1.0	55	5	21	44	51	5
Bk1	50–80	10YR4/6	2 m abk	c2sc	7.8	0.5	50	34	19	30	54	16
Bk2	80–107	10YR4/6	1 m abk	c2sc	7.8	0.4	44	35	18	23	63	14
Ck	107–150	10YR4/6	m	f1sc	7.7	0.2	42	34	15	18	69	13

Table 7.32 Morphological and physicochemical properties of a Typic Haploxeralf developed in Gorganrud Watershed, Golestan Province (Khormali and Ajami 2011)

Hor.	Morphologic properties			pH	OC (%)	CCE (%)	CEC ($cmol^{+}kg^{-}$)	Particle size distribution		
	Depth (cm)	Color (moist)	Structure					Clay	Silt	Sand
Fine-silty, mixed, calcareous, thermic, Typic Haploxeralfs										
A	0–15	10YR4/4	m1 gr	7.31	2.96	6.5	29.2	20.8	71.2	8
Bt	15–45	10YR4/6	f1 abk	7.49	2.24	21.8	15.5	32.6	57.9	9.5
Bw1	45–80	10YR4/6	f1 abk	7.62	0.43	22	22.6	21.7	70.8	7.5
Bw2	80–120	10YR4/6	f1 abk	7.58	0.29	17.5	18.6	19.2	70.8	10
Bk	120–150	10YR5/6	f1 abk	7.6	0.13	22	13.8	15.8	74.2	10

Table 7.33 Physicochemical properties of a Ultic Hapludalf developed in Lahijan area, Gilan Province (Rezaei et al. 2012)

Hor.	Depth (cm)	pH ($CaCl_2$)	CEC (meq 100^{-1} gr)	K (mg/kg)	P (mg/kg)	N	BS	OC	Fed	Clay	ρb ($grcm^{-1}$)
						%					
Ultic Hapludalfs											
A	0–25	4.44	18.6	147.42	8.7	0.254	39	3.25	0.51	21.1	1.32
Bt1	25–57	4.47	16.5	76.44	4.9	0.058	36	1.75	0.58	15.1	1.31
Bt2	57–84	4.45	26.79	132.48	6.8	0.038	33	1.44	0.64	49.2	1.34
Bt3	84–110	4.77	23.95	92.82	6.3	0.026	40	0.27	0.63	47.1	1.36
C	110–150	4.89	17.73	80.34	4.5	0.002	40	0.04	0.41	27.3	1.4

[a]*CCE* Calcium Carbonate Equivalent, measured by ammonium acetate

Table 7.34 Selected morphological and physicochemical properties of a Mollic Hapludalf developed in Royan area, Mazandaran Province (Emadi et al. 2012)

Hor.	Morphologic properties				pH	OC (%)	CCE (%)	CEC ($cmol^{+}kg^{-}$)	Tex	Particle size % distribution		
	Depth (cm)	Color (moist)	Structure	Effervescence n						Clay	Silt	Sand
Fine, mixed, Thermic, Mollic Hapludalf												
A	0–14	10YR3/2	1 m sbk	–	6.2	2.5	11	55.5	CL	39	35	25
Bt1	14–74	7.5YR3/4	3 m abk	–	6.4	2.1	24	48.2	C	58	38	4
Bt2	74–109	7.5YR3/3	3c abk	–	6.7	0.9	24	55.0	C	62	34	4
BC	109–150	7.5YR6/6	1f abk	e	7.2	0.2	12	33.2	Sic	45	41	14

Table 7.35 Selected morphological and physicochemical properties of a Typic Hapludalf developed under forest in Golestan Province (Khormali et al. 2012)

Hor.	Morphologic properties				pH	OC (%)	SP	CCE (%)	CEC ($cmol^{+}kg^{-}$)	Particle size distribution		
	Depth (cm)	Color (moist)	Structure	Calcite nodules						Clay	Silt	Sand
Typic Hapludalf												
A	0–13	10YR2/2	2 m gr	–	6.4	4.8	70	7	33	32	58	10
Bw	13–30	7.5YR4/6	3 m sbk	–	6.4	0.6	67	9	21	33	53	14
Bt1	30–62	5YR4/6	3 m abk	–	7.4	1.4	53	6	31	52	37	11
Bt2	61–103	5YR4/5	2 m abk	–	7.2	0.2	58	8	19	38	51	11
Bk	103–125	10YR4/5	1 m abk	m2sc	7.4	0.1	46	24	37	28	60	12
Ck	125–165	10YR5/6	m	f1sc	7.6	0.1	40	21	22	21	65	14

Diffuse distribution of iron hydroxides (100–300 μm) and the occurrence of moderately to well impregnated typical ferruginous nodules (40–600 μm) are signs of well-developed hydromorphism (Stoops 2003). According to Stoops (2003), the typical ferruginous nodules have a pedogenic origin. The high amount of calcium carbonate in Bk horizons can be attributed to the later re-calcification processes (Khormali and Ajami 2011; Khormali 2003). This suggestion is confirmed by observing calcite coatings and hypocoatings along chambers and channels in Fig. 7.34.

7.7.3 Ultisols (*Alisols*)

These soils occuring a very limited area of about 5000 hectares under native forest vegetation in the rugged and undulating mountainous areas of the Caspian Sea region in Gilan and Mazandarn provinces.

These soils do not generally have potential to be used for crop production due to their physiographic position and steep slopes and are generally under natural forest vegetation. They usually have well-developed and prominent

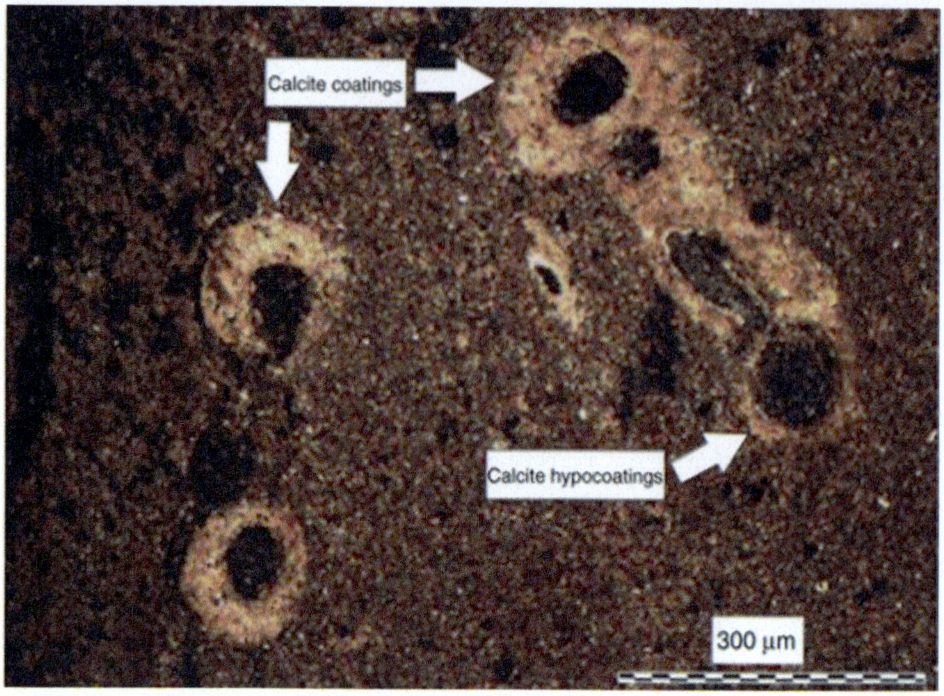

Fig. 7.34 Calcite coatings and hypocoatings along chambers and channels of a Udalf soil in Northern Alborz Mountain (Khormali et al. 2012)

agillic horizon below the ochric epipedon. In the surface of most of the soils, the O horizon consisting of slightly decomposed to highly decomposed organic materials such leaves, twigs, moss, and lichens is usually observed. The base saturation below 180 cm of the mineral surface is usually less than 35%. Very limited field survey studies have been carried out on these soils, and as a result, little information is available on the extent of these soils in the forest areas of the Caspian Sea.

Hapludults and Haplohumults have been formed in small and scattered areas as inclusion with Inceptisols (Dystrudepts) on mountainous landscapes in Gilan and west of Mazandaran. These soils exhibit a typical elluvial and illuvial horizons. They are generally formed under the forest on landforms that are at least late Pleistocene in age. According to pedostratigraphic correlation and results of luminescence age estimates, the uppermost layers of the loess deposits covering the Alborz Mountain foothills in the Caspian Sea region were accumulated during the last glacial under dry and probably colder climatic conditions than today (see Chap. 8 and Kehl et al. 2005). This enhanced weathering causes mineral alteration and the dominance of iron and aluminum oxides. The presence of the iron oxides causes the A horizon of these soils to be stained red. Leaching causes these soils to have low quantities of basic cations.

7.7.4 Inceptisols

These soils are found under udic, xeric, and aquic moisture regimes in the Caspian Sea region in association with other soils. They cover an area of about 895,000 ha or 18.5% of the total soils in the region (Table 7.26). They are the third widespread soils in the region and mostly used for production of various crops and fruit trees. The main subgroups of Inceptisols are Xerepts, Udepts, and Aquepts.

Xerepts (*Cambisols*) are developed on relatively stable landforms in an area of about 322,000 ha or 36% of the Inceptisols. They are mostly found in Golestan Province in association with Xerolls. The soils have generally ochric epipedon and cambic (Haploxerepts) or calcic subsurface horizons (Calcixerepts) within 100 cm of the soil surface (Fig. 7.35).

Aquepts (*Gleysols*) in Mazandaran and Gilan provinces are mostly under paddy cultivation. They cover a total area of about 160,000 ha or 18% of the Inceptisols in the region. They are found mostly in association with aquolls or aqualfs.

Endoaquepts cover an area of about 220,000 hectares. They are mostly under rice cultivation and continuous water stagnation and saturation process as indicated by redoximorphic features (Fig. 7.36). In Endoaquepts, the soils are saturated with groundwater fluctuation from bottom to top

Fig. 7.35 Calcixerepts developed in association with Calcixerolls in east of Caspian Sea region in Golestan Province

Fig. 7.36 A Typic Endoaquept soil profile and field under continuous rice cultivation in Mazandaran Province (Courtesy by Dr A. Heidari)

(Endosaturation). However, in Epiaquepts, the saturation is not caused by groundwater, but it forms by permanent surface water percolation. The Episaturation may take place by anthric saturation in which soil saturation occurs due to flood irrigation. Paddy soils in provinces of Gilan and Mazandaran are mostly classified in Endoaquepts, but there are some small areas of such paddy soils in inland provinces (Isfahan, Khuzestan, Fars, and others) that are mostly classified in Epiaquepts.

Morphology and Physicochemical properties: In northern paddy fields, the soil color has a grayish hue denoting the strongly reducing condition; soil structure is massive in ochric epipedon and cambic horizon. The paddy soils of Mazandaran province (Raheb and Heidari 2011) show very

Table 7.36 Selected properties of a paddy soil (Endoaquept) developed in Ramsar area, Mazandaran Province, (Raheb and Heidari 2011)

Hor.	Depth (cm)	Color (moist)	Sand	Silt	Clay	OC	pH	ECe	CEC	CCE (%)	K^+
			%					(dS/m)	($cmol^+kg^-$)		Soluble
Fine, mixed, thermic, Typic Endoaquept											
Apg	0–16	10YR3/1	10.5	36	53.5	4.5	7.7	0.1263	24.8	4.2	2.7
Bg1	16–40	2.5Y3/1	32.4	34	33.6	0.4	7.7	0.1163	24.7	2.2	2.3
Bg2	40–100	2.5Y3/1	30.4	34	35.6	0.6	7.7	0.1113	23.4	9.6	2.3

CCE Calcium Carbonate Equivalent

distinct morphological, physical, and chemical properties from those of non-paddy soils.

Table 7.36 shows selected properties of a paddy soil (Typic Endoaquept) developed in Mazandaran Province in Ramsar area which has low altitude of 20 m below sea level. The field observations confirmed the presence of shallow water table and poor drainage conditions that are responsible for reducing the soils condition. The mean annual precipitation in the region is about 1250 mm, and the mean annual air temperature is 15.8 °C.

Udepts are another prevalent Inceptisols in the region, which are mostly developed in Gilan Province. They cover a total area of about 410,000 ha or 46% of the Inceptisols in the region (Table 7.26). Dystrudepts and Eutrudepts are the most dominant great groups. Eutrudepts have a base saturation (by NH4OAc) of 60% or more in one or more horizons at a depth between 25 and 75 cm from the mineral soil surface or directly above a root-limiting layer that is at a shallower depth (Soil Survey Staff 2014).

Table 7.37 shows selected properties of a Oxyaquic Dystrudept formed on undulating and hilly landforms in Gilan Province, Lahigan area under a tea plantation. The climate is humid with the mean annual precipitation of about 1300 mm and the mean minimum and maximum annual temperatures of 2.8° and 19.5 °C, respectively. The mean annual humidity is 77.5%, and the mean annual potential evapotranspiration is 884 mm. The soil moisture and temperature regimes are udic and thermic, respectively. Tea is mainly cultivated on these undulating hilly landforms. The major geological formations are composed of thick sedimentary and metamorphic rocks of Tertiary and Quaternary periods.

7.7.5 Entisols

They are the second most prevalent soils after the Mollisols and cover an area of about 1.14 million hectares or about 24% of the total soils of the region. They are developed in four soil moisture regimes of xeric, udic, aridic, and aquic on unstable landscapes or recent deposits. The main suborders of the Entisols are Orthents, Fluvents and Aquents.

Orthents (*Regosols*) are the most widespread and cover an area of about 770,000 ha or 67% of all Entisols in the region (Table 7.26). Udorthents are the most dominant great group and cover about 0.5 million ha or 44% of the Entisols. These soils are mainly under natural vegetation and are mostly developed on the sloping landforms and steep slope in the west of the region (Table 7.3).

Fluvents (*Fluvisols*) occur on the recent alluvial deposits in the udic, xeric, and aridic moisture regimes near the rivers and are mainly used for agricultural production under rainfed or irrigated systems. They cover a total area of about 315,000 ha or 28% of the Entisols (Table 7.26).

Aquents (Glysols) cover a small area of 55,000 ha or about 5% of all Entisols in the region. They are not generally suitable for crop production unless they are drained. Most of these soils, however, are used for paddy cultivation in Gilan and Mazandaran.

Table 7.37 Physicochemical properties of a Dystrudept developed under tea cultivation in Lahijan, Gilan Province (Rezaei et al. 2012)

Hor.	Depth (cm)	pH ($CaCl_2$)	CEC (meq 100^{-1} gr)	K (mg/kg)	P (mg/kg)	N	BS	OC	Fed	Clay	ρb ($grcm^{-1}$)
						%					
Oxyaquic Dystrudept											
Ap	0–25	3.6	27.5	44.88	6.4	0.24	13	3.78	0.84	19.5	1.16
C1	25–60	3.96	13.4	47.78	3.4	0.034	28	0.12	0.41	24.1	1.29
C2	60–90	3.95	10.8	44.39	3.2	0.052	34	0.12	0.4	20.1	1.30
2Cg1	90–120	3.9	22.8	120.9	3.7	0.004	54	0.35	0.59	36.2	1.32
2Cg2	120–150	4.07	19.7	91.69	3.6	0.004	66	0.31	0.45	38.2	1.31

Fig. 7.37 Aridisols (Aqusalids/Solonchaks) shown in the left and Entisols (Torriorthents/Regosols) shown in the right developed under aridic soil moisture regime in Turkmen Sahara, northern part of the Golestan Province (by M H Roozitalab)

7.7.6 Aridisols

These soils have developed in the Caspian Sea region only at the northern part of the Golestan Province that has an aridic soil moisture regime. These soils cover an area of about 530,000 ha or 11% of the soils of the region. The main suborders of Aridisols in the region are Salids, Calcids, and Cambids (Fig. 7.37). The morphological and physicochemical properties of these soils have been defined the soils of the arid region, Sect. 7.5.

Many areas of Cambids and a limited area of Calcids in the region are under irrigated agriculture, but Salids are not generally suitable for crop production due to high salinity and low groundwater level.

7.8 Soils of the Tropical Semiarid Region

As indicated earlier, the ustic soil moisture regime occurs in the soils of warm and tropical semiarid region of the South and Southwest of the country, close to coastal arid regions of the Persian Gulf. This moisture regime covers an area of about 10% of the country, mainly in parts of Khuzestan, Ilam, "Kohgiluyeh and Boyer-Ahmad", Hormozgan, and Fars provinces where ustic moisture regime and thermic or hyperthermic soil temperature regimes are dominant and rainfall occurs mostly during the warm winter and spring seasons.

The total area of soils with ustic moisture regime is estimated to be around 5.83 million hectares or 6% of the total area of the soils in the country. These soils are generally formed in the landscapes bordered to soils with aridic soil moisture regime. The main two soil orders, which cover more than 99% of the soil areas, are Entisols and Inceptisols.

The low amount of precipitation and high amount of evapotranspiration is not inducive for development of other soil orders. There are only small areas of Mollisols (Ustolls/Calcic Kastanozem) and Alfisols (Ustalfs/Calcic Luvisols) formed mainly on stable landforms and calcareous parent materials (Tables 7.3 and 7.38).

Due to steep slopes, undulating landscapes or lack of water resources, majority of the soils in the region with ustic soil moisture regime are currently under native vegetation or scattered forest trees. Small areas (about 10%) of the soils are currently used for dry-farming of wheat and barley in places where rainfall is enough and slope and other soil

Table 7.38 Soil orders and suborders developed in ustic soil moisture regime

Orders	Suborders	Area/1000 ha	Soil area %
Alfisols	Ustalfs	9	0.1
Entisols	Orthents	3582	61.4
	Fluvents	92	1.6
Inceptisols	Ustepts	2117	36.3
Mollisols	Ustolls	36	0.6
	Total soils	5836	100

properties are appropriate (Figs. 7.38, 7.39, and 7.40). Water shortage is usually the main constraint for many places, hence irrigated agriculture is not feasible.

7.8.1 Entisols

These soils have only an ochric epipedon, are the most widespread soils in the region, and constitute around 63% of the soil areas. Ustorthents (*Regosols*) are the most prevalent Entisols and cover a total area of about 61% of the soil area (Table 7.38). They are mostly course textured and developed on steep or undulating landscapes on recent deposits. These soils are not generally suitable for cultivation of crops or fruit trees.

7.8.2 Inceptisols

These soils are the second most important soils developed mostly on stable landforms of calcareous or gypsiferous materials. They may have cambic, calcic, or gypsic horizons. Some soils have petrocalcic or petrogypsic horizons. They constitute around 36% of the total soils of the region (Table 7.38). The main great groups of the Inceptisols as shown in Table 7.3 are Haplustepts (*Cambisols*) and Calciustepts (*Calcisols*) .

Calciustepts and Haplustepts occur in a large area of the region in Khuzestan, Fars, Ilam, Lorestan, and Kohgiluyeh and Boyer-Ahmad" provinces on sloping/undulating piedmont fans, old gravelly alluviums of bajada and glacis, plateaus, river alluvial plains. Calciustepts have mainly

Fig. 7.38 A Inceptisol (Lithic Haplustept) developed on limestone under oak trees in Southern Zagros Mountain near Kazeroon, Fars Province (By M H Roozitalab)

Fig. 7.39 Dry-farming of rapeseed and wheat on soils (Calciustepts/Calcisols) developed in ustic moisture regime in Gachsaran Research Station in Kohgiluyeh and Boyer-Ahmad Province (Courtesy of Dr. M. Roustaii)

Fig. 7.40 Dryland wheat varietal trial on an Inceptisol (Typic Calciustept/Calcisol) in Shirvan Chardavol Research Farm, Ilam Province (Courtesy of Dr. M. Roustaii)

developed on calcareous and gypsiferous parent materials (Abbaslou et al. 2013; Owliaie 2012; Owliaie et al. 2006; Khormali 2003). Gypsum and calcium carbonate accumulate in such soils, according to their relative solubility, at the depth in which the leachate rests. These soils with petrocalcic or petrogypsic horizons have also been reported.

Morphology: Some morphological properties of a selected profile formed in Fars Province are shown in Table 7.39. Soil matrix colors vary according to their location. Soils with the heavier texture have darker colors. The calcic, gypsic, and salic are the main morphological pedofeatures in soils of Hormozgan Province (Abbaslou et al. 2013). Calcic features were observed in the field as powdery pockets, nodules, or concretions in subsurface horizons. In addition, for gypsum the features are pendants and pockets of gypsum crystal (Abbaslou et al. 2013). The soils have generally medium to fine texture in subsurface horizons but according to micromorphological studies, illuviation process was not taken place. However, a lithological discontinuity occurs in river alluvial plain, suggesting that the soil at this site has formed on deposit recently laid over older sediment.

Physicochemical properties: The physicochemical properties of a Calciustept profile from Hormozgan province are given in Table 7.40. The soil textures in most regions of south coastal province (Bushehr and Hormozgan) are sandy loam, silt loam, loamy sand, silt, sand, loam, and silty clay loam, with low clay content (Owliaie 2012).

Micromorphological features: The overall microstructure of the Calciustepts ranged between weakly and well-separated sub-angular blocky. Since soils are highly calcareous, the b-fabric is mostly calcitic crystallitic, except in some soils with argillic horizons, where it can be speckled when calcite depletion occurs as well as granostriated b-fabric. Within the soils of more humid areas, spots, coating, hypo-coating, diffusive rings, and concentrations of ferromanganese formations were common.

Micromorphological observations confirmed that the calcite features are as medium-to-coarse typic lithogenic nodules (200–6000 μm in size) with a sharp boundary as well as calcite pendants beneath small gravels or previously formed calcite nodules (Fig. 7.41). These soils exhibited coarser texture with larger voids, hence secondary carbonate

Table 7.39 Morphological properties of a selected Calciustept developed near Fasa, Fars Province (Khormali 2003)

Horizon	Depth (cm)	Color (moist)	Structure[a]	Consistence[a] (moist)	Boundary[a]	Other components[+]
Fine-loamy, carbonatic, thermic, Typic Calciustept						
Ap	0–30	10YR4/3	M	fr	cs	Few to common fine roots
Bk1	30–60	7.5–10YR4/4	m ± (c1abk)	vfi	gs	Common irregular lime powdery pockets, few fine roots
Bk2	60–135	7.5–10YR4/4	m ± (c1abk)	fi		Few irregular lime powdery pockets

[a]Symbols used according to abbreviation given in Soils Survey Manual. + All soils are calcareous throughout and ± Indicates primary and secondary structure

Table 7.40 Physicochemical properties of a Calciustept formed in Hormozgan Province (Abbaslou et al. 2013)

Horizon	Sand	Silt	Clay	pH	EC (dSm^{-1})	SAR	CEC ($cmol^{+}\ kg^{-1}$)	Gypsum	CCE	OM
	%							%		
Torreric Calciustept										
A	59	39	2	7.76	6.4	1.0	10	2.15	39.2	0.75
Bk1	52	44	4	7.89	5.0	0.03	11	3.4	56.3	0.59
C	47	49	4	7.94	5.1	0.03	9	3.9	61.0	0.55

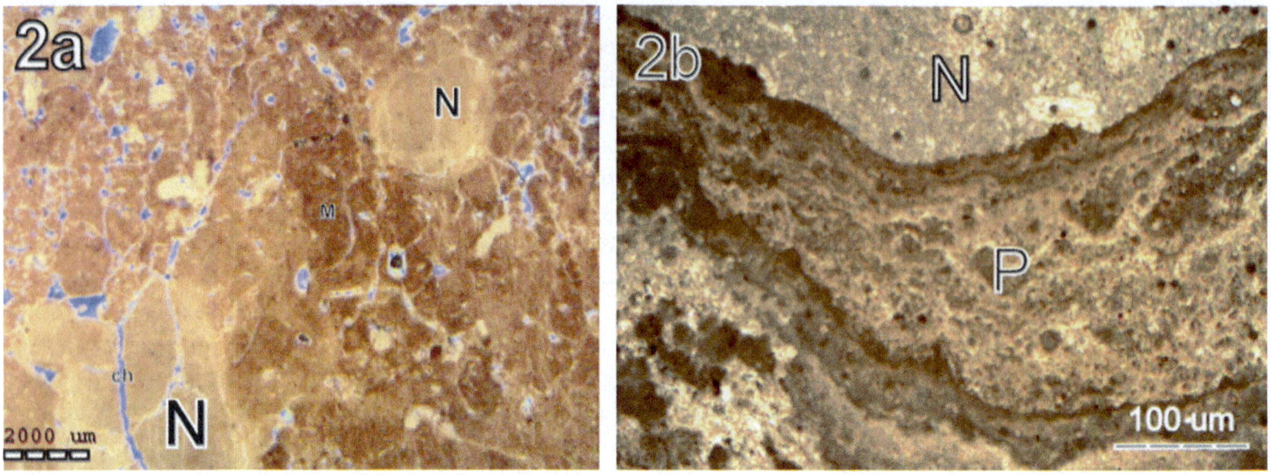

Fig. 7.41 Large lithogenic calcite nodules (N) in soil matrix with crystallic b-fabric (2a, xpl), Laminar pendant of calcite (P) formed beneath a calcite nodule (N) (2b, xpl) (Owliaie 2012)

precipitates in interpedal and/or intrapedal voids as large sparry crystals. Sparry calcite has been attributed to both precipitations from supersaturated solutions and micrite recrystallizations. Large voids are the main sites for calcite precipitation, as they dry more rapidly than smaller voids and are usually in more direct contact with lower atmospheric CO_2 concentration (Owliaie 2012).

In these soils, secondary calcium carbonate occurred as infilling with sparitic cytomorphic calcite (Fig. 7.42 3a), and typic calcite nodules as well as micritic calcite coating along channels. No clay coating was observed in the thin sections of the Calciustept soils. The high carbonate content throughout the soils must inhibit the translocation of clays because of its flocculation.

7.8.3 Khuzestan Plain

Khuzestan plain with a surface area of about 3 million hectares, which is a southeastern extension of the Mesopotamian plain, located in the arid and semiarid tropic regions of the southwest of Iran. The region is a very important food-producing area of the country. It borders in the west to Iraq and in the south to the Persian Gulf. The plain is

Fig. 7.42 Cytomorphic calcite crystals (C) in root channel (3a xpl), SEM image of rhombohedral calcite (3b) in micritic nodule of a Calciustept formed in Kohgiluyeh and Boyer-Ahmad Province (Owliaie 2012)

excessively hot and dry in the summer and mild in the winter.

The annual rainfall ranges from 300 to 400 mm and usually occurs during late fall and winter. The amount of rainfall increases in the North and East of the plain and in the upper terraces close to the Zagros Mountain, where ustic moisture regime is partially prevalent and dry farming of wheat and barley is common. Soils of the plain are mainly developed on river alluvial materials. Soils are generally productive and fertile in the northern and eastern part of the plain. Soils are highly saline and calcareous in the south and southwestern areas and in the lowlands with low drainage potential.

Khuzestan plain benefits from relatively secured surface water resources originated from several rivers such as Dez, Karun, and Karkheh that are regulated by several big dams built on the up streams. Almost 30% or about 30 BCM of the total surface water resources of the country run downstream into the Khuzestan plain. The hot prevailing aridic and ustic moisture regimes also allows for irrigated cultivation of crops (Fig. 7.43) such as wheat, barley, sugarcane, sugar beet, sorghum, oilseeds, melons, and vegetables and trees such as date palms, citrus, olive, and other fruits on the various soils (Chorom et al. 2009). In 2015, total agricultural land under crop cultivation (irrigated and dry farming) and orchards in the whole Khuzestan Province were respectively about 820,000 and 35,000 hectares (MOJA 2015). The irrigated land in Khuzestan constitutes about 10% of the total irrigated land of the country.

Soils are mostly classified in order of Entisols and Aridisols. However, Inceptisols have developed in ustic moisture regime. The main soil suborders are Fluvents, Orthents, Calcids, Cambids, Salids, and Ustepts. The main diagnostic subsurface horizons are cambic, calcic, gypsic, and salic horizons. The main soil great groups are Calciustepts, Haplustepts, Ustorthents, Ustifluvents, Torrifluvents, Torriorthents, Haplocambids, Haplocalcids, Haplosalids, and Aquisalids. High salinity, slow infiltration, and low internal drainage are the main constraints of the soil resources in the southwest of the plain. These soils are mostly classified in Aquisalids and are not generally suitable for crop production.

Since 2014, after several years of study, an irrigated and drainage project is being implemented on the soils in the northwest of the plain (Dashte Azadegan). The main purpose is to expand the irrigated land using the water resources regulated by the Karkheh River Dam. The initial aim of the project was to increase the irrigated land by about 550,000 hectares. But due to recent drought and shortage of the water resources, the area of the proposed irrigated land might be reduced. The soils are mainly saline and fine textured (Torrifluvents or Salic Fluvisols) . They need to be reclaimed and drained for proper crop, soil, and water management (Fig. 7.44).

7.9 Soil Clay Mineralogy

A nationwide review of the soil mineralogy shows that chlorite, illite, smectite, palygorskite, kaolinite, and inter-stratified illite-smectite are the major clay minerals of soils and rocks belong to Lower Cretaceous, Upper Cretaceous, and Paleocene, Oligo-Eocene, Oligo-Miocene, Miocene, and Pliocene-Pleistocene (Khormali 2003; Abbaslou and Abtahi 2007; Hojati et al. 2013). The clay minerals in the soils are mostly originated from the parent materials. Kaolinite is present only in Cretaceous sedimentary rocks, and it has higher occurrence in Lower Cretaceous samples comparing to the Upper Cretaceous ones. Smectite is absent in Lower Cretaceous rocks while it is seen in the Upper Cretaceous and late Paleocene and all other Tertiary and Quaternary rocks. However, it shows a lower occurrence in Oligo-Miocene rocks. Unlike kaolinite, palygorskite is absent in Cretaceous and late Paleocene rocks but is seen in Eocene and all other younger formations. The highest occurrence of this fibrous clay mineral is observed in

Fig. 7.43 Irrigated sugarcane plantations on a fertile soil (Haplocambids) in northeast of the Khuzestan plain (Courtesy by Dr. A Landi)

Fig. 7.44 Drainage and irrigation systems are being installed in saline soils (Typic Torrifluvents) of Dashte Azadegan, west of Khuzestan plain (Courtesy by Dr. N. Ebrahimi)

Oligo-Miocene samples. Illite and chlorite are present almost in all rocks. Few inter-stratified illite-smectite has only been in sediments of Eocene-Oligocene. Clay minerals, condition in which they form, develop and evolve and their relative abundance may provide information on climate, burial diagenesis, or reworking. Since formation of clay minerals needs special physicochemical and climatic conditions, their presence in sediments or soils can give insight toward the paleoclimatic evolution and its reconstruction. There are some studies in this regard in the Tethys region and Iran (Khademi and Mermut 1998; Farpoor and Krouse 2008).

The climatic evolution of the Tethys in southern Iran, as inferred from clay mineralogy of some parent rocks of Fars Province is shown in Fig. 7.45. Presence of high kaolinite and the absence or little occurrence of chlorite, smectite, palygorskite, and illite in the Lower Cretaceous sediments is in accordance with warm and humid climate of that period. However, lower amounts of kaolinite and the occurrence of smectite in Upper Cretaceous indicate a gradual shift from warm and humid to more seasonal climate. Kaolinite disappearance and presence of some palygorskite and smectite in the late Paleocene sediments indicate the increase in aridity. This absence of kaolinite in late Paleocene shows that the southern Iran has not experienced the Late Paleocene Thermal Maximum (LPTM) period with warm and humid climate. The same results were also reported for the Tethyan Sea environment of central Iran (Khademi and Mermut 1998; Farpoor and Krouse 2008). Based on their studies, authigenically formed palygorskite and sepiolite are predominant in the lacustrine sediments of the post-Tethyan era (Oligo-Miocene epoch). Older formations, including Cretaceous limestones and sandstones and Jurassic shale, contain only traces of detrital palygorskite and in contrast, they are dominated by kaolinite.

Illite and chlorite: Illite and chlorite are two commonly observed clay minerals occurring mainly in areas of steep relief where active mechanical erosion limits soil formation. Their abundance in the soils is largely due to their presence in parent rocks. They are dominant clay minerals of the coarse clay fraction (Owliaie et al. 2006). In most of the soils, illite shows a general decreasing trend with depth. Several processes may account for this decrease, including aeolian deposition of this mineral, enhanced physical weathering of biotite grains in the near-surface horizons due to large daily and seasonal fluctuations in temperature and moisture and to limited lessivage in the aridic climate (Boettinger and Southard 1995). Accumulation of higher smectite in Bt horizons might also be due to leaching of transformed illite to Bt horizons. Since illite is largely inherited into the soils from the surrounding sedimentary rocks, with increasing depth, formation and stability of smectite are favored from either soil solution, due largely to longer periods of moisture availability for chemical weathering and dampened temperature fluctuations relative to near-surface horizons.

Chlorite is also inherited to the studied soils, as well as illite, but it does not show any regular decreasing pattern with depth. Chlorite and illite were reported as the major clay minerals of Entisols of alluvial and colluvial fans of the southern Iran. These clay minerals are considered as the precursors for the pedogenic formation of other clay minerals in arid and semiarid regions (Khademi and Mermut 1998, Farpoor and Krouse 2008). Moreover, the presence of inter-stratified minerals in pedons occurring in plains and downslopes can be considered as an intermediate stage of

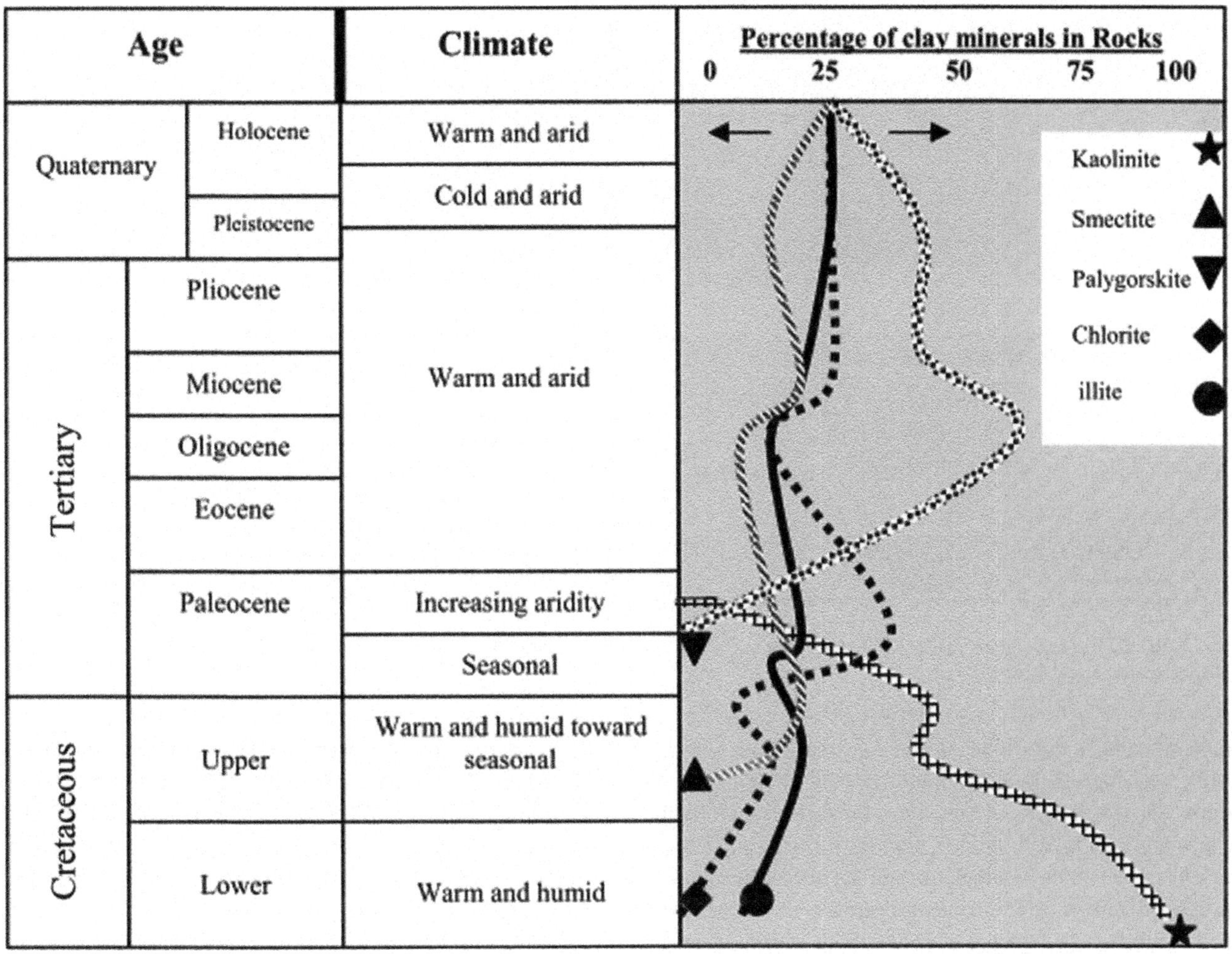

Fig. 7.45 Climatic evolution of Tethys area of southern Iran based on the clay mineralogy of the studied parent rocks. Arrows indicate the variable occurrence of clay minerals in soils (Khormali 2003)

this transformation. Inter-stratified minerals of chlorite-smectite or illite-smectite are observed in clay fraction of soils occurring in plains and downslope positions due to further weathering and transportation from upslope to downslope (Abbaslou and Abtahi 2007). Studies in soils developed on igneous rocks of Binaloud zone of eastnorth Iran, confirmed the formation of chlorite, smectite, mica, and different inter-stratified clay minerals of mica/chlorite, chlorite/smectite, some neoformed, and some inherited.

Kaolinite: Studying the origin and formation of kaolinite minerals showed that precipitation from solution of kaolinite requires acid conditions with moderate silica activity and lesser amounts of base cations. This clay mineral forms mainly from feldspars and mica alteration, in conditions of low K^+/H^+ activity ratio. These conditions occur in warm and humid tropical and subtropical climates. Kaolinite is mainly detected in clay fraction of pedons located near the Cretaceous outcrops (Khormali and Abtahi 2003). Thus, it can be concluded that the presence of this clay mineral in the soils of arid and semiarid region is due mainly to inheritance from surrounding parent rocks. Figure 7.46 shows the hexagonal kaolinite in representative pedon. Traces of this mineral are detected in the formations of younger than Cretaceous.

Smectite and palygorskite: Smectite constitutes the major portion of the clay minerals in well-drained Alfisols, some poorly drained Mollisols, and some Calcixerepts with high precipitation (Khormali and Abtahi 2003). In well-drained soils with increasing soil moisture (as expressed in P/ET°), smectite shows an increasing trend. Gharaee and Mahjoory (1984) also reported neoformation of smectite under saline and alkaline conditions with high concentrations of Si, Mg, and Al in southern Iran. However, neoformation cannot solely explain the general increasing trend of this mineral with increasing soil available moisture in relatively well-drained soils of the northwestern parts.

Increasing soil moisture, and consequently a relative leaching environment for the release of K^+ from micaceous

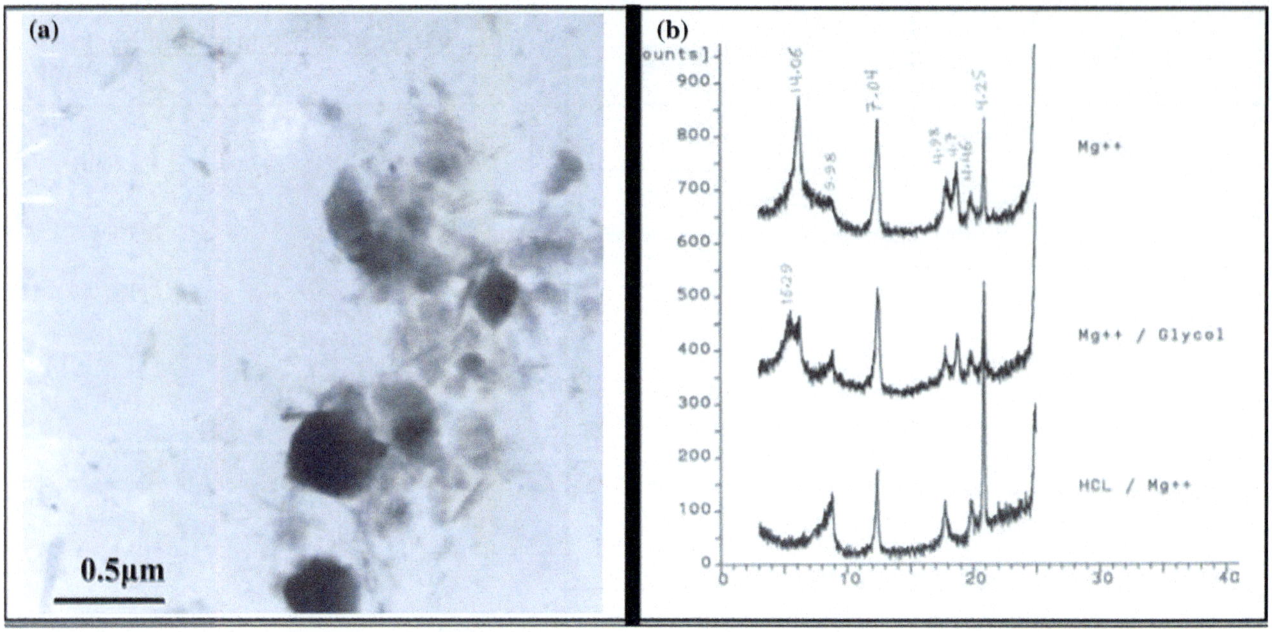

Fig. 7.46 a TEM micrograph and b XRD patterns of carbonate-free clay fraction (Bt) showing hexagonal kaolinite and very few palygorskite fibers (Khormali and Abtahi 2003)

minerals and mainly illite, in the calcareous soil with high Mg^{++} and high Si mobility might provide favorable conditions for formation of smectite through transformation especially at the soil surface, and its subsequent lessivage results in its higher accumulations in the deeper horizons. As discussed earlier, illite content decreases with increasing depth, unlike smectite, which shows a rather increasing trend. Moreover, the presence of inter-stratified illite-smectite in some pedons can be an intermediate stage of this transformation. In contrast to illite, transformation of chlorite into smectite has not been reported.

Palygorskite is the other possible precursor mineral for smectite in the arid and semiarid environments (Nadimi and Farpoor 2011, Moazallahi and Farpoor 2012). Singer (1989) noted that the frequent association of smectite and palygorskite is a result of proximity in their stability fields, but does not necessarily imply a solid-phase transformation of one into another. Figure 7.47 shows there is a reverse correlation between palygorskite and smectite content in the studied soils in Iran. Smectite forms the dominant clay mineral of the argillic horizons in P/ET° of more than about 0.4, but in the argillic horizons of the arid regions (lower P/ET°), palygorskite is the dominant clay mineral (Hojati et al. 2013; Hashemi et al. 2013; Razavipour 2009).

Climate or soil moisture has strong influence in distribution and formation of smectite in different areas. Khormali (2003) reported that smectite in less weathered C-horizons of some soils is most probably of detrital origin. Regarding the smectite in the argillic horizons of the Argids suborder or argillic horizon of arid regions, most evidences suggest that these minerals are Pleistocene relicts related to a more humid climate and to elluvial accumulation in the soil. The smectite is regarded as being of relict origin when viewed from the standpoint of current pedogenic conditions. The relative abundance and size of smectite increase in the deeper horizons indicating that its weathering is accelerated in the surface soil. Moreover, it is dominant in the fine clay portion. Therefore, it may be concluded that smectite is also the major illuvial clay mineral in the argillic horizons (Khormali 2003).

As Khormali and Abtahi (2003) reported in Fars Province, in contrast to smectite, palygorskite shows a decreasing trend with increasing soil available moisture, or in other words, increases with increasing aridity. The highest

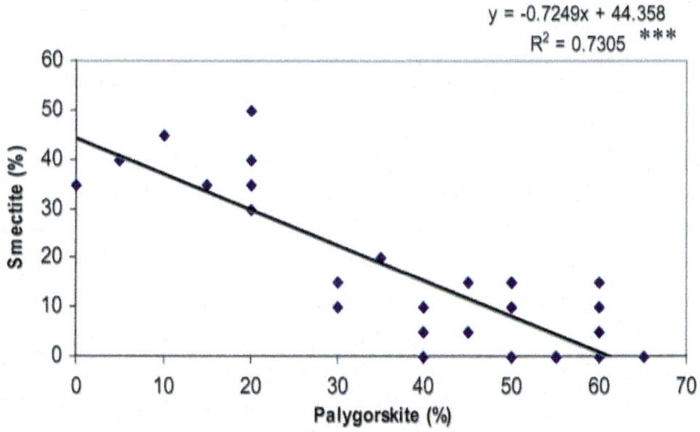

Fig. 7.47 Palygorskite (%) versus smectite (%) in the studied pedons (Khormali 2003)

occurrence of this fibrous mineral happens in the more arid areas of the south, southeast, and north, and it is either in trace amounts or nil in soils of the northwestern parts where the available moisture is high in soil (P/ET° $\geq$ 0.4). Presence of palygorskite and sepiolite in the Tertiary deposits indicates that the sedimentary environment of the study region during the Tertiary was suitable for the formation of palygorskite and sepiolite (Khademi and Mermut 1998; Farpoor and Krouse 2008). Several researchers reported that palygorskite was formed pedogenically in the soils of Iran (Gharaee and Mahjoory 1984; Khademi and Mermut 1998; Farpoor and Krouse 2008). An environment rich in Si and Mg with high pH but poor in Al and Fe is favorable for the formation of palygorskite and sepiolite (Singer 1989).

It is commonly agreed that the presence of gypsum (Nadimi and Farpoor 2011; Moazallahi and Farpoor 2012), calcrete or caliche, saline, and alkaline groundwater table would favor the in situ formation of palygorskite from soil solution. Palygorskite in arid soils has been stated to have two main origins; (1) inheritance from parent materials and (2) pedogenic formation through neoformation or transformation of 2:1 types clay minerals. Well-preserved long palygorskite bundles of plateau soils of southern arid regions formed directly in the late Tertiary sedimentary rocks containing a lot of palygorskite are most likely of geogenic or inherited origin (Fig. 7.48). This was also indicated by Khademi and Mermut (1998) and Farpoor et al. (2012) in central Iran. In a representative pedon, it is observed that palygorskite constitutes the major portion of the fine clay especially in argillic horizon. Presence of palygorskite in alluvial or colluvial soils could also be related to inheritance from the parent rocks (detrital). As mentioned earlier, gypsum would favor the in situ formation of palygorskite from soil solution. Gypsum like other evaporite minerals has been precipitated after drying of the shallow saline and alkaline lakes of the Tertiary.

In areas with relatively high P/ET°, palygorskite decreases, whereas, smectite content of the soils increases, indicating gradual disappearance of palygorskite and possible transformation into smectite. Therefore, it can be concluded that the critical available moisture as expressed by P/ET° is about 0.4, above which palygorskite is highly unstable and weathers mainly to smectite. However, presence of specific conditions such as high gypsum or shallow saline groundwater would prevent this transformation as in case of Iranian central playas. As Hashemi et al. (2013) reported, the highest content of palygorskite (>50%) in the clay fraction in the gypsiferous soils was observed in: (1) the aridic moisture regimes, (2) the upper part of alluvial fans and piedmont plain and in plateaus, and (3) the materials with high contents of gypsum and carbonate, such as Bk, By, and Bym horizons. Not only the amount of palygorskite but also its morphology is different in soils with different soil moisture regimes. Transmission electron micrographs indicated that with increasing aridity, palygorskite bundles contained longer and thinner fibers. Results showed that smectite/(illite + chlorite) ratio increased with the increase in moisture, and the highest ratio was found in the xeric moisture regime.

Moazallahi and Farpoor (2012) have reported from Kerman Province that there is some clear relationship between soil properties and parent material, topography, and climate. They found that smectite, illite, kaolinite, and chlorite clay minerals were formed in almost all studied soils. Palygorskite mineral was found in soils of the arid area as the dominant clay mineral, but due to increasing soil moisture in the upper part of the transect, palygorskite was not formed

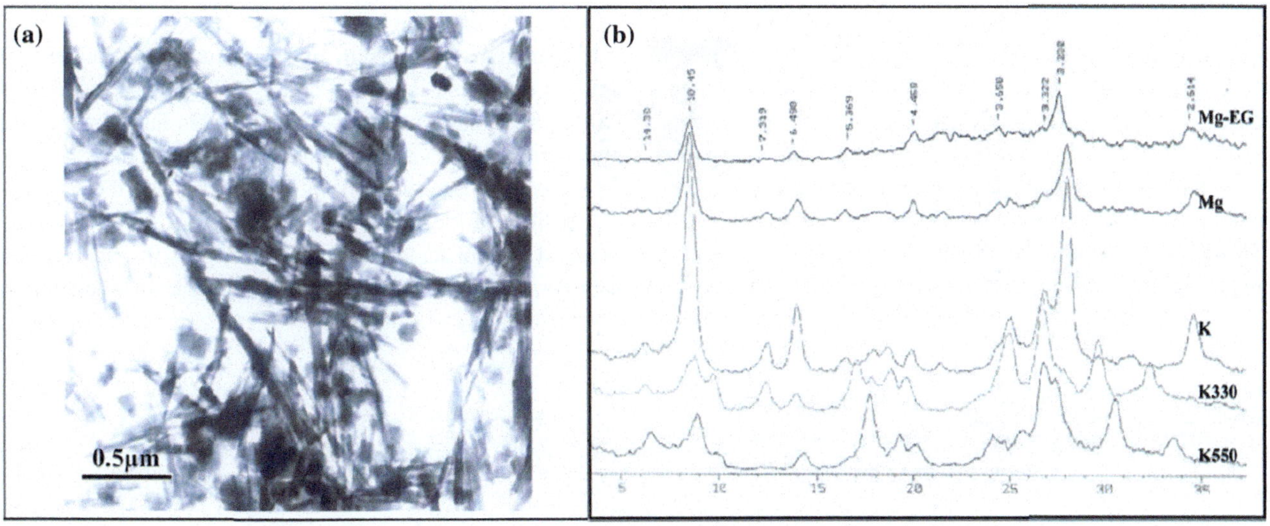

Fig. 7.48 **a** TEM micrograph and **b** XRD patterns of fine clay fraction in a pedon (Btk), showing well-preserved long palygorskite bundles (Khormali 2003)

and/or transformed to smectite. The same major formation pathways of chlorite, illite, and kaolinite are reported by Abbaslou et al. (2013) in gypsiferous, calcareous, and saline soils of Hormozgan province. Neoformation of palygorskite was the major mechanism for the occurrence of this clay mineral in the studied soils; these results are found in gypsiferous soils of southern Tehran.

Sepiolite is rarely detected in soils by XRD, mainly because of its dissolution under NaOAC treatment in pH 4.5 for the removal of carbonate. Under acidic conditions, sepiolite is destroyed more rapidly than palygorskite because of its magnesium-containing composition and the larger size of its structural microchannels. Another explanation for rare occurrence of sepiolite in soil environment is its lower stability under the presence of enough Al in soil, which is more favorable for the palygorskite formation.

Vermiculite: Occurrence and stability of vermiculite in calcareous and silica-rich soils have not been well-documented (Boettinger and Southard 1995). Trace amount of vermiculite is reported in calcareous soils of southern or central Iran (Khademi and Mermut 1998). They have detected only rare occurrence of this clay mineral mainly in northwestern parts of Fars Province where there is higher available moisture in soils ($P/ET^{\circ} > 0.4$). According to literature, mica in Aridisols can weather to vermiculite via direct simple transformation. However, this cannot explain the rare occurrence of this mineral in the highly calcareous soils of southern and/or central Iran.

7.10 Constraint Facing Prime Agricultural Land

As indicated earlier, the total land of Iran is about 165 million hectares from which the total soil area is about 96 m ha or 58% of the total land. The total non-soil areas and miscellaneous land types are about 68 m ha or 42% of the total land area. In 2015, the total cultivated land under irrigated, dryland, and fallow agricultural system were about 14 m ha (MOJA 2015). Since 1963, the total annual irrigated land (crop and orchards) has increased from 3.0 to 8.3 and cultivated dryland from 3.4 to 5.7 million hectares (Dewan and Famouri 1964, MOJA 2015). On the other hand, the total agricultural land of the country has reduced from about 18.8 in 1963 to about 16.5 million hectares in 2014 (SCI 2014), mainly due to land-use change and encroachment of urban and industrial expansion (Table 7.41).

As indicated in Table 7.41, the fertile and productive agricultural land resources of the country are very limited as only about 14 m ha or 8% of the total lands were under irrigated and dry-farming system in 2015. Mesgaran et al. (2016) using a relatively large geospatial datasets reported that based on soil properties, topography and climatic variables only 4.2 m ha (2.6%) of the total land have very good to good suitability and 12.8 m ha (7.9%) have medium suitability for sustainable crop production. Therefore, the total land with very good to medium suitability for crop production is roughly estimated to be around 17 m ha or 10.5% of the total land area. They reported the remaining lands were either have poor to very poor suitability or are unsuitable for crop production (Fig. 7.49). In this study, among the soil and terrain attributes considered, low soil organic carbon, steep slope, and high soil sodium content were identified as the most frequent factors limiting the suitability of lands for cropping.

The average yields of crops or land productivity under dryland farming system in the highlands of the semiarid regions of the country are generally low because of low annual precipitation during the crop growth and high frequency of drought. Soils are also facing severe water erosion on steep slopes due to improper tillage practices and monocropping system of cereals (see Chap. 12). In 2015, while the total land under dryland agriculture was 5.7 m ha or 40% of the total land area under cultivated crops and orchards, the share of dry-farming system to the total agricultural production was less than 8% (MOJA 2015). On the other hand, irrigated agriculture is responsible for about 90% of total water consumption compared to 8% for domestic and 2% for industrial use.

In summary, expansion of agricultural land beyond the present cultivated land area is not feasible in the future due to the increasing limitation of the suitable land for the crop production and scarcity of water resources in the vast areas of the country. In addition, the productivity of the soils and availability of the water resources will further be adversely affected by impact of the population growth and incipient climate change. Therefore, the sustainable use and proper

Table 7.41 General kinds of agricultural land use in 1963 and 2015 (SCI 2014, MOJA 2015)

Kind of agricultural land use	1963 m ha	2015 m ha
Annual irrigated cropland	2.3	6.0
Orchards (Irrigated)	0.7	2.3
Dryland farming (crops and orchards)	3.4	5.7
Land under fallow	12.4	2.5
Total agricultural land	18.8	16.5

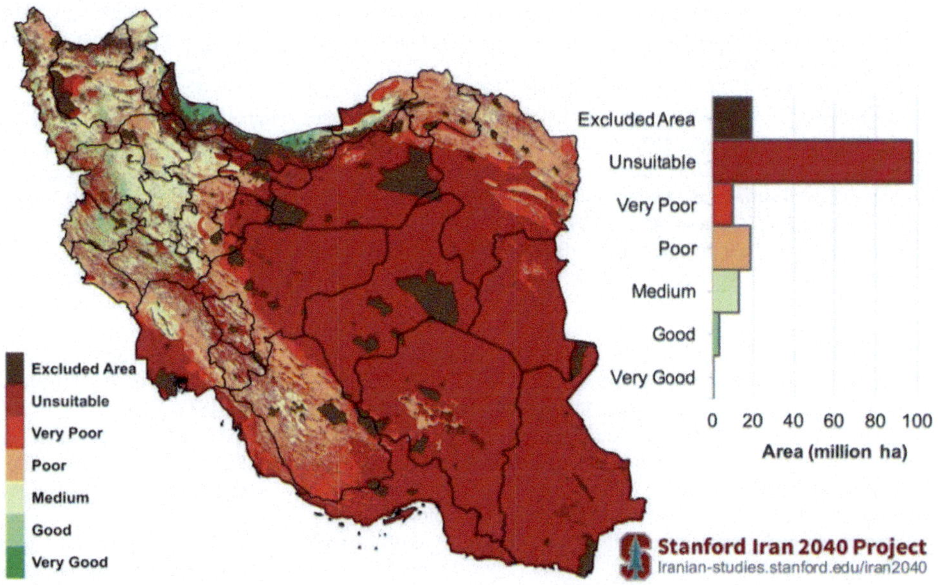

Fig. 7.49 A general land suitability map showing various classes for crop production in Iran based on soil properties, topography, and climatic variables (Mesgaran et al. 2016)

management of the soil resources should be a high priority for the government and landowners.

Acknowledgements The authors would like to express their appreciation and thanks to Dr. Kamran Eftekhari, Head of the Soil Genesis and Land Classification Department of Soil and Water Research Institute (SWRI) for his collaboration and providing the high-resolution maps.

References

Abbaslou H, Abtahi A (2007) Origin and distribution of clay minerals in calcareous, gypsiferous, saline soils and sediments of Bakhtegan Lake Bank, Southern Iran. Iran Agric Res 25(2):70–86

Abbaslou H, Abtahi A, Peinado FJM, Owliaie H, Khormali F (2013) Mineralogy and characteristics of soils developed on Persian Gulf and Oman Sea Basin, Southern Iran: Implications for soil evolution in relation to sedimentary parent material. Soil Sci 178(10):568–584

Abtahi A (1977) Effect of a saline and alkaline ground water on soil genesis in semiarid southern Iran. Soil Sci Soc Am J 41:583–588

Bakhshipour R (1993) Genesis and mineralogy of some soils in Lahijan area, Gilan, Iran, MSc Dissertation, Faculty of Agriculture, Isfahan University of Technology

Banaei MH (1998) Soil moisture and temperature regimes map of Iran. Soil and Water Institute, Tehran, Iran

Banaei MH (2005) Soil resources of Iran and their potential, In: Banaei MH, Moameni A, Bybordi M, Malakouti MJ (ed) The soils of Iran (in Farsi), Soil and water research Ins. Tehran. Iran, pp 161–212

Boettinger JL, Southard RJ (1995) Phyllosilicate distribution and origin in Aridisols on a granitic pediment, Western Mojave Desert. Soil Sci Soc Am J 59:1189–1198

Chorom M, Baghernejad M, Jafari S (2009) Influence of rotation cropping and sugar cane production on the clay mineral assemblage. Appl Clay Sci 46:385–395

De Pauw E, Ghaffari A, Ghasemi VR (2004) Agroclimatic zones map of Iran, explanatory notes. ICARDA

Dewan ML, Famouri J (1964) The soils of Iran. FAO, Rome

Emadi M, Baghernejad M, Bahmaniar MA, Morovvat A (2012) Changes in soil inorganic phosphorous pools along a precipitation gradient in northern Iran. Int J Forest, Soil Erosion 2(3):143–147

Esfandiarpour I, Salehi MH, Karimi A, Kamali A (2013) Correlation between soil taxonomy and world reference base for soil resources in classifying calcareous soils: a case study of arid and semi-arid regions of Iran. Geoderma 197–198:126–136

Farpoor, MH (1995) Soil geomorphology relationship in Godar-e-Kabk area, Chaharmahal and Bakhtiari Province, Iran. MSc dissertation, Isfahan University of Technology

Farpoor MH, Krouse HR (2008) Stable isotope geochemistry of sulfur bearing minerals and clay mineralogy of some soils and sediments in Loot Desert, central Iran. Geoderma 146:283–290

Farpoor MH, Neyestani M, Eghbal MK, Esfandiarpour I (2012) Soil–geomorphology relationships in Sirjan playa, south central Iran. Geomorphology 138:223–230

Gharaee HA, Mahjoory RA (1984) Characteristics and geomorphic relationships of some representative Aridisols in southern Iran. Soil Sci Soc Am J 48:115–119

Hashemi SS, Baghernejad M, NajafiGhiri M (2013) Clay mineralogy of gypsiferoussoils under different soil moisture regimes in Fars Province, Iran. J Agric Sci Technol 15:1053–1068

Heidari A, Mahmoodi Sh, Roozitalab MH, Mermut AR (2008) Diversity of clay minerals in the vertisols of three different climatic regions in Western Iran. J Agric Sci Technol 10:269–284

Hojati S, Khdemi H, Fazcano A, Ayubi S, Landi A (2013) Factors affecting the occurrence of Palygorskite in central Iranian soils developed on Tertiary sediments. Pedosphere 23(3):359–371

Kehl M, Sarvati R, Ahmadi H, Frechen M, Skowronek A (2005) Loess paleosol-sequences along a climatic gradient in northern Iran. Eiszeit Gegenw 55:149–173

Khademi H, Mermut AR (1998) Source of palygorskite in gypsiferous Aridisols and associated sediments from central Iran. Clay Miner 33:561–578

Khademi H, Mermut AR (2003) Micromorphology and classification of Argids and associated gypsiferous Aridisols from central Iran. Catena 54:439–455

Khormali F (2003) Mineralogy, micromorphology and development of the soils in arid and semiarid regions of Fars Province, southern Iran. Ph.D. dissertation, Soil Science Department of Shiraz University

Khormali F, Abtahi A (2003) Origin and distribution of clay minerals in calcareous arid and semi-arid soils of Fars Province, southern Iran. Clay Miner 38:511–527

Khormali F, Ajami M (2011) Pedogenetic investigation of soil degradation on a deforested loess hill slope of Golestan Province, northern Iran. Geoderma 167–168:274–283

Khormali F, Kehl M (2011) Micromorphology and development of loess-derived surface and buried soils along a precipitation gradient in northern Iran. Quatern Int 234:109–123

Khormali F, Nabiollahi K (2009) Degradation of Mollisols in western Iran as affected by land use change. J Agrical Sci Tech (JAST) 11:363–374

Khormali F, Abtahi A, Mahmoodi Sh, Stoops G (2003) Argillic horizon development in calcareous soils of arid and semiarid regions of southern Iran. Catena 53:273–301

Khormali F, Ghorbani R, Omrani RA (2005) Variations in soil properties as affected by deforestation on loess-derived hill slopes of Golestan Province, northern Iran. Sociedade and Natureza, Uberlândia, special Issue, pp 440–445

Khormali F, Abtahi A, Stoops G (2006) Micromorphology of calcic features in highly calcareous soils of Fars Province, southern Iran. Geoderma 132:31–45

Khormali F, Ghergherechi S, Kehl M, Ayoubi S (2012) Soil formation in loess-derived soils along a sub humid to humid climate gradient, northeastern Iran. Geoderma 179–180:113–122

Mahjoory RA (1975) Clay mineralogy, physical and chemical properties of some soils in arid regions of Iran Soil Science Society of America Proceedings 39:1157–1164

Mahmoodi Sh, Heidari A, Masihabadi MH, Stoops G (2007) Soil-landscape relationship as indicated by micro-morphological data on selected soils from Karaj basin, Iran. J Agric Sci Technol 9:153–164

Manafi Sh (2009) Mineralogical and micro-morphological investigation on arid and semiarid soils and possibility of using these data for reconstruction of paleoclimate in some parts of southern Alborz (Qazvin). Ph.D. dissertations, Faculty of Soil and Water Engineering Department of Soil Science, University of Tehran

Manafi Sh, Mahmodi Sh, Sarmadian F, Heidari A, Poch RM (2008) Paleoclimatic significance of secondary calcium carbonate coatings in some arid and semiarid soils in southern Alborz, Takestan-Iran. In: Xiubin He (eds) Proceedings of 13th international conference on soil micromorphology. Chengdu, China

Mesgaran M, Madani k, Hashemi H, Azadi P (2016) Evaluation of land and precipitation for agriculture in Iran, Stanford -Iran 2040 Project, working paper 2

Milani PM, Eftekhari K, Fatehi SH, Sepahvand M, Turkelboom F, De Pauw E, Bruggeman A (2010) Semi-detailed soil surveys of Merek and Honam watersheds, Karkheh river basin, Iran. ICARDA, Aleppo, Syria. vi 79 pp

Ministry of Jihad-e Agriculture -MOJA (2015) Annual agricultural statistics—crops and fruit trees (Vol. 1 and 3), Tehran, Iran

Moazallahi M, Farpoor MH (2012) Soil genesis and clay mineralogy along the xeric—aridic climo-toposequence in south central Iran. J Agricul Sci Technol 14:683–696

Moghiseh E (2005). Study of Physico—chemical, mineralogical, micromorphological characteristics and classification of some rich-gypsum soils (Gypseous) in Bam area (southeastern Iran). MSc Thesis, Department of Soil Science Eng. Tehran University

Moghiseh E, Heidari A (2012) Polygenetic saline gypsiferous soils of the Bam region, southeast Iran. J Soil Sci Plant Nutr 12(4):729–746

Nadimi M, Farpoor MH (2011) Genesis and clay mineralogy of soils on different geomorphic surfaces in Mahan-Joupar area, central Iran. Arab J Geoscience 6:825–833

Owliaie HR (2012) Micromorphology of pedogenic carbonate features in soils of Kohgiluyeh, Southwestern Iran. J Agr Sci Tech 14:225–239

Owliaie HR, Abtahi A, Heck RJ (2006) Pedogenesis and clay mineralogical investigation of soils formed on gypsiferous and calcareous materials, on a transect southwestern Iran. Geoderma 134:62–81

Raheb A, Heidari A (2011) Clay mineralogy and its relationship with potassium forms in some paddy and non-paddy soils of northern Iran. Aust J Agricul Eng 2(6):169–175

Razavipour SM (2009) An investigation of soil evaluation along a toposequence in Sedeh and Comfirooz plains (Fars province, Iran). MSc dissertation, College of Agriculture and Natural Resources, Faculty of Agricultural Engineering and Technology, Department of Soil Science Engineering. Tehran University

Rezaei N, Roozitalab MH, Ramezanpour H (2012) Effect of land use change on soil properties and clay mineralogy of forest soils developed in the Caspian Sea Region of Iran. J Agr Sci Tech 14:1617–1624

Roozitalab MH, Serghini H, Keshavarz A, Eser V, de-Pauw E (2013) Elements of research strategy and priorities for sustainable agricultural development of highlands in Central. West Asia and North Africa, ICARDA

Sarmadian F (1997) Soil genesis, classification and land suitability assessment in different climates of Gorgan and Gonbad in North of Iran (in Persian), Ph.D. dissertation, Soil Sci. Dept, Tehran University, Karaj, Iran

Singer A (1989) Palygorskite and sepiolite group minerals. In: Dixon JB, and Weed SB (eds.) Minerals in soil environment. Soil Sci Soc Am J, pp 829–872

Soil Institute of Iran (1970) Manual of multipurpose land classification, joint project with FAO/UNDP, Publication No. 212

Soil Institute of Iran (1979) Manual of land classification for irrigation, third revised edition, publication No. 205

Soil Survey Staff (2014) Keys to soil taxonomy, United States Department of Agriculture, Natural Resources Conservation Service, 12th edition, USA

Soil and Water Research Institute, SWRI (2000) Soil resources and use potentiality map of Iran (1:1000, 000). Tehran, Iran

Soil and Water Research Institute-SWRI (2015) Dominant soil order map of Iran, Karaj, Iran

Statistical Center of Iran, SCI (2014) Results of detailed census on agriculture in 2014 (in Persian), Tehran, Iran

Stoops G (2003) Guidelines for analysis and description of soil and regolith thin section, handbook for soil thin section description. SSSA Inc., Madison, Wisconsin, USA

Toomanian, N (1995) Origin of gypsum and evolution of gypsiferous soils in northwest area of Isfahan. Msc thesis, Isfahan University of Technology. Isfahan, Iran. 219 pages

Toomanian N (2007) Landscape evolution, pedodiversity and mapping of some pedogenic attributes of soils in central Iran. Ph.D. dissertation. Isfahan University of Technology, Isfahan, Iran

Toomanian N (2011) Gypsiferous soils (specifics and usages). Pelk Publication Agency, Tehran

Toomanian N, Jalalian A, Eghbal MK (2001) Genesis of gypsum enriched soils in north-west Isfahan, Iran. Geoderma 99:199–224

Toomanian N, Jalalian A, Eghbal MK (2003) Application of the WRB (FAO) and US taxonomy systems to gypsiferous soils in northwest Isfahan, Iran. J Agric Sci Technol 5:51–66

Yusefifard M (2012) Evolution of soils developed on some igneous rocks, Ormie-Dokhtar strip, north-west Iran. Ph.D. dissertation, Isfahan University of Technology, Isfahan, Iran

Zamanian K (2005) Study of Fashand-Hashtgerd soils and investigation on the mechanisms of petrocalcic horizon formation in these soils. MSc dissertation, College of Agriculture and Natural Resources, Faculty of Agricultural Engineering and Technology. Tehran University, Iran

Author Biographies

Dr. M.H. Roozitalab is a distinguished soil scientist with wide range of scientific and working experiences in Iran and abroad, and he received his MSc (1974) and Ph.D. (1978) from Oklahoma State University, Stillwater, Oklahoma, USA. His major research and teaching activities at Tehran University has focused mainly on soil genesis and classification and clay mineralogy. He has contributed to the advancement of soil science in Iran as President of Soil Science Society for 17 years from 1992 to 2009 and Director General of Soil and Water Research Institute from 1983 to 1995. Dr. Roozitalab also served as Deputy Head of Agricultural Research, Education and Extension Organization (AREEO) in Iran from 1995 to 2006. He has contributed to strengthening the regional and global agricultural research collaboration as President of Association of Agricultural Research Institutions in the Near East and North Africa, AARINENA (1998–2000), Coordinator of Regional Highland Research Network of ICARDA (2008–2011), and Chairman of the Global Forum on Agricultural Research (GFAR) from 2002 to 2006.

Dr. Norair Toomanian is an associate professor of Soil Science in Soil and Water Research Department, Isfahan Agricultural and Natural Resources Research and Education Center, AREEO, Isfahan, Iran. His current research interests are digital soil mapping, suitability mapping, and databasing of Iranian soils. He has been teaching graduate students at various universities in Iran and has supervised many theses and dissertations.

Dr. Vahid Reza Ghasemi Dehkordi is senior researcher at Soil and Water Research Institute (SWRI), Iran. His current research interests focus on geomatics, geostatistics, and spatial modeling, using similarity analysis to evaluate soil and environment characteristics and potential, impact of climate change on soil and environment, agro-climatic zoning (ACZ), and agro-ecological zoning (AEZ).

Dr. Farhad Khormali is a professor of pedology at Gorgan University of Agricultural Sciences and Natural Resources, Iran. He is also a committee member of the Soil Science Society of Iran as well as editor in chief of the *Journal of Soil Management and Sustainable Agriculture.* His main research interests are soil genesis, loess–paleosols studies, micromorphology, clay mineral weathering, and soil quality and soil carbon storage.

Paleosols and Past Climate Change

8

Martin Kehl and Farhad Khormali

Abstract

Paleosols in Iran have received less attention than modern soils, although Quaternary deposits and exposures therein are widespread. The chapter gives a short introduction into the general nature of buried and relict paleosols and on methodological aspects how to recognize, characterize and date paleosols. Examples are given for Pre-quaternary and Quaternary paleosols in Iran. The focus is on paleosols in loess-soil sequences in Northern Iran, which have been studied in more detail. These loess exposures contain paleosols of various development degrees separated by layers of more or less unweathered sediment. Strongly developed argillic horizons of paleo-Luvisols occur as well as very weakly developed brown horizons reflecting syngenetic soil formation. The loess-paleosol sequences are excellent geological archives of climate change, where paleosols mark periods of increased humidity and layers of unweathered loess dry phases. The paleosols most probably correlate with interglacial and interstadial periods of the last glacial cycles suggesting that climate change in Northern Iran was in phase with well-known cycles recorded in the temperate zone of the northern hemisphere. The high potential of paleosols as stratigraphic markers and geological archives of climate change and landscape evolution in Iran is still widely unchallenged.

Keywords

Nature of paleosols • Quaternary • Loess • Pedostratigraphy • Climate archive

8.1 Introduction

While the state of knowledge on modern soils of Iran is advanced, comparatively little information is available on the distribution and properties of paleosols in Iran and adjacent countries. Paleosols are soils formed at land surfaces of the geological past (Ruhe 1956). They may represent excellent stratigraphic markers in geological context and provide important information on paleo-environmental conditions, because the constellation of soil forming factors such as climate, vegetation or topography can often be deduced from paleosol properties. The formation and preservation of paleosols are closely linked to changes in geomorphodynamic conditions, i.e., changes between stable and unstable land surfaces, providing sufficient time for soil formation on the one hand and burial under protecting sediment covers on the other hand. These changes are often driven by climate change hence sediment-soil sequences may provide excellent records of climate change. Here, we give a brief overview on occurrences of paleosols in Iran in order to demonstrate their general properties and high potential as indicators of climate change and landscape evolution. The focus is on paleosols derived from Quaternary loess and loess-like sediments, i.e., deposits of eolian dust or those showing similar sediment properties including silt as the dominant grain size, homogenous structure, lack

M. Kehl (✉)
University of Cologne, Cologne, Germany
e-mail: kehlm@uni-koeln.de

F. Khormali
Agricultural Sciences and Natural Resources,
Gorgan University, Gorgan, Iran

M.H. Roozitalab et al. (eds.), *The Soils of Iran*, World Soils Book Series,
https://doi.org/10.1007/978-3-319-69048-3_8

of bedding, forming stable walls, including carbonate and fossils of land snails etc. Considerable progress has been made in the study of these deposits in Iran during the last decades. Loess forms thick deposits in the North, in particular in the Iranian Loess Plateau and along the Alborz Mountains northern foothills. Other deposits of loess are patchy or consist of dust additions in thin cover beds. Loess-like sediments cover large areas in the Central Iranian highlands for instance in depressions of large basins. Besides representing detailed terrestrial archives of climate and past environmental change, loess itself and the loess-derived modern and ancient soils are of large value for land use, related to the high storage capacity for plant-available water, high nutrient contents and good workability. Loess-derived soils in Iran thus belong to the most fertile lands of the country, which depending on climate conditions and topography are used for grazing, rain-fed or irrigated agriculture as well as forestry as perfectly demonstrated in Golestan Province.

8.2 Nature of Paleosols

The general properties, definition and classification of paleosols have been dealt with in detail previously (e.g., Catt 1990; Yaalon 1995; Retallack 2001; Mason and Jacobs 2007), and a comprehensive discussion would be beyond this chapter. The following remarks shall provide general background information on the descriptions below. Paleosols have either been buried under younger sediments (buried or fossil soils), remained at the land surface during periods of climate change (relict soils) or have been partly exhumed after erosion of overlying sediments. Buried paleosols can easily be distinguished from the modern soil if sediment strata with limited pedogenetic overprint are separating them, which is often the case in unconsolidated sediments and sediment-soil sequences of the Quaternary.

While close similarities between the modern soil and buried paleosols may exist, as is often the case for soils developed from similar parent materials during the present and previous interglacials, there may still be some general differences. Firstly, diagenetic processes including mineralization of organic matter, compaction and loss of soil structure, iron loss by microbial reduction, color changes due to dehydration/recrystallization of hydrous pedogenic minerals, or recalcification from overlying strata may have altered the buried soil. Post-burial alteration may even result in lithification, which is often encountered in Pre-Quaternary paleosols. Secondly, although not necessarily apparent in the field, buried soils may have experienced several phases of soil formation at former land surfaces. Sediment covers on top of the paleosol could have been eroded or the cover was too thin to substantially protect the buried soils from later soil formation. In these cases, welded or composite paleosols may have formed over one or multiple cycles of climate change. These paleosols can attain several meters in thickness. Thick buried or relict soils may also form by soil formation under more or less continuous dust input or deposition of fluvial deposits. These soils are called syngenetic, accretionary or upbuilding (e.g. Palmer 2007). It is difficult or sometimes impossible to define, when their formation started or ended, in contrast to paleosols formed by top-down pedogenesis on a pristine or pre-weathered substrate. Thirdly, past periods of soil erosion may have considerably truncated the soils at former land surfaces. Paleosols observed today often consist of Bt or Bk horizons, which are more resistant against erosion than top soils. Preservation of complete profiles is the exception rather than the rule and the classification of paleosols is often limited to the horizon level. Fourthly, the climatic conditions and related vegetation and fauna, as well as the paleosols themselves may not have modern analogues in the area of interest, which often holds for interstadial soils formed during short phases of climate amelioration within glacial stages. Examples are the Tundra gley of Central European loess-soil sequences formed above permafrost or the weakly developed but thick accretionary CBb horizons in Iranian loesses. Finally, the duration of paleosol formation may considerably differ from that of modern soils.

Relict soils contain features that have not formed under present-day climatic conditions, implying that they experienced at least one significant change in climate, hence can be defined as polygenetic. Identification of relict soils may be difficult, because (i) morphological imprints of past climates will only be recognizable if the diversity of morphological features in modern soils is well known and (ii) the onset of modern soil formation is often difficult to determine if the timing and magnitude of climate change are not well known. Exhumed paleosols represent a special type of relict soils. The time of exhumation can be deduced if the soil can be traced below layers of younger sediments and if deposition of these sediments can be dated using geochronological methods.

8.3 Recognition, Characterization and Geochronological Dating of Paleosols

In the field, paleosols can be distinguished from sediments by changes in color, intensity of hydromorphic mottling, grain size, development of soil structure, presence of root traces or krotovina, and pedogenic enrichments in soluble salts, gypsum, carbonates, iron and manganese hydroxides or silicate clay. These changes are sometimes subtle and in particular weakly developed paleosols are often not readily identified in the field. It is of high importance to carefully clean field exposures and to provide detailed field descriptions using pedological field manuals or those suggested for the description of paleosols (e.g., Catt 1990). An important observation is, whether the soil horizon identified consists of redeposited pre-weathered soil material, i.e., can be considered a soil sediment, or if it has formed in situ. Sharp lower contacts with underlying strata provide evidence of soil sediments, but pedogenic processes such as carbonate leaching and reprecipitation may result also in sharp boundaries between Bw or Bt horizons and an underlying Bk. Close inspection of carefully cleaned profiles will help to avoid misinterpretations. Sampling in the field should be as detailed as possible. High-resolution sampling along vertical columns is the basis for detecting sedimentary discontinuities within paleosol horizons and abrupt or gradual transitions from soil horizons to sandwiching sediment layers. Laboratory analyses may involve the whole spectrum of methods available to the soil scientist. However, care has to be taken in interpreting the data, because post-burial alterations may have strongly changed the properties acquired during soil formation. In the past two decades, the use of rock magnetism and geochemical ratios has proven beneficial for characterizing the development degrees of paleosols and stable isotope ratios of carbon and oxygen used to infer information on paleoclimate. In order to reconstruct pedogenic processes, the use of micromorphology is indispensable.

Several geochronological methods are available to determine the age of paleosols and the duration of former soil forming phases. The latter can be assessed if buried paleosol horizons can be sandwiched, i.e., if age estimates can be provided for over- and underlying strata. This approach implies that the time determined for parent material deposition and burial of the paleosol is identical to the onset and end of soil formation, respectively. In case of accretionary soils, this assumption is not valid. In case of top-down pedogenesis, it has to be considered that the time of sediment accumulation within the C horizon can considerably differ from that of the substrate in which the paleosol formed. The different dating techniques have various limitations. Down to an age of about 40,000 years before the present day (40 ka), radiocarbon dating on bulk samples containing humus, or on pieces of charcoal, bone and pollen extracted from soil horizons is feasible. Age estimates down to about 250 ka or even older may be gained by luminescence dating of the time of sediment deposition. This approach has been used to date loess-soil sequences in Northern Iran (Frechen et al. 2009; Karimi et al. 2011; Lauer et al. 2017a, b).

Other approaches are to apply paleomagnetic techniques including polarity changes or intensity variations of the earth magnetic field or to wiggle-match fluctuations in climate proxies such as soil color, pedogenic iron etc. recorded in sediment-soil sequences with climate proxy data established for well-dated geological archives of paleo-environmental change including ice cores, marine sediments or long limnic records. Both techniques, however, yield relative age estimates and must be combined with numerical dating, otherwise it is impossible to provide reliable age estimates, because complete terrestrial records are rarely to be found and stratigraphic gaps are the rule rather than the exception.

The paleosols can be used as stratigraphic markers to correlate sediment-soil sequences of different eolian, fluvial or colluvial deposits or for correlation with climatic phases known from other climate archives. The legitimate assumption of this pedostratigraphical approach is that weakly developed paleosols—such as Ab, CBb and Bwb horizons in Iranian loess sequences—were often formed during interstadial periods, whereas the well-developed ones (e.g., BAtb and Btb horizons) formed during interglacials of the climate record. In loess covered areas of Northern Iran, where a Bt horizon is expressed in the modern soil, the first strongly developed buried Bt horizon can most probably be correlated with the last interglacial soil, a terrestrial equivalent to Marine Isotope Stage (MIS) 5e. The chronostratigraphic position of the second, third and following Btb may represent the penultimate and older interglacials, but age control for this assumption is still lacking. The counting of Btb-horizons from the top can only provide a first estimate, since loess deposits often represent discontinuous sediment records due to erosional discontinuities.

8.4 Pre-quaternary Paleosols in Iran

During the Devonian, large plants began to colonize the continents and roots and root-like structures became more common. If the presence of large plants is considered a prerequisite for soil formation and root traces a key characteristic of soils, paleosols have formed in terrestrial systems since post-silurian times. Using the wider concept of paleosols of Retallack (2001), even Precambrian deposits may contain paleosols. Pre-Quaternary rocks of Iran are also known to include paleosols. For example, Aharipour et al. (2010) described red calcretes, yellowish calcretes, or dolocretes formed in clastic sequences of the lower Devonian Padeha Formation in the Eastern Alborz Mountains. The authors identified alluvial fan, distal fan, and palustrine/lacustrine environments, in which the paleosols formed. Moussavi-Harami et al. (2009) report on several calcrete horizons in Lower Cretaceous fluvial deposits of the Shurijeh formation in the KopehDagh range, which at least partly have a pedogenic origin. In a study of the Miocene Upper Red formation in the Southern Alborz, Ballato et al. (2010) sampled pedogenic carbonates to study changes in stable carbon and oxygen isotopes as indicators of climate change. Recently, a thin red paleosol (10–15 cm) has been described in the Chehel-Kaman Formation of the Central Kopet-Dagh basin, which probably delineates the Paleocene/Eocene boundary (Rivandi et al. 2013).These examples may illustrate that Pre-Quaternary paleosols can be important stratigraphic markers and paleo-environmental indicators in terrestrial sequences of Iran. They may be difficult to notice in the field, because, after burial, extensive diagenetic alteration may have taken place (e.g., Driese 2009).

8.5 Quaternary Paleosols in Iran

In general, paleosols can be expected in all kinds of unconsolidated Quaternary sediments, which experienced subaerial exposure and stable geomorphodynamic conditions for a sufficiently long period of time to promote soil formation. The preservation of paleosols is promoted in low-energy depositional environments such as floodplains or on elevated land terraces and plateaus where eolian or low-energy fluvial deposition occurred but fluvial erosion, mass wasting or eolian deflation was limited. If conditions for soil preservation and sediment build-up are favorable, extensive sediment-soil sequences may develop over time. This was the case for loess deposits in Northern Iran reaching several decameters in thickness and containing a large number of weakly to strongly developed paleosols.

8.5.1 Lower to Middle Pleistocene

In the Iranian loess plateau, bivalve conglomerates of the Akchagyl formation mark the final Tertiary dated to about 3 mio years ago (e.g., Forte and Cowgill 2013). The conglomerate is unconformably covered by a series of silty to clayey reddish-brown terrestrial sediments (Fig. 8.1), identified as loess and dated to the lower Pleistocene ($\sim$2.4–1.8 Ma ago) as based on palaeomagnetic data (Wang et al. 2016). These deposits show a strong overprint of soil formation and have been interpreted as a thick pedocomplex (Taheri et al. 2017). Diagenetic alteration is indicated by continuous gypsum infillings between large soil prisms and in fissures. The upper

Fig. 8.1 Reddish-brown paleosols in silty to clayey deposits exposed in a limestone quarry near the town of Agh Band, where the underlying bivalve conglomerate of the Akchagyl formation (Pliocene) is exploited (not shown). In the lowest series, a comparatively thick (2 m) well-developed paleosol is visible. In series 2, the light colored sediment is intercalated by a large number of thin weakly developed paleosols which are perfectly aligned. This indicates several short periods of soil formation interrupted by phases of sediment accumulation. Toward the top, a thick weakly developed paleosol testifies syngenetic soil formation. Series 3 consists of at least three strongly developed paleosols, the lower two being separated by a thin layer of sediment. These soils are strongly enriched in gypsum. Series 1–3 reflect changing dominance of soil formation and sediment accumulation, which could be related to varying eolian inputs. Typical loess (series 4) disconformably covers the soils of series 3. For more information on this section see Taheri et al. (2017) and Wang et al. (2016)

part (series 3 in Fig. 8.1) has an erosional upper boundary and is covered by about 10 m of loess, resembling loess of the nearby key sections of Agh Band where a middle to upper Pleistocene age was determined using luminescence dating (Frechen et al. 2009; Wang et al. 2017; Lauer et al. 2017a, b). The erosional unconformity between the red clay and overlying loess thus probably represents a long stratigraphic gap.

The lower part of several loess profiles in Northern Iran contains reddish-brown, strongly developed paleosols, which consist of Btb- or Bwb-horizons (lower series in Figs. 8.2, 8.3, and 8.4) often showing recalcification from overlying strata. These paleosols can be observed at the section at Rustamabad, located in the Sefid-Rud valley (Kehl 2010), and in several sections situated at the northern foothills of Alborz Mountains, for instance, at loess sections near Neka (Fig. 8.2; Kehl et al. 2005) and Toshan (Vlaminck et al. 2016; Lauer et al. 2017a). Little is known about these presumably middle to early Pleistocene paleosols. They are often welded and polygenetic, and if light colored sediment layers separate them, these are heavily overprinted by pedogenesis. However, it is most likely that these paleosols developed under forest covers similar to the present-day deciduous forest. In comparison with the modern soils under forest (e.g., Khormali et al. 2012), the older interglacial paleosols rarely show dark A-horizons, exhibit a stronger development of blocky to prismatic soil structure (Fig. 8.3 left), are richer in clay and have a more reddish soil color. Interestingly, some of these old paleosols are distorted, probably resulting from tectonic activity (Fig. 8.4). Another interesting observation is that at the section at Now Deh two weakly developed brown paleosols (Bw horizons over CBk horizons) are intercalated within the loess deposit of presumably MIS 6. These paleosols provide evidence for Middle Pleistocene interstadials in Northern Iran (Kehl et al. 2005; Frechen et al. 2009). Recent evidence for two interstadials of MIS 6 in Northwestern Iran deduced from the $\delta^{18}O$ record of Lake Urmia (Stevens et al. 2012) corroborates this notion.

While tracing paleosols along exposures, separate paleosol horizons may unify to form a composite or welded paleosol. In Northern Iran, this has been observed for example on the presumably MIS 7 paleosols at the section at Now Deh (Kehl et al. 2005) and at several sections located in brickeries near the city of Gorgan. On the other hand, a single paleosol may fan out into separate horizons, as observed at the section of Rustamabad (Kehl 2010). Both phenomena render the pedostratigraphic correlation of different sections and establishment of a pedostatigraphic framework for the Lower to Middle Pleistocene difficult. As based on "counting Btb horizons from the top" and on

Fig. 8.2 Loess quarry near the town of Neka exposes several series of paleosols. In the lower half of the exposure, at least five strongly developed reddish-brown paleosols can be distinguished (series 1, ~8 m high). The slight curvature of the soils delineates former land surfaces and locally shows former incision of gullies. This sequence is capped by a dark ABt-horizon and two well-developed paleosols of dark color (series 2; ~3.5 m high). These three soils probably represent the pedocomplex of MIS 5. Equivalents of this pedocomplex can be found in many loess exposures along the Alborz Mountains foothills, which include the sections at Neka 1, Toshan, Now Deh and Mobarakabad (Kehl et al. 2005; Shahriari et al. 2017; Vlaminck et al. 2016). The upper series (~5 m high) includes at least three weakly developed brown paleosols and the modern soil at the surface. At about 3 m below surface, the loess color changes from yellowish brown to grayish reflecting a higher intensity of pedogenic alteration toward the top

Fig. 8.3 Two paleosols in the loess exposure of Fig. 8.2. Left is a presumably Middle Pleistocene, welded paleosol of series 3 with clayey texture, prismatic structure, clay coatings and secondary calcite. Horizon boundaries as well as the lower boundary of the paleosol are difficult to identify, because the loess itself has a strong overprint by pedogenic alteration (spate is 40 cm high). On the right, the well-developed ABw-horizon in series 2 represents an interstadial paleosol of the last Glacial (presumably MIS 5a). The sharp lower contact is caused by enrichment of underlying loess with secondary carbonate (Bk horizon). The gradual transition toward the overlying loess reflects continuously increasing dust input caused by aridification of the landscape. This loess forms the lowest member of series 3

luminescence age estimates for the overlying loess deposits (Frechen et al. 2009; Lauer et al. 2017a), the Btb-horizons below the first buried Btb can be tentatively correlated with MIS 7, 9 and 11 or older interglacials. At least five interglacial soils of this type are exposed in the lower loess series near the town of Neka (Fig. 8.2), indicating, that the loess deposits along the Alborz Mountains foothills may provide a detailed stratigraphy of the Middle Pleistocene.

Observations in many exposures of alluvial fans and fluvial terraces suggest that these deposits also contain paleosols. For example, the fluvial terraces of Sefid-Rud show intercalations of gravel beds, several meters thick, with loess strata including brown paleosols (Fig. 8.5) and road cuts in the alluvial fan within the city of Tehran expose thick gravel sequences with paleosol horizons. In fluvial environments, changes between sediment accretion and erosion can be related to an intrinsic variation of channel courses. In these dynamic systems, erosion of paleosols may take place at the same time like paleosol formation. This leads to limited lateral continuity of paleosols and potentially large gaps in the stratigraphic records. Therefore, chronostratigraphic estimates for these soils are difficult.

8.5.2 Upper Pleistocene and Holocene

The Late Quaternary includes the last interglacial/glacial cycle and the Holocene, i.e., MIS 5 to MIS 1. Loess deposits formed in Northern Iran during this time are considerably thicker than those accumulated within previous interglacial/glacial cycles are. The deposits contain a detailed sequence of paleosol horizons (Table 9.1), some of which can be correlated over large longitudinal distances along the loess covered northern foothills of the Alborz Mountains. The sequences start with the last interglacial or Eemian soil (MIS 5e), which consists of strongly developed ABtb horizons near the town of Neka (Fig. 8.2; Kehl 2010) or Btb horizon, as, for example, at the section at Toshan (Vlaminck et al. 2016; Lauer et al. 2017a; Fig. 8.6). Toward the top, about 3–4 m of slightly weathered loess follows in which two well-developed interstadial paleosols have been preserved. These soils consist of ABw- or Bwb-horizons over Bkbs (e.g., Fig. 8.3 right) and are clearly expressed in many loess sections. As based on luminescence age estimates and pedostratigraphic arguments, they probably formed during interstadials of MIS 5, i.e., MIS 5c and 5a, whereas intense

Fig. 8.4 Vertical distortion of the fifth reddish-brown paleosol (Btb) counted from the top at a road cut near the town of Rustamabad, Sefid-Rud valley. The two uppermost Btb-horizons are not shown in the picture. Most probably, the distortion was caused by tectonic activity, which also affected the silty and partly stony slope deposits overlying the paleosol. The third Btb appears to be unaffected, hence distortion occurred before formation of this soil. Reproduced from Kehl (2010, changed)

accumulation of dust took place during the stadials 5d and 5b and after MIS 5a. The three paleosols thus form a pedocomplex (Catt 1998) of MIS 5, which serves as a stratigraphic marker for correlating the loess-paleosol sequences at Neka 1 and Now Deh (Frechen et al. 2009; Kehl 2010) as well as Mobarakabad and Neka 2 (Kehl and Khormali 2014; Ghafarpour et al. 2016; Shahriari et al. 2017). Interestingly, the upper transition from paleosol horizons into overlying loess is often gradual (Fig. 8.3 right), as clearly shown by the lithological characteristics of the section at Toshan (Fig. 8.6). This suggests that many paleosols in Northern Iranian loess were not truncated before

Fig. 8.5 Fluvial deposits exposed in the valley of Sefid-Rud River near the town of Rustamabad. In between two gravel beds, about 3 m of silty and sandy deposits are intercalated. This unit includes two brown paleosols. The upper one was truncated during accumulation of the upper gravel bed and only a thin remnant of this soil is preserved www.schweizerbart.de/9783443090241

the onset of dust accumulation. The MIS 4 loess deposit at Toshan is the most "typical loess" of this section. It shows coarse texture, low lightness values (L*) and low redness rating (Fig. 8.6). Toward the top, two weakly developed paleosols at 3–9.5 m depth were detected in the field. In addition, the loess of MIS 3 and MIS 2 at Toshan shows a high degree of pedogenic alteration reflected by frequent calcite concretions, comparatively dark color and elevated redness ratios pointing to syngenetic soil formation during the time of dust accumulation. In the Sefid-Rud valley, sandy loess at the section at Saravan contains two well-developed Ahb-horizons formed during MIS 3 (Kehl 2010).

In Northeastern Iran, Btk horizons were identified within several loess sections (Karimi et al. 2009) possibly dating back to the last interglacial (Karimi et al. 2011). Recently, Karimi et al. (2013) demonstrated that, if taking the high amounts of gypsum and carbonates into account, magnetic susceptibility clearly increases in one of these Btk horizons compared to the parental sediments and overlying layers. The increased magnetic susceptibility of paleosol horizons may represent a valuable indicator of soil development degree (see also Frechen et al. 2009; Ghafarpour et al. 2016) and possibly climatic conditions in the past. A new study of Shahriari et al. (2017) on biomarkers extracted from paleosols of four loess-soil sequences sheds light on the vegetation of the past enabling distinction between semi-arid and forest biomes.

Besides loess, Late Quaternary paleosols are also known from other deposits. Antoine et al. (2006) reported on a well-developed buried paleosol within alluvial deposits of the Caspian lowlands about 25 km southwest of Babolsar. Two radiocarbon ages indicate that the soil is younger than 28,486 ± 190 cal BP and older than 12,119 ± 82 cal BP. Several studies were carried out on surface soils in the central Iranian Plateau and in the Zagros Mountains. The presence of clay coatings and even argillic horizons in arid regions was related to increased rainfall and edaphic moisture during the past (Khademi and Mermut 2003; Khormali et al. 2003; Ayoobi et al. 2003; Farpoor et al. 2004). Similarly, layered pendents of calcite in coarse-textured soils of the xeric moisture regime probably represent relicts of more humid climate (Khormali et al. 2006). In addition, the stable carbon isotope compositions of gypsum hydration water (Khademi et al. 1997) and pedogenic carbonate (Khademi and Mermut 1999) indicate phases of increased humidity in the past.

In the alluvial plain of the Kor River in Southern Iran, a well-developed paleosol probably formed in loess-like sediments between 27 and 21 ka, as indicated by luminescence age estimates of the underlying and overlying deposits. Several younger weakly developed paleosols were identified in these sediments (Kehl et al. 2009). It is also evident that colluvial deposits on limestone slopes of the Zagros Mountains bear paleosol horizons and that grike-fillings contain argillic horizons formed during moister periods of

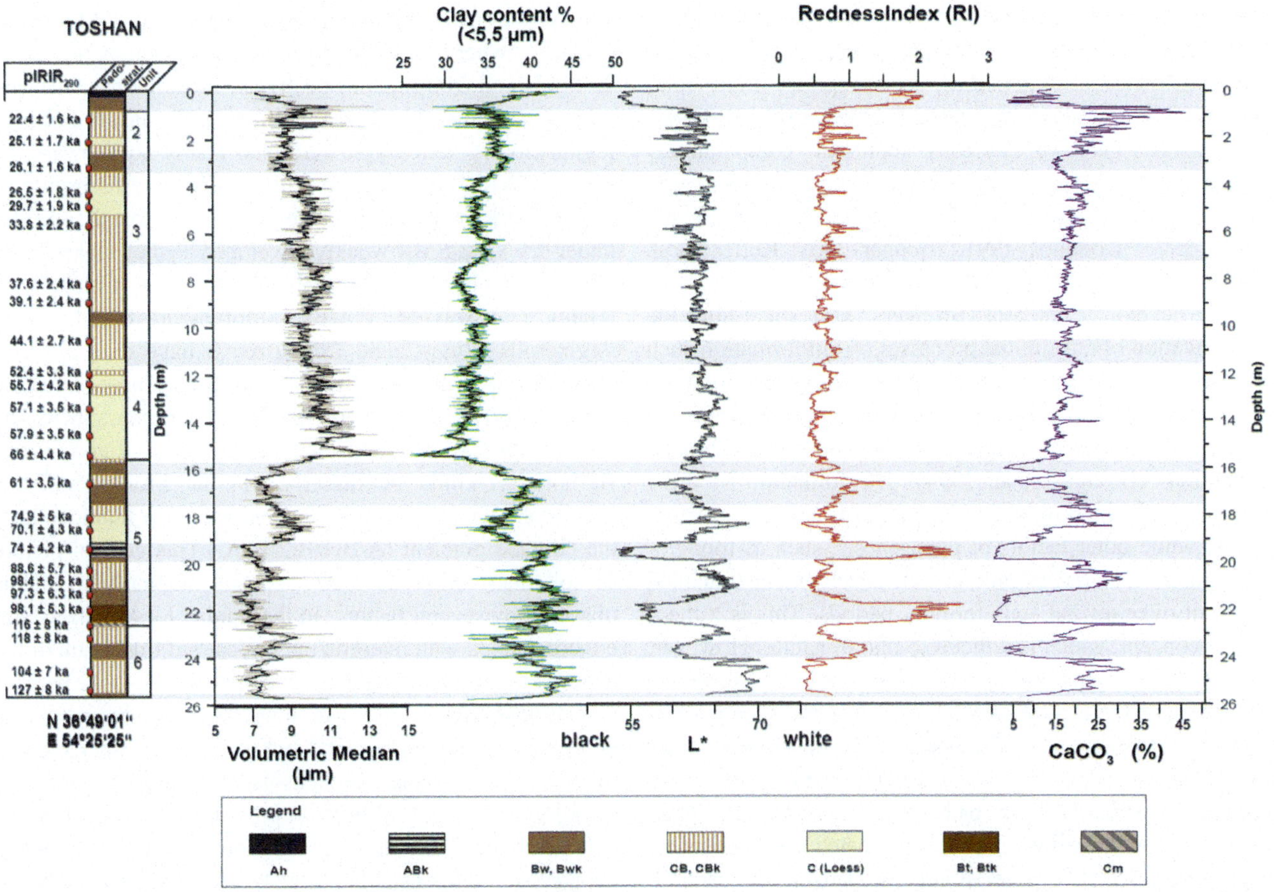

Fig. 8.6 Major stratigraphic units and pedostratigraphic column of the loess-soil sequence at Toshan. The units roughly correlate with MIS 6 to MIS 1 as indicated by results of luminescence dating. The grain size records show finest facies and highest clay content in Btb-horizons and in pedogenically altered loess of the lower part of the sequence. Weakly developed paleosols detected in the field at 3–9.5 m depth below surface show slight increases in the clay fraction in comparison with over- and underlying loess strata. The sediment color reflects boundaries between more strongly developed soils and loess in the lower part, as well as minor changes within units 4 and 3, where weakly developed soils are present. The $CaCO_3$ content strongly varies from almost carbonate free in the Bt-horizon at 22 m depth and more than 40% in CBk horizons. In the field, the most typical loess was detected at 15.75 m depth, reflected in high median grain size, moderate L*-values and low redness ratio. Data were taken from Vlaminck et al. (2016) and Lauer et al. (2017a)

the past. In addition, buried paleosols were found in playa deposits near Isfahan (Ayoobi et al. 2003) and in the Yazd-Ardakan plain (Mehrjardi et al. 2009). However, estimates of the timing and duration of pedogenic processes are often insufficient.

8.6 Paleosols as Indicators of Climate Change in Iran

Proxy data such as the $\delta^{18}O$ or Ca^{2+} concentration in Greenland ice cores for temperature and dust content of the atmosphere, respectively, provide detailed records of climate change on orbital or millennial time scales during the Last Glacial. Rasmussen et al. (2014) defined 26 Greenland stadials and interstadials between the last and present interglacials (Eemian and Holocene) including subdivisions with shorter changes in climate. Many of these fluctuations can be traced in marine cores, which reach further down in the geological past than ice cores (e.g., Lisiecky and Raymo 2005). The question remains, which cycles and amplitudes of climate change occurred on land and how terrestrial ecosystems reacted to these conditions. Stacked sequences of sediments and paleosols are excellent archives to at least partly answer these questions. Loess deposits are particularly useful because the accumulation of loess involves a series of climate-controlled processes including the production of silt-sized particles, their deflation, eolian transport and deposition as well as syn- and post-depositional transformations including soil formation (e.g., Pécsi 1990; Gardner and Rendell 1994; Pye 1995; Smalley1995; Wright 2001). Loess layers with no obvious pedogenic overprint can be correlated with stadial phases, reflecting dry (and cold) conditions during dust accumulation under a sparse vegetation cover. Paleosol horizons represent comparatively moister (and warmer) conditions and steppe or forest vegetation during interstadials or interglacials (e.g., Fink and Kukla 1977; Dodonov 1991; Bronger 2003; Rutter et al. 2003). However, the loess itself has to be understood as a weakly developed paleosol (Bronger 2003), since after the accumulation of dust, initial processes of soil formation such as formation of calcite bridges between grains, precipitation of iron hydroxides and humus accumulation took place.

Climaphytomorphic soils are best suited for paleoclimatic deductions (Bronger et al. 1994). The formation of these soils was governed by climate conditions and related vegetation, while other factors of pedogenesis, such as topography, parent material or time of soil development remained constant over several soils forming periods. This is a theoretical concept, which is difficult to find in nature. However, loess-paleosol sequences in plateau position far above the groundwater table may provide such records. Comparison with a climosequence of modern soils developed under known climate conditions is necessary in order to derive climate proxy data from paleosols. Khormali and Kehl (2011) and Khormali et al. (2012) presented pedons derived from loess, which can serve for comparison. In general, horizon differentiation and degree of soil development increase from the dry northern areas of the Turkmen steppe toward the more humid areas of the Alborz Mountains foothills, where mean annual rainfall amounts to about 200 mm and 900 mm, respectively. Accordingly, the soil moisture regime (SMR) ranges from aridic in the northern regions to dry xeric, typic xeric and finally udic toward the South (SWRI 2000). In the aridic SMR, an A/C-profile of Typic Torriothents (Haplic Regosols) developed, whereas under dry xeric SMR weakly developed Typic Haploxerepts (Haplic Cambisols) with A/Bw/C and Typic Calcixerepts (Haplic Calcisols) with A/Bk/C-profiles are found. Calcic Argixerolls (Luvi-Calcic Chernozems) have developed an A/Bt/Bk/C-profile (see Chaps. 6 and 7). These soils are common in the typic xeric SMR. Under udic SMR, the intensity of clay illuviation is more pronounced and several Bt horizons could be discerned in the Mollic or Typic Hapludalfs (Haplic Luvisols) developed at the humid side of the climatic gradient (A/Bt1/Bt2/Bk/C).This general picture of soil distribution on loess deposits in Northern Iran is obscured by several factors. In the Iranian loess plateau, soil erosion related with overgrazing has caused truncation of the upper soil horizons such that wide areas in the dryer parts of the climosequence have lost their soils and the raw dust deposit is exposed at the land surface instead. Since erosion may also have truncated paleosols, the mere presence of a certain type of paleosol horizon provides information on the paleo climate (Table 8.1). An attempt was made to compare the paleosol horizons at the three key loess sections at Neka, Now Deh and Agh Band with horizons of the modern soils (Table 8.2). This comparison shows that the horizons of

Table 8.1 Paleosol horizons in Northern Iranian loess and their paleoclimatic implications

Horizon	Main characteristics	Pedological interpretation	Paleoclimatic implications
C	Yellowish brown to yellow orange "unweathered" loess; massive structure	Eolian sediment with incipient soil formation	Comparatively dry and cold climate (during stadials)
CBb	Light brown color; weakly developed subangular blocky structure; root channels, signs of bioturbation, contains primary carbonates, diffuse lower boundary toward Bk or C	Incipient soil resembling a weakly developed Kastanozem or Calcisol, syngenetic soil formation	Slight increase in edaphic moisture (+ temperature rise) during weak (and/or short) interstadials, dry steppe vegetation
CByb	Similar to CBb horizons but enrichment of gypsum as concretions, veins or large crystals	Same as above, but with gypsum enrichment	Same as above
Bwb	Brown color, moderately well-developed subangular blocky structure, root channels, free of primary carbonate, diffuse lower boundary toward Bk	Resembling degraded A-horizons of well-developed Kastanozems or Calcisols with low amounts of humus	Comparatively moist (and/or long) interstadial or dry interglacial, steppe vegetation
ABtb	Dark brown to brownish black color, well-developed subangular to angular blocky structure, partly nut shaped aggregates; some clay skins, sharp lower boundary toward Bk	Phaeozem or degraded Tschernosem	Moist and warm climatic conditions of interglacials with forest steppe or deciduous forest
Btb	Strong brown to reddish-brown, well-developed angular blocky structure, common clay skins, Mn and humus coatings; sharp lower boundary toward Bk	Bt-horizons of Luvisols	Moist and warm climatic conditions of interglacials with deciduous forest
Bkb	Light brown rich in secondary carbonate as concretions or pseudo mycelia, sharp upper boundary, massive structure, partly cemented	Subsurface horizon below Bwb or Btb	

Table 8.2 Modern soils and paleosols at the sections at Neka, Now Deh and Agh Band (Khormali and Kehl 2011). The correlation with MIS is based on pedostratigraphical considerations and luminescence age estimates (Kehl 2009; Frechen et al. 2009)

	Neka	Now Deh	Agh Band[a]
Estimated annual precipitation	750 mm	600 mm	350 mm
Modern soils			
– Subsoil horizons	Bt	BAw, (BAt)	Bk
– Soil units (WRB)	Luvic Phaeozem	Calcic Chernozem	Haplic Calcisol
Paleosol horizons of			
– MIS 5a, 5c	Bw(t)b	CBkb, Bwkb	CByb, Bwyb
– MIS 5e	ABtb	Btb	Bwyb
– MIS 6	not found	CBkb, Bwkb	not found?
– MIS 7	Btkb	Btb, Bwkb	not found?

[a] A new profile near the former section at Agh Band (Kehl et al. 2005) was studied by Lauer et al. 2017b and Wang et al. 2017. It contains a weakly and two moderately well-developed paleosols. Preliminary age estimates suggest that the most strongly developed soil (Bwyb) is older than expected from previous work (Frechen et al. 2009) and could belong to MIS 7 (Lauer et al. 2017b).

interglacial paleosols reflect a climatic gradient similar to that of the modern soils.

Interestingly, more reddish paleosols are typical of the lower strata of loess-paleosol sequences, whereas gray and brown soils are found in the upper parts indicating a trend of aridisation during the in Northern Iran. Shorter aridisation trends are represented by gradual transitions in sediment color and soil structure from paleosol horizons into overlying loess layers. These transitions indicate that at the end of soil forming periods dust accumulation continuously increased and finally outbalanced soil formation. The balance between these two mechanisms changed repeatedly during the Quaternary most probably driven by moisture changes. It appears likely that during interglacials top-down pedogenesis dominated while during transitional periods after interglacials sensustricto and interstadials syngenetic or upbuilding soil formation was common. Therefore, Iranian loess-paleosol sequences provide excellent geological archives of Late Quaternary climate change.

Overall, earlier hypotheses on repeated Holocene cycles of loess accumulation and soil formation in Northern Iran (e.g., Bobek 1937; Ehlers 1971) must be rejected as based on luminescence dating (Frechen et al. 2009; Lauer et al. 2017a) and pedostratigraphic reasoning. There is convincing evidence that, in Northern Iran, loess deposition and paleosol formation closely followed the climate changes of interglacials, interstadials and stadials of the Northern Hemisphere. In Central and Southern Iran, however, less information is available on the timing of soil forming periods and their relation with past changes in the circulation systems of the Westerlies or the Monsoon. It is to be expected that future studies on sediment-paleosol sequences of Iran will considerably contribute to our understanding of climate change in Western Asia.

8.7 Conclusions and Outlook

There is large potential to improve our knowledge on the properties, genesis, stratigraphic significance and paleoclimate implications of paleosols in Iran. Future investigations could benefit from the large number of exposures available in Quaternary deposits. Pedological characterization and interpretation of paleosols in Iran are still at the beginning and the use of proxy data such as stable isotopes, rock magnetism, or biomarkers for estimating climatic conditions and related vegetation covers during past soil forming periods is quite promising. However, along with this kind of paleo climatic deductions, more absolute dating of paleosols is needed for sound correlation of paleosols with other records of past climate change and landscape evolution. Future stratigraphic investigations would benefit from paleomagnetic measurements to identify the polarity reversal at the Brunhes-Matuyama boundary or magnetic excursions including the Laschamps or Blake events in loess-soil sequences of Northern Iran. This could provide independent estimates for the age of paleosols. Most information on paleosols in Iran and other places has been gathered by the study of spatially confined exposures. It is an ongoing challenge to describe the catenary relationships of paleosols or their relation with physiographic units widely used in the mapping of modern soils.

References

Aharipour R, Moussavi MR, Mosaddegh H, Mistiaen B (2010) Facies features and paleoenvironmental reconstruction of the Early to Middle Devonian syn-rift volcano-sedimentary succession (Padeha Formation) in the Eastern-Alborz Mountains, NE Iran. Facies 56:279–294

Antoine P, Bahain J-J, Berillon G, Asgari-Khaneghan A (2006) Tufcalcaire et séquense alluviale en context etectoniqueactif: la formation de Baliran (Province du Mazadaran, Iran). Quaternaire 17 (4):321–331

Ayoobi S, Jalalian A, Eghbal MK (2003) Evidences of climate change recorded in Quaternary paleosols from Isfahan and Emam-Gheis (Chaharmahal and Bakhtiary). J Sci Technol Agric Natur Resour 7 (in Persian)

Ballato P, Mulch A, Landgraf A, Strecker MR, Dalconi MC, Friedrich A, Tabatabaei SH (2010) Middle to late Miocene Middle Eastern climate from stable oxygen and carbon isotope data, southern Alborz mountains, N Iran. Earth Planet Sci Lett 300:125–138

Bobek H (1937) Die Rolle der Eiszeit in Nordwestiran. Z Gletscherk 25:130–183

Bronger A (2003) Löss als Produkt der Verwitterung und Bodenbildung. Mitteilgn Dtsch Bodenkundl Gesellsch 102:437–438

Bronger A, Bruhn-Lobin N, Heinkele T (1994) Micromorphology of paleosols-genetic and paleoenvironmental deductions: Case studies from central China, south India, NW Morocco and the Great Plains of the USA. Dev Soil Sci 22:187–206

Catt JA (1990) Paleopedology manual. Quat Int 6:1–95

Catt JA (1998) Report of the Working Group on definitions used in paleopedology. Quat Int 51(52):84

Dodonov AE (1991) Loess of Central Asia. GeoJournal 24:185–194

Driese SG (2009) Paleosols, Pre-Quaternary. In: Encyclopedia of paleoclimatology and ancient environments, Springer, Netherlands, p 748–751. 10.1007/978-1-4020-4411-3_178

Ehlers E (1971) Südkaspisches Tiefland Nordiran und Kaspisches Meer. Beiträge zu ihrer Entwicklungsgeschichte im Jung- und Postpleistozän. Tübinger GeogrStud 44, Tübingen

Farpoor MH, Khademi H, Eghbal MK, Krouse RH (2004) Mode of gypsum deposition in southeastern Iranian soils as revealed by isotopic composition of crystallization water. Geoderma 121:233–242

Fink J, Kukla GJ (1977) Pleistocene climates in Central Europe: At least 17 interglacials after the Olduvai event. Quat Res 7:363–371

Forte AM, Cowgill E (2013) Late Cenozoic base-level variations of the Caspian Sea: A review of its history and proposed driving mechanisms. Palaeogeogr Palaeoclimatol Palaeoecol 386:392–407

Frechen M, Kehl M, Rolf C, Sarvati R, Skowronek A (2009) Loess chronology of the Caspian Lowland in Northern Iran. Quat Int 128:220–233

Gardner R, Rendell H (1994) Loess, climate and orogenesis: implications of South Asian loesses. Z Geomorph NF 38:169–184

Ghafarpour A, Khormali F, Balsam W, Karimi A, Ayoubi S (2016) Climatic interpretation of loess-paleosol sequences at Mobarakabad and Aghband, Northern Iran. Quat Res 86:95–109

Karimi A, Khademi H, Kehl M, Jalalian A (2009) Distribution, lithology and provenance of peridesert loess deposits in northeastern Iran. Geoderma 148:241–250

Karimi A, Frechen M, Kehl M, Khademi H (2011) Chronostratigraphy of loess deposits in Northeast Iran. Quat Int 234:124–132

Karimi A, Khademi H, Ayoubi S (2013) Magnetic susceptibility and morphological characteristics of a loess–paleosol sequence in northeastern Iran. Catena 101:56–60

Kehl M (2009) Quaternary climate change in Iran - the state of knowledge. Erdkunde 63:1–17

Kehl M (2010) Quaternary loesses, loess-like sediments, soils and climate change in Iran. Relief, Boden, Paläoklima, 24, Schweizerbart, Stuttgart

Kehl M, Frechen M, Skowronek A (2009) Nature and age of Late Quaternary basin fill deposits in the Basin of Persepolis, Southern Iran. Quat Int 196:57–70

Kehl M, Khormali F (2014) Excursion guide. Int Symp Loess, Soils and Climate Change in Southern Eurasia, Gorgan Univ of Agricultural Sciences and Natural Resources

Kehl M, Sarvati R, Ahmadi H, Frechen M, Skowronek A (2005) Loess paleosol-sequences along a climatic gradient in Northern Iran. Eiszeitalter u Gegenwart 55:149–173

Khademi H, Mermut AR (1999) Submicroscopy and stable isotope geochemistry of carbonates ad associated palygorskite in Iranian Aridisols. Eur J Soil Sci 50:207–216

Khademi H, Mermut AR, Krouse RH (1997) Isotopic composition of gypsum hydration water in selected landforms from Central Iran. ChemGeol 138:245–255

Khademi H, Mermut AR (2003) Micromorphology and Classification of Argids and Associated Gypsiferous Aridisols from Central Iran. Catena 54:439–455

Khormali F, Abtahi A, Stoops G (2006) Micromorphology of Calcic Features in Highly Calcareous Soils of Fars Province, Southern Iran. Geoderma 132:31–45

Khormali F, Abtahi A, Mahmoodi S, Stoops G (2003) Argillic horizon development in calcareous soils of arid and semiarid regions of southern Iran. Catena 53:273–301

Khormali F, Ghergherechi S, Kehl M, Ayoubi S (2012) Soil formation in loess-derived soils along a subhumid to humid climate gradient, Northeastern Iran. Geoderma 179–180:113–122

Khormali F, Kehl M (2011) Micromorphology and development of loess-derived surface and buried soils along a precipitation gradient in Northern Iran. Quat Int 234:109–123

Lauer T, Frechen M, Vlaminck S, Kehl M, Lehndorff E, Shahriari A, Khormali, F (2017a) Luminescence-chronology of the loess palaeosol sequence Toshan, Northern Iran—A highly resolved climate archive for the last glacial-interglacial cycle. Quat Int 429: 3–12

Lauer T, Vlaminck S, Frechen M, Rolf C, Kehl M, Sharifi J, Lehndorff E, Khormali F (2017b) The Agh Band loess-palaeosol sequence—A terrestrial archive for climatic shifts during the last and penultimate glacial–interglacial cycles in a semiarid region in northern Iran. Quat Int 429:13–30

Lisiecki LE, Raymo ME (2005) A Pliocene-Pleistocene stack of 57 globally distributed benthic $\delta^{18}O$ records. Paleoceanography 20: PA1003

Mason JA, Jacobs PM (2007) Nature of Paleosols. Encyclopedia of Quaternary Science. UK, Elsevier, Oxford, pp 2086–2095

Mehrjardi R, Mahmoodi S, Heidari A, Akbarzadeh A (2009) Micromorphological evidences of climatic change in Yazd region. Iran. J Mt Sci 6:162–172

Moussavi-Harami R, Mahboubi A, Nadjafi M, Brenner RL, Mortazavi M (2009) Mechanism of calcrete formation in the Lower Cretaceous (Neocomian) fluvial deposits, northeastern Iran based on petrographic, geochemical data. Cretac Res 30:1146–1156

Palmer AS (2007) Pedostratigraphy. Encyclopedia of Quaternary Science. UK, Elsevier, Oxford, pp 2847–2855

Pécsi M (1990) Loess is not just the accumulation of dust. Quat Int 7 (8):1–21

Pye K (1995) The nature, origin and accumulation of loess. QuatSci Rev 14:653–667

Rasmussen SO, Bigler M, Blockley SP, Blunier T, Buchardt SL, Clausen HB, Cvijanovic I, Dahl-Jensen D, Johnsen SJ, Fischer H, Gkinis V, Guillevic M, Hoek WZ, Lowe JJ, Pedro JB, Popp T, Seierstad IK, Steffensen JP, Svensson AM, Vallelonga P, Vinther BM, Walker MJC, Wheatley JJ, Winstrup M (2014) A stratigraphic framework for abrupt climatic changes during the Last Glacial period based on three synchronized Greenland ice-core records: refining and extending the INTIMATE event stratigraphy. Quat Sci Rev 106:14–28

Retallack GJ (2001) Soils of the Past. An Introduction to Paleopedology.2nded, Blackwell, Oxford

Rivandi B, VahidiniaM, Nadjafi M, Mahboubi A, Sadeghi A (2013) Biostratigraphy and sequence stratigraphy of paleogene deposits in central kopet-dagh basin (NE of Iran). J Geol Res, vol 2013. Article ID 892198. 10.1155/2013/892198

Ruhe RV (1956) Geomorphic surfaces and the nature of soils. Soil Sci 82:441–455

Rutter NW, Rokosh D, Evans ME, Little EC, Chlachula J, Velichko AA (2003) Correlation and interpretation of paleosols across European Russia and Asia over the last interglacial-glacial cycle. Quat Res 60:101–109

Shahriari A, Khormali F, Bläsing M, Vlaminck S, Kehl M, Frechen M, Karimi A, Lehndorff E (2017) Biomarkers in modern and buried soils of semi-desert and forest ecosystems of northern Iran. Quat Int 429: 62–73

Smalley IJ (1995) Making the material: The formation of silt-sized primary mineral particles for loess deposits. QuatSci Rev 14:645–651

Stevens L, Djamali M, Andrieu-Ponel V, de Beaulieu J-L (2012) Hydroclimatic variations over the last two glacial/interglacial cycles at Lake Urmia Iran. J Paleolimnol 47:645–660

SWRI - Soil and Water Research Institute Iran (2000) Soil resources and use potentiality map of Iran (1:1 000 000), Tehran

Taheri M, Khormali F, Wang X, Amini A, Wei H, Kehl M, Frechen M, Chen F (2017) Micromorphology of the lower Pleistocene loess in the Iranian Loess Plateau and its paleoclimatic implications. Quat Int 429:31–40

Vlaminck S, Kehl M, Lauer T, Shahriari A, Sharifi J, Eckmeier E, Lehndorff E, Khormali F, Frechen M (2016) The loess-soil sequence at Toshan (Northern Iran): Insights into late Pleistocene climate change. Quat Int 399:122–135

Wang X, Wei HT, Taheri M, Khormali F, Danukalova G, Chen FH (2016) Early Pleistocene climate in western arid central Asia inferred from loess-palaeosol sequences. Sci Rep 6:20560. doi:10.1038/srep20560

Wang X, Wei H, Khormali F, Taheri M, Kehl M, Frechen M, Lauer T, Chen F (2017) Grain-size distribution of Pleistocene loess deposits in northern Iran and its palaeoclimatic implications. Quat Int 429:41–51

Wright J (2001) "Desert" loess versus "glacial" loess: Quartz silt formation, source areas and sediment pathways in the formation of loess deposits. Geomorphology 36:231–256

Yaalon DH (1995) The soils we classify. Essay review of recent publications on soil taxonomy. Catena 24:233–241

Author Biographies

Dr. Martin Kehl obtained his Ph.D. on soil erodibility and specialized in soil geography, palaeosols and Quaternary research. His major interests are how past climate change and landscape evolution may be deduced from sediments and intercalated palaeosols. Since the year 2000, he has been investigating sediment-soil sequences in Iran in cooperation with soil scientists and geographers of different Iranian universities and with German colleagues.

Dr. Farhad Khormali is a professor of pedology at Gorgan University of Agricultural Sciences and Natural Resources, Iran. He is also a committee member of the Soil Science Society of Iran as well as editor in chief of the *Journal of Soil Management and Sustainable Agriculture*. His main research interests are soil genesis, loess-paleosol studies, micromorphology, clay mineral weathering and soil quality and soil carbon storage.

Agroecological Zones

9

Eddy De Pauw, Vahid R. Ghasemi Dehkordi, and Abdolali Ghaffari

Abstract

Although Iran consists almost entirely of drylands, it has a high diversity in agro-ecologies and agricultural systems, which has major consequences for agricultural land use planning and environmental management. The key determinants of this diversity are climate, landscapes, soils and water for irrigation. An agro-ecological zones (AEZs) approach is able to capture adequately the ecological diversity as it impacts the potential performance of agriculture and provides a suitable integrated framework for country-level agricultural planning. This chapter describes the method used to integrate national-level datasets on climate, terrain, soils and land use into an agro-ecological zones map of Iran consisting of 688 mapping units. Of these, 92 agro-ecological zones cover 80% of Iran, with the remainder 596 AEZs representing 'niche' agricultural environments. A policy implication is that agricultural research and planning need to promote diverse farming and production systems, each fitting well into each particular environment.

Keywords

Agricultural Environments • Climate • Soils • Land use

9.1 Introduction

Iran, with a total land area of 1,648,195 sq.km,[1] consists almost entirely (99%) of drylands.[2] At the same time its total population, using figures of 2014[3] is nearly 81 million, of which almost 30% lives in rural areas. Even accounting for the fact that about 8 Mha of agricultural land is irrigated, the extent of the dryland areas, which all experience water deficits to a larger or smaller degree, is staggering, certainly in comparison to the population that needs to be fed.

In this undoubtedly dramatic but oversimplified picture, the fact is often overlooked that, within its overall dryland context, Iran also has a very high diversity in agro-ecologies and agricultural systems. Climate is the first determinant of agro-ecological diversity. Using the UNESCO classification system for Arid Zones (UNESCO 1979), by the simple combination of major aridity and temperature regimes, 28 different agro-climatic zones were differentiated in Iran (De Pauw et al. 2004). Within this highly heterogeneous climatic setting, more diversity in biophysical environments is created by differences in landscapes, soils, geological substrata, surface water and groundwater resources. In dryland areas, irrigation development is the single most important factor in creating artificial agro-ecological niches, areas where natural

[1]Source http://en.wikipedia.org/wiki/Iran.

[2]Drylands defined as areas having an aridity index (ratio annual precipitation/potential evapotranspiration) of less than 0.75.

[3]Source http://www.census.gov/population/international/data/idb/region.php?N=%20Results%20&T=13&A=separate&RT=0&Y=2014&R=-1&C=IR.

E. De Pauw (✉)
Digital Atlas Development, ICARDA, Aleppo, Syria
e-mail: edepauw@pavocarto.be

V.R. Ghasemi Dehkordi
Soil and Water Research Institute (SWRI), Karaj, Iran

A. Ghaffari
Agriculture and Natural Resources Research and Education Center of West Azerbaijan, West Azerbaijan, Iran

M.H. Roozitalab et al. (eds.), *The Soils of Iran*, World Soils Book Series,
https://doi.org/10.1007/978-3-319-69048-3_9

conditions and production systems show abrupt differences with their surroundings.

This implies that it is not enough to simply look at general precipitation levels to understand a dryland area and provide appropriate guidance for sustainable land utilization. This certainly has major consequences for environmental management: ignoring the complex nature of diverse agricultural dryland environments complicates biodiversity management, crop production intensification or diversification, land use optimization or combating land degradation.

What country-level agricultural planners need is an integrated spatial framework that (i) is able to capture at the level of a country the ecological diversity as it impacts the potential performance of agriculture, (ii) is simple enough to be understood by non-specialists but at the same time allows greater technical detail to be seamlessly incorporated and (iii) for the sake of efficiency and cost minimization, uses the best national datasets available, with possibilities for updating.

It is argued that such suitable integrated framework for country-level agricultural planning can be provided by an *agro-ecological zones (AEZs)* approach. At the national level, the main use of an AEZ-based spatial framework would be to serve as a tool for rapid identification of different biophysical environments and characterization of resources and constraints to assist agricultural research planning and policy development.

The AEZ concept is not new. Agro-ecological zones have been used in different regions or countries, and at different scales, for a variety of agricultural planning purposes. Agricultural planning covers many charges and can be undertaken at different levels. It ranges from the physical location of research stations, the introduction of particular crops, cultivars and technologies to suit the conditions in different areas, the allocation of water resources to agriculture, fertilizer recommendations, policies and regulations for rural land use, inputs and technology subsidies etc. When planners are more aware of the SWOTs (strengths, weaknesses, opportunities and threats) of geographically well-defined areas, their planning can also be more effective.

The term 'agro-ecological zones' has no uniform meaning in the literature. Depending on the source it has different connotations, sometimes it refers to zones with similar climate for agricultural use, on other occasions it has been used to delineate areas with defined precipitation amounts. The Food and Agriculture Organization of the United Nations (FAO), followed later by the International Institute for Applied Systems Analysis (IIASA), has a 30-year legacy of studies based on a concept of 'agro-ecological zones' as spatial entities that delineate areas with different production potential (or 'suitability') for specific crops (e.g. FAO 1978; Fischer et al. 2012). For this reason, it is important to clearly define what the concept stands for in this study.

The term *agro-ecological zones* is used here in a broader meaning of *integrated* and more or less homogeneous spatial units in which available water resources, climate, terrain and soil conditions combine to create *unique* environments, which can be associated with distinct farming systems and land use and settlement patterns. By integrating the key components of the agricultural environments, an agro-ecological zones map offers a bird's-eye view of internal diversity, agricultural potential and constraints that decision-makers find easier to understand than a stack of single-theme maps. Agro-ecological zones are therefore in concept related to but are not similar to *agro-ecosystems*, which refer more to farming systems within an ecological setting, as the following definition implies (Smit et al. 1998).

> An agro-ecosystem is the basic unit of study in agro-ecology, and is somewhat arbitrarily defined as a spatially and functionally coherent unit of agricultural activity and includes the living and nonliving components involved in that unit as well as their interactions.

Whereas in the past the manual integration of multidisciplinary spatial information was a major handicap for producing a reliable AEZ map, geographic information systems (GIS) technology, through its seamless overlaying capabilities of thematic datasets, makes this now perfectly practicable. Collaborative research in North Africa and West Asia, through partnerships between ICARDA[4] and different national agricultural research systems, particularly in Iran, Morocco, Syria and Turkey, has confirmed the feasibility of rapidly defining agro-ecological zones by the combination of climatic, land use/land cover, terrain, soil and other data using GIS procedures.

Whereas agro-ecological zones can certainly be defined at different geographical levels, from the global to the subnational, in this chapter the AEZ-mapping methodology is illustrated through a case study that covers all of Iran. This case study aims in the first place to demonstrate how new added value can be created by integrating existing high-quality single-theme national datasets.

9.2 Methodology

In theory, a map of agro-ecological zones can be created by simple overlaying of climatic, land use/land cover, terrain and soils data. However, without additional processing this is not practical since in the case of Iran, the direct overlaying of the most relevant thematic layers leads to 16,054 individual mapping units. For this reason, the following three major GIS approaches were applied for generating the Iran AEZ map:

[4]International Center for Agricultural Research in Dry Areas, Amman, Jordan.

- generalization of raster themes related to climate, land use/land cover, terrain and soils through reclassification;
- creating an integrated AEZ layer through overlaying the reclassified single-theme layers;
- harmonization and removal of 'spurious' mapping units, created by the overlaying process.

Four themes were considered the most relevant in order to generate the AEZ map:

- agro-climatic zones
- land use/land cover groups
- landform groups
- soil management domains.

The data sources and processing steps involved with each of these themes are explained in the following sections.

9.2.1 Generalization

9.2.1.1 Generalizing the Climate Theme

The climate theme was generalized using the Agro-climatic Zones Map of Iran. As documented by De Pauw et al. (2004), the Agro-climatic Zones Map of Iran is already a fairly integrated interpretation of climate, in the sense that it incorporates variables related to water supply and demand and temperature. The map was developed using the following procedure:

- spatial interpolation of meteorological station data to generate continuous surfaces of mean monthly precipitation and temperature;
- conversion of temperature data into estimates of potential evapotranspiration (PET) and generation of continuous surfaces of mean monthly PET;
- classification of the precipitation and PET surfaces using their annual ratio (aridity index) into *moisture regime zones* (Table 9.1) and of the temperature surfaces into zones of *winter types* (Table 9.2) and *summer types* (Table 9.3), in accordance with the criteria and class thresholds as implemented in the UNESCO classification system for arid regions (UNESCO 1979);
- integration of the climatic surfaces related to moisture regime and winter and summer temperature regimes into agro-climatic zones.

From Iran, 126 stations were used with a precipitation record length of at least 20 years and 590 stations with a temperature record length of at least 5 years. The database also included some precipitation and temperature data from neighbouring countries, leading to a total database of 244 precipitation stations and 627 temperature stations. For the spatial interpolation of the meteorological station data the 'thin-plate smoothing spline' method of Hutchinson (1995), as implemented in the ANUSPLIN software (Hutchinson 2007), was used.

Mean monthly PET estimates according to the Penman-Monteith method (1998) were obtained by a two-step procedure:

- estimation of PET from temperature according to the Hargreaves method (Choisnel et al. 1992);
- estimation of PET Penman-Monteith from PET-Hargreaves through regression.

Generalization of the Agro-climatic Zones Map of Iran was done on the basis of the conversion procedure of Table 9.4.

The symbols for the agro-climatic zones in the first column of Table 9.4 are the ones used by De Pauw et al. (2004). The symbols for the generalized agro-climatic zones are the same ones as used in the Generalized Agro-Climatic Zones Map of Iran.[5]

9.2.1.2 Generalizing the Land Use/Land Cover Theme

The source data were provided by the land use/land cover map of Iran.

The original 19 classes of the latter map were generalized in accordance with Table 9.5.

9.2.1.3 Generalizing Topography

The most important attributes of topography are the elevation and the slope. Due to the lapse of air temperature that takes place with increasing altitude, the effect of absolute elevation is basically a temperature effect, which is already taken into consideration through the climatic classification explained in Sect. 5.2.1.

To represent slopes accurately, high-resolution topographical maps or a digital elevation model, such as SRTM, is required. However, to get an overview of landforms over a huge country like Iran, slope patterns are too small scale to be practical, hence the need to adopt a simple classification of landforms. For these reasons, generalized landform categories were generated[6] using SRTM30, a global DEM with 30 arc-second (approximately 1 km) resolution.[7]

This was achieved by using a proxy for slope classes based on the *maximum elevation difference between*

[5]Link https://www.dropbox.com/s/7cyre7bpr1jba2y/ACZ_generalized_v2.tif?dl=0.
[6]Link https://www.dropbox.com/s/31t81tizcxl1cau/Landforms_generalized.tif?dl=0.
[7]Link http://dds.cr.usgs.gov/srtm/version2_1/SRTM30/srtm30_documentation.pdf.

Table 9.1 Classes for the moisture regime

Moisture regime	Hyper-arid (HA)	Arid (A)	Semi-arid (SA)	Sub-humid (SH)	Humid (H)	Per-humid (PH)
Aridity index	<0.03	0.03–0.2	0.2–0.5	0.5–0.75	0.75-1	>1

Table 9.2 Classes for the winter type

Winter type	Warm (W)	Mild (M)	Cool (C)	Cold (K)
Mean temp. coldest month	>20 °C	>10 °C	>0 °C	≤ 0 °C

Table 9.3 Classes for the summer type

Summer type	Very warm (VW)	Warm (W)	Mild (M)	Cool (C)
Mean temp. warmest month	>30 °C	>20 °C	>10 °C	≤ 10 °C

Table 9.4 Generalization of Iran Agro-climatic Zones Map

Full ACZ symbol	Generalized ACZ	Description
HA-M-VW; HA-C-VW	ACZ1	Hyper-arid climates with mostly warm or very warm summers and warm or mild winters; the overwhelming constraint is lack of moisture
A-M-VW; A-M-W	ACZ2	Arid climates with mild or warm winters and warm or very warm summers; winter coldness is not an ecological constraint
A-C-VW; A-C-W; A-C-M; A-K-W; A-K-M	ACZ3	Arid climates with cool or cold winters and warm or very warm summers; winter coldness is an ecological constraint
SA-M-VW; SA-M-W	ACZ4	Semi-arid climates with mild or warm winters and warm or very warm summers; winter coldness is not an ecological constraint
SA-C-VW; SA-C-W; SA-K-W	ACZ5	Semi-arid climates with cool or cold winters and mostly warm summers; winter coldness is an ecological constraint
SA-C-M; SA-K-M;	ACZ6	Mostly semi-arid climates with mostly cold winters and mild summers; winter coldness is an ecological constraint as well as short thermal growing periods
SH-C-VW; SH-C-W; SH-K-W; SH-K-M	ACZ7	Sub-humid climates with cool or cold winters and mild or warm summers; winter coldness is an ecological constraint
H-C-W; H-K-W; H-K-M	ACZ8	Humid climates with cool or cold winters and mild or warm summers; winter coldness is an ecological constraint
H-K-C; PH-K-C	ACZ9	Humid or per-humid climates with cold winters and cool summers; winter coldness is an ecological constraint as well as short thermal growing periods
PH-C-W; PH-K-W; PH-K-M	ACZ10	Per-humid climates with cool or cold winters and mostly mild or warm summers

Table 9.5 Generalization of Iran Land Use-Land cover categories

LULC-Iran	Generalized land use/land cover classes
Irrigated farming	Irrigated cropland
Dry farming; scattered dry farming	Rainfed cropland
Non suitable range; range; range habitat	Rangelands
Forest; afforestation	Forest cover
Sand dunes/lands; saline soils; playa; rock outcrops; bare lands; desert; urban	Non-agricultural land cover
Orumieh Lake; lake/reservoir; wetlands; Gorgan Gulf	Water bodies and wetlands

Table 9.6 Conversion of USDA Soil Taxonomy subgroups into new soil management groups (SMGs)

Soil Taxonomy subgroups (abbreviations)	Soil management group
Water; XBE; XBL; XDL; XKL; XLF; XPL; XRW; XSD; XSP; XUR; XWA; XWB	Undifferentiated non-soil areas
ADH,T;AXH,K; AXH,T; EVD,M; EVD,T;EVU,T; EVX,M; EVX,T; IMH,F; IMH,T; IOD,T; IOE,F; IOE,T; IOU,C; IOU,F; IOU,K; IOU,T; IOU,V; IOX,C; IOX,F; IOX,G; IOX,T; IOX,V; MDA,K; MDA,T; MDH,F; MDH,T; MDK,T; MUK,T; MXA, K; MXA,T; MXH,F; MXH,K; MXH,T; MXK,T; MXK,V; VXH,R; VXK,E; VXK,T	Agricultural soils, without major soil-related limitations
ADH,L; DGH,L	Soils with potential for agricultural use, with depth limitations
AUH,D; DGH,I; DGH,T; DGH,U; DGH,X; DGK,T; DGK,U; DGK,X; DGP,T; DGP,X; DKH,Q; DKH,T; DKH,U; DKH,X; DMH,F; DMH,I; DMH,N; DMH,Q; DMH,T; DMH,U; DMH, UF; DMH,X; DMQ,X; DRH,U; DRN,X; EVT,Q; EVT,T; EVT, U; EVT,X; EVU,D; IOE,DY; IOU,D; IOU,DK; IOU,G; VXK,A	Soils with potential for agricultural use, requiring irrigation
AXH,Q; IOD,Q; IOE,Q; IOX,Q; MDA,Q; MDH,Q; MXH,Q; MXK,Q	Soils with potential for agricultural use, requiring drainage
AQE,M;AQE,T; DGH,Q; EQE,M; EQE,T; EQI,T; EQV,A; EQV,M; EQV,T; IQE,T; IQH,T; IQM,T; MQE,F; MQE,I; MQE,T; MQK,T; XMA	Poorly drained soils of wetlands and floodplains
EST,T; EST,U; EST,X; ESU,D; ESU,T	Sandy soils
AQN,T; DGH,S; DMH,S; DMH,SX; DMH,SY; DSH,G; DSH, PG; DSH,Q; DSH,T; DSQ,G; DSQ,T	Sodic and/or saline soils
DGH,P; DKH,LX; EOD,L; EOT,L; EOT,LU; EOT,LX; EOU, L; EOX,L; IOD,L; IOU,L; IOX,L; MR,L; MR,T; MXH,L; XLF; XRM; XRO	Rock outcrops and very shallow soils
EOD,T; EOI,U; EOT,T; EOT,U; EOT,X; EOT,XL; EOU,D; EOU,Q; EOU,T; EOV,D; EOX,Q; EOX,T; SD	Non-agricultural soils
ADH,HU; IOD,B; UHH,T	Soils with high acidity and/or low nutrient status

The Soil Taxonomy abbreviations are explained in Table 9.7

neighbouring pixels. The latter was calculated using the Range function in the Spatial Analyst module of ArcGIS software with subsequent classification.

Generalized landform classes were obtained using the following rules:

- *Plains and plateaux*: maximum elevation difference between neighbouring pixels 0–50 m
- *Hills*: maximum elevation difference between neighbouring pixels 50–300 m
- *Mountains*: maximum elevation difference between neighbouring pixels >300 m

9.2.1.4 Generalizing the Soils Theme

The source data of soil information were provided by the 1:1,000,000 scale Soil Map of Iran (Banaei et al. 2000). The mapping units of this map consist of soil associations, classified according to Soil Taxonomy (USDA 2014).

Together with topography, soils are the main factor of spatial variability: the Soil Map of Iran contains 161 soil taxonomic units, distributed over 375 different soil associations. The reduction of this vast variability by regrouping was essential in order to establish eco-zones which are not over-fragmented.

Moreover, the highly technical detail of soil taxonomic units was inappropriate for direct application to agricultural management and therefore required a 'translation' step into more management-oriented categories.

Data reduction and interpretation were undertaken by the following two-step approach:

Step 1 conversion of the Soil Taxonomy units into broader groupings that are relevant to their broad management properties ('soil management groups', SMGs). This regrouping is summarized in Table 9.6.

Step 2 establishing the main combinations of the soil management groups ('soil management domains', SMDs)

Through this process, the Soil Map of Iran was converted by reclassifying each soil association, consisting of the newly formed soil management groups, into a map of '*Soil Management Domains*'.[8] This has a double benefit: (i) data generalization and (ii) making a link from soil classification to soil management.

The 57 SMDs thus identified are described in the Legend of the Map of Soil Management Domains.[9]

9.2.2 Integration of Thematic Layers

Once the component layers had been established, the first phase of generating the agro-ecological zones consisted of simple overlaying in a GIS procedure that retains all characteristics and attributes of the component generalized themes.

Unfortunately direct overlaying of information layers with different levels of accuracy or spatial precision tends to propagate the errors that were already present in the component datasets. Generalization of these datasets, as explained in Sect. 2.1, was one way to reduce, but not completely eliminate the errors resulting from overlaying. Two additional steps were needed to further reduce possible errors and, in the same processes, reduce the number of mapping units:

- harmonization and removal of inconsistencies
- Removal of spurious mapping units

9.2.2.1 Harmonization and Removal of Inconsistencies

Especially between the land use/land cover and soil map inconsistencies could appear. In order to harmonize the soil and land use/land cover maps, special decision rules were introduced that aim at giving priority to the most accurate information.

- If land use/land cover = irrigated cropland the landform and soils are not specified: it is assumed that for this intensive land use to be feasible, landform and soils cannot constitute major limitations, and are therefore of less relevance in the mapping. The SMDs are then not differentiated.
- If land use/land cover = rainfed cropland, it is assumed that for this land use to be feasible the soils cannot constitute major limitations. The SMDs are then not differentiated, but the landform can still imply different levels of suitability and therefore still needs to be specified.
- If land use/land cover = water bodies or wetlands, the SMDs are not considered relevant and are therefore not differentiated.
- Elimination of inconsistencies between landforms and soil management domains: if landform = mountains, plausible SMDs are: 0–6; 14–15; 23–24; 42–50; 54. If other SMDs occur in combination with mountains, they are considered implausible and replaced by the undifferentiated SMD.
- Elimination of inconsistencies between climate and soil types: in more humid climates one does not expect SMDs typical for drier climates: if ACZ = 6–9 then SMD 4, 42, 43 are replaced by SMD 0 (undifferentiated).

9.2.2.2 Removal of Spurious Mapping Units

Using overlay processing on the raster datasets causes sometimes small dispersed clusters (ranging from a few pixels to several hundred) of one class to appear inside another class. 'Spurious' mapping units are defined as homogeneous mapping units that occupy groups of maximum 10 pixels. These are a major source of fragmentation of the map and their removal does not affect the general pattern. To solve this problem of over-fragmentation an automated cleanup procedure was applied using GIS functions, available in standard GIS software, to absorb the 'orphaned' pixels into their nearest neighbours.

9.3 Result

Direct overlaying of the generalized thematic layers (agro-climatic zones, land use/land cover, landforms and soil management domains) generated 1919 mapping units, an impractical number to encompass in a single legend and to display, and moreover including too many very small mapping units. After application of harmonization rules and elimination of inconsistencies the corrected AEZ map still contained 1010 different mapping units. After removal of the spurious mapping units the AEZ map could be finalized with a total number of 688 mapping units, still a large number, but manageable through appropriate legend design. The final AEZ map[10] can be downloaded from the link in footnote.

A very generalized version of the AEZ map is included as Fig. 9.1.

The relationship between number of AEZ, from the largest to the smallest, and area covered is illustrated in Fig. 9.2. The full AEZ list, with descriptions of their component themes, can be downloaded from the provided link.[11]

[8]https://www.dropbox.com/s/2hg37cen3ubl69b/SMD.tif?dl=0.
[9]https://www.dropbox.com/s/cxbas2v186o09p0/SMD_Full%20Legend.tif?dl=0.
[10]https://www.dropbox.com/s/s283930gcc1yqla/AEZ-layers_A0_Final_v2.tif?dl=0.
[11]https://www.dropbox.com/s/ivwvg64530o1ml7/AEZ_Iran_full-list.xlsx?dl=0.

Table 9.7 Abbreviations used for Soil Taxonomy subgroups

Abbreviation	Soil Taxonomy subgroup	Abbreviation	Soil Taxonomy subgroup	Abbreviation	Soil Taxonomy subgroup
ADH,HU	Ultic Hapludalfs	EOT,U	Ustic Torriorthents	IOX,L	Lithic Xerochrepts
ADH,L	Lithic Hapludalfs	EOT,X	Xeric Torriorthents	IOX,Q	Aquic Xerochrepts
ADH,T	Typic Hapludalfs	EOT,XL	Xerertic Torriorthents	IOX,T	Typic Xerochrepts
AQE,M	Mollic Endoaqualfs	EOU,D	Aridic Ustorthents	IOX,V	Vertic Xerochrepts
AQE,T	Typic Endoaqualfs	EOU,L	Lithic Ustorthents	IOX,V	Vertic Xerochrepts
AQN,T	Typic Natraqualfs	EOU,Q	Aquic Ustorthents	IQE,T	Typic Endoaquepts
AUH,D	Aridic Haplustalfs	EOU,T	Typic Ustorthents	IQH,T	Typic Halaquepts
AXH,K	Calcic Haploxeralfs	EOV,D	Aridic Ustorthents	IQM,T	Typic Humaquepts
AXH,Q	Aquic Haploxeralfs	EOX,L	Lithic Xerorthents	MDA,K	Calcic Argiudolls
AXH,T	Typic Haploxeralfs	EOX,Q	Aquic Xerorthents	MDA,Q	Aquic Argiudolls
DGH,I	Vitrandic Haplogypsids	EOX,T	Typic Xerorthents	MDA,T	Typic Argiudolls
DGH,L	Lithic Haplogypsids	EQE,M	Mollic Endoaquents	MDH,F	Fluventic Hapludolls
DGH,P	Leptic Haplogypsids	EQE,T	Typic Endoaquents	MDH,Q	Aquic Hapludolls
DGH,Q	Aquic Haplogypsids	EQI,T	Typic Psammaquents	MDH,T	Typic Hapludolls
DGH,S	Sodic Haplogypsids	EQV,A	Aquandic Fluvaquents	MDK,T	Typic Calciudolls
DGH,T	Typic Haplogypsids	EQV,M	Mollic Fluvaquents	MQE,F	Fluventic Endoaquolls
DGH,U	Ustic Haplogypsids	EQV,T	Typic Fluvaquents	MQE,I	Fluvaquentic Endoaquolls
DGH,X	Xeric Haplogypsids	EST,T	Typic Torripsamments	MQE,T	Typic Endoaquolls
DGK,T	Typic Calcigypsids	EST,U	Ustic Torripsamments	MQK,T	Typic Calciaquolls
DGK,U	Ustic Calcigypsids	EST,X	Xeric Torripsamments	MR,L	Lithic Haprendolls
DGK,X	Xeric Calcigypsids	ESU,D	Aridic Ustipsamments	MR,T	Typic Haprendolls
DGP,T	Typic Petrogypsids	ESU,T	Typic Ustipsamments	MUK,T	Typic Calciustolls
DGP,X	Xeric Petrogypsids	EVD,M	Mollic Udifluvents	MXA,K	Calcic Argixerolls
DKH,LX	Lithic Xeric Haplocalcids	EVD,T	Typic Udifluvents	MXA,T	Typic Argixerolls
DKH,Q	Aquic Haplocalcids	EVT,Q	Aquic Torrifluvents	MXH,F	Fluventic Haploxerolls
DKH,T	Typic Haplocalcids	EVT,T	Typic Torrifluvents	MXH,K	Calcic Haploxerolls
DKH,U	Ustic Haplocalcids	EVT,U	Ustic Torrifluvents	MXH,L	Lithic Haploxerolls
DKH,X	Xeric Haplocalcids	EVT,X	Xeric Torrifluvents	MXH,Q	Aquic Haploxerolls
DMH,F	Fluventic Haplocambids	EVU,D	Aridic Ustorthents	MXH,T	Typic Haploxerolls
DMH,I	Ustertic Haplocambids	EVU,T	Typic Ustifluvents	MXK,Q	Aquic Calcixerolls
DMH,N	Vitrandic Haplocambids	EVX,M	Mollic Xerofluvents	MXK,T	Typic Calcixerolls
DMH,Q	Aquic Haplocambids	EVX,T	Typic Xerofluvents	MXK,V	Vertic Calcixerolls
DMH,S	Sodic Haplocambids	IMH,F	Fluventic Humudepts	SD	Sand Dune
DMH,SX	Sodic Xeric Haplocambids	IMH,T	Typic Humudepts	UHH,T	Typic Haplohumults
DMH,SY	Sodic Ustic Haplocambids	IOD,B	Umbric Dystrochrepts	VXH,R	Chromic Haploxererts
DMH,T	Typic Haplocambids	IOD,L	Lithic Dystrochrepts	VXK,A	Aridic Calcixererts
DMH,U	Ustic Haplocambids	IOD,Q	Aquic Dystrochrepts	VXK,E	Entic Calcixererts
DMH,UF	Ustifluventic Haplocambids	IOD,T	Typic Dystrochrepts	VXK,T	Typic Calcixererts
DMH,X	Xeric Haplocambids	IOE,DY	Dystric Fluventic Eutrochrepts	Water	Water
DMQ,X	Xeric Aquicambids	IOE,F	Fluventic Eutrochrepts	XBE	Beaches
DRH,U	Ustic Haplargids	IOE,Q	Aquic Eutrochrepts	XBL	Badlands

(continued)

Table 9.7 (continued)

Abbreviation	Soil Taxonomy subgroup	Abbreviation	Soil Taxonomy subgroup	Abbreviation	Soil Taxonomy subgroup
DRN,X	Xeric Natrargids	IOE,T	Typic Eutrochrepts	XDL	Dune Lands
DSH,G	Gypsic Haplosalids	IOU,C	Calcic Ustochrepts	XKL	Kalut
DSH,PG	Leptic Haplosalids	IOU,D	Aridic Ustochrepts	XLF	Lava Flow
DSH,Q	Aquic Haplosalids	IOU,DK	Aridic Lithic Ustochrepts	XMA	Marsh
DSH,T	Typic Haplosalids	IOU,F	Fluventic Ustochrepts	XPL	Playa
DSQ,G	Gypsic Aquisalids	IOU,G	Gypsic Ustochrepts	XRM	Rocky Mountain
DSQ,T	Typic Aquisalids	IOU,K	Calcic Ustochrepts	XRO	Rock Outcrop
EOD,L	Lithic Udorthents	IOU,L	Lithic Ustochrepts	XRW	River Wash
EOD,T	Typic Udorthents	IOU,T	Typic Ustochrepts	XSD	Sand Dune
EOI,U	Vitrandic Troporthents	IOU,V	Vertic Ustochrepts	XSP	Salt Plug
EOT,L	Lithic Torriorthents	IOX,C	Calcixerollic Xerochrepts	XUR	Urban
EOT,LU	Lithic Ustic Torriorthents	IOX,F	Fluventic Xerochrepts	XWA	Water
EOT,LX	Lithic Xeric Torriorthents	IOX,G	Gypsic Xerochrepts	XWB	Water
EOT,T	Typic Torriorthents				

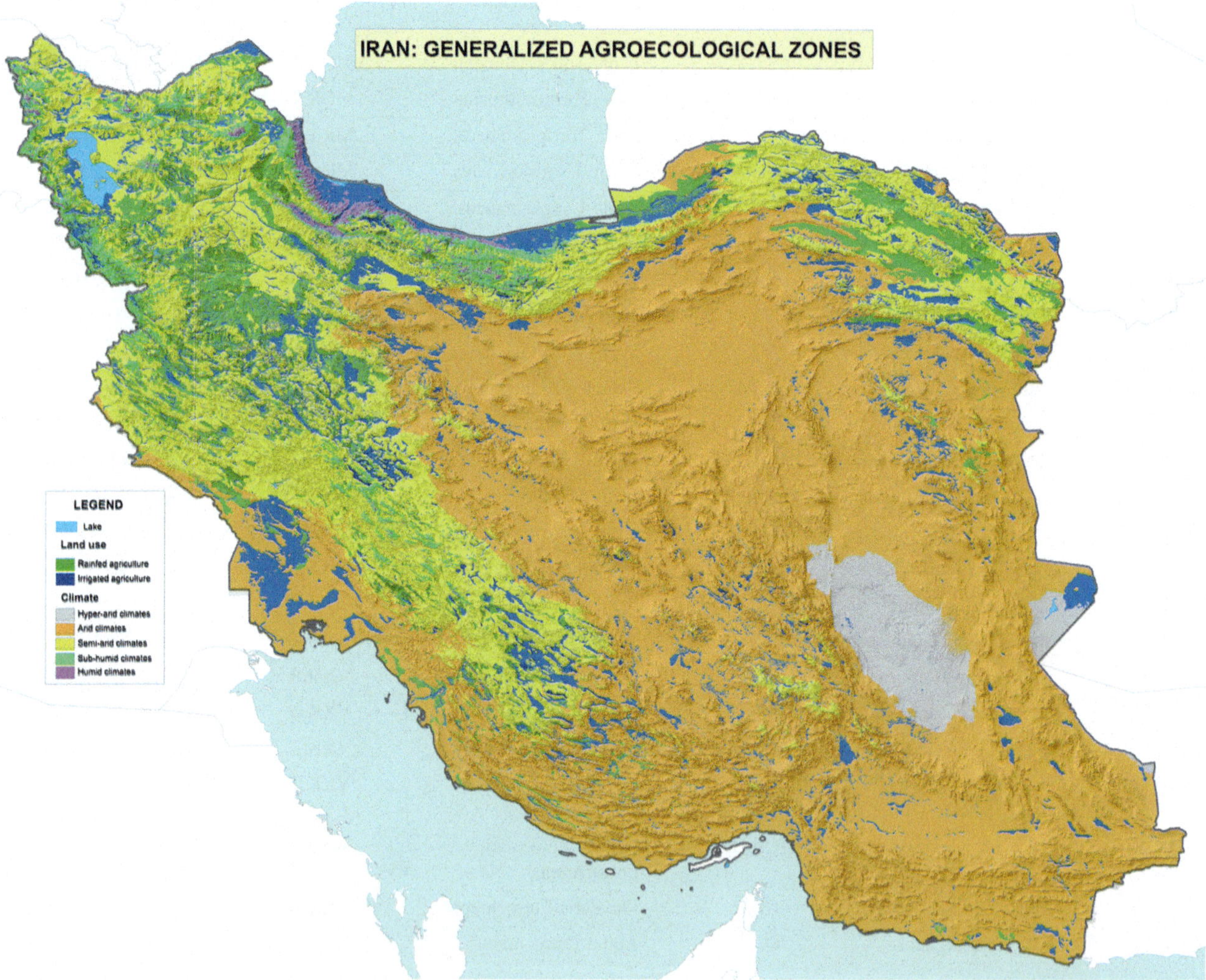

Fig. 9.1 Generalized agro-ecological zones map of Iran

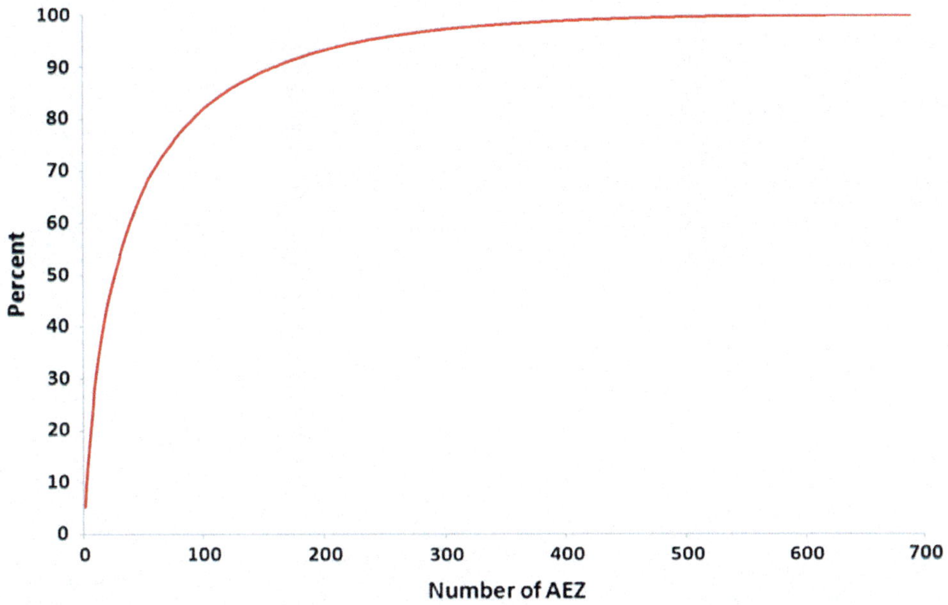

Fig. 9.2 Cumulative area (in per cent) of Iran covered by number of AEZ

Table 9.8 Ten largest AEZs in Iran

AEZ code	% of Iran	Climatic zone	Land use	Landform	Soil management domain
23243	5.34	Arid climates with cool or cold winters and warm or very warm summers; winter coldness is an ecological constraint	Rangelands	Hills	Predominantly rock outcrops and very shallow soils
42200	4.15	Semi-arid climates with cool or cold winters and mostly warm summers; winter coldness is an ecological constraint	Rainfed cropland	Hills	Undifferentiated soils
23123	3.04	Arid climates with cool or cold winters and warm or very warm summers; winter coldness is an ecological constraint	Rangelands	Plains and plateaus	Mainly soils with potential for agricultural use, requiring irrigation, associated with non-agricultural soils
13243	2.66	Arid climates with mild or warm winters and warm or very warm summers; winter coldness is not an ecological constraint	Rangelands	Hills	Predominantly rock outcrops and very shallow soils
44343	2.55	Semi-arid climates with cool or cold winters and mostly warm summers; winter coldness is an ecological constraint	Non-agricultural land use	Mountains	Predominantly rock outcrops and very shallow soils
43243	2.51	Semi-arid climates with cool or cold winters and mostly warm summers; winter coldness is an ecological constraint	Rangelands	Hills	Predominantly rock outcrops and very shallow soils
25100	2.39	Arid climates with cool or cold winters and warm or very warm summers; winter coldness is an ecological constraint	Non-agricultural land use	Plains and plateaus	Undifferentiated soils
43343	2.39	Semi-arid climates with cool or cold winters and mostly warm summers; winter coldness is an ecological constraint	Rangelands	Mountains	Predominantly rock outcrops and very shallow soils
23223	2.38	Arid climates with cool or cold winters and warm or very warm summers; winter coldness is an ecological constraint	Rangelands	Hills	Mainly soils with potential for agricultural use, requiring irrigation, associated with non-agricultural soils
41000	1.99	Semi-arid climates with cool or cold winters and mostly warm summers; winter coldness is an ecological constraint	Irrigated cropland	Plains and plateaus	Undifferentiated soils

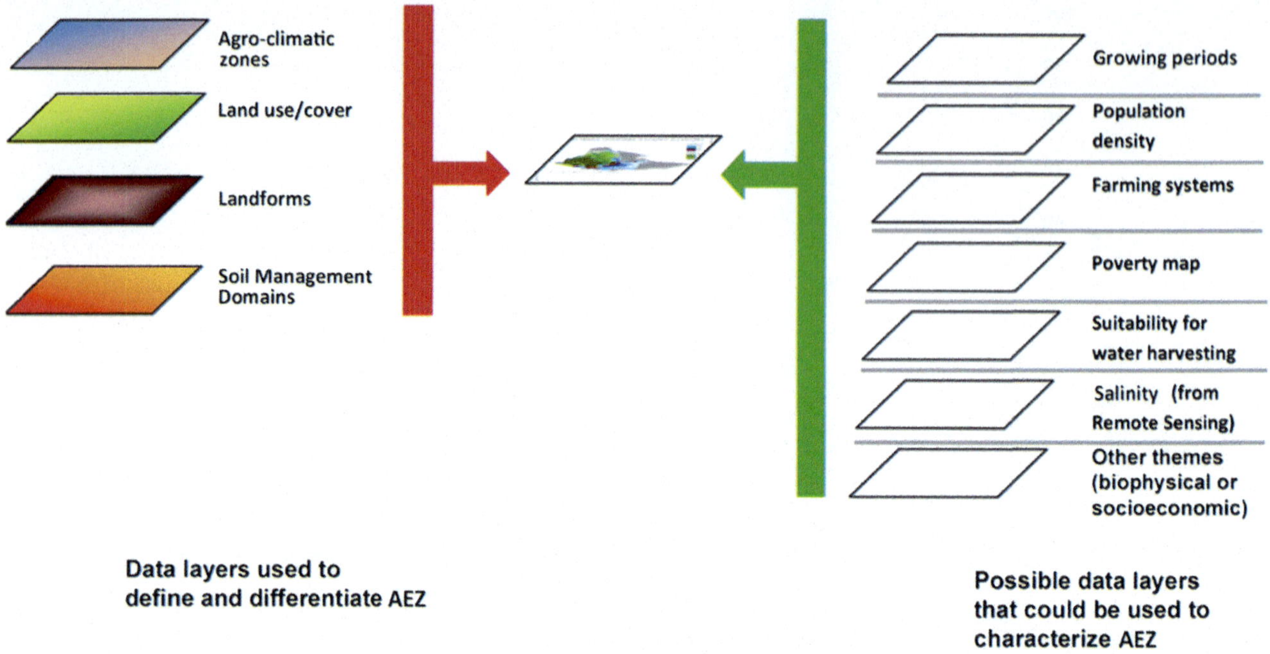

Fig. 9.3 Using thematic spatial datasets to characterize different AEZ

The ten largest AEZs (Table 9.8) cover only about 30% of Iran. Twenty-six AEZs are needed to cover half of Iran, and 92 AEZs are required to cover 80% of the country. The remaining 596 AEZs, with areas below 0.24% of Iran, can be considered 'niche' agricultural environments. Of these niche environments, a further 379 AEZs occupy individually more than 100 km^2.

As procedures were already included in the methodology to simplify the four overlay themes and to remove spurious units created by the overlaying, and considering the wide classes within the climate, land use, terrain and soil themes, the very large number of AEZ bears testimony to the great diversity in agricultural environments in Iran.

A very effective way to characterize the different AEZ is by means of histograms that provide the relative importance of each thematic class inside each zone. These so-called *characterization tables* can be prepared for any spatial dataset that is considered relevant and sufficiently reliable to provide meaningful information. The spatial datasets could cover themes related to the quality of the natural resource base, such as salinity, growing periods, land suitability, but also socio-economic themes, such as poverty, farming systems, population and others. Several example themes are shown in Fig. 9.3.

In fact, any theme that can be presented as spatial data can be used, as long as the particular theme is relevant to the research questions to be addressed, sufficiently reliable and at a level of detail that makes sense to differentiate by AEZ. In the AEZ description table[12], the particular example is given of the characterization of the AEZ in terms of the percentages they contain of specific classes of the duration of the first[13] and second[14] growing period, as limited by both moisture availability and temperature.

9.4 Conclusion

In order to reduce the great complexity of agricultural environments, this paper provides a generalized approach for defining agro-ecological zones using GIS procedures. It is based on the combination of climatic, land use/land cover, terrain and soil patterns, which constitute *defining* elements of each AEZ. Each zone can be further characterized through additional datasets, using the capabilities of GIS. The GIS procedures can be applied to a wide range of scales, subject to data availability at the required level of detail for the integration to be meaningful. The datasets are combined in an overlaying procedure of different biophysical frameworks, each one characterized separately through its own specific attributes.

[12]https://www.dropbox.com/s/ivwvg64530o1ml7/AEZ_Iran_full-list.xlsx?dl=0.

[13]https://www.dropbox.com/s/icjys3o1vunf17n/LGPmt1_Iran_v2.tif?dl=0.

[14]https://www.dropbox.com/s/5rwyeue7jo4etc7/LGPmt2_Iran_v2.tif?dl=0.

This approach is useful to define areas that can be considered relatively homogeneous in their biophysical characteristics and can thus serve as a first basis for area-specific agricultural (research) planning. Characterization of the identified AEZ in terms of themes relevant for specific planning purposes is an essential step.

For agricultural planning the idea of synthesis maps is useful as they provide to non-specialists greater clarity than thematic layers, but they need to include also socio-economic information, which is more difficult to spatialize. A farming systems framework appears the most useful and easy to combine with an AEZ framework. Experience from other countries indicates that this approach has much potential to provide an adequate spatial basis for agricultural planning and project development, but that it requires at the same time a substantial investment in the generation, integration and updating of local and scientific knowledge on rural environments.

The diversity of agricultural environments in Iran is already hinted at by the generalized AEZ map in Fig. 9.1. The detailed AEZ map is able to convey the full extent of agro-ecological diversity, within an overall dryland context, created by the adaptation of land use systems to site-specific climatic, terrain and soil conditions. It goes against the conventional, but erroneous, 'wisdom' that drylands can only support a limited number of agricultural systems.

A policy implication from the observed diversity in agricultural environments in Iran is that agricultural research and planning, in order to optimize each zone's agricultural potential and productivity, will fully need to take into consideration this diversity. Promoting diverse farming and production systems, each fitting well into each particular environment, is the way to go and an excellent justification for increased investment in agricultural research and extension.

References

Banaei MH et al (2000) Soil Map of Iran 1:1,000,000. GIS ESRI coverage, Soil & Water Research Institute (SWRI), Tehran, Iran

Choisnel E, de Villele O, Lacroze F (1992) Une approche uniformisée du calcul de l'évapotranspiration potentielle pour l'ensemble des pays de la Communauté Européenne. Centre Commun de Recherche, Commission des Communautés Européennes, EUR 14223, 178 pp. (Link: http://bookshop.europa.eu/en/une-approche-uniformis-e-du-calcul-de-l-evapotranspiration-potentielle-pour-l-ensemble-des-pays-de-la-communaut-europ-enne-pbEUNA14223/downloads/EU-NA-14-223-FR-C/EUNA14223FRC_001.pdf;pgid=y8dIS7GUWMdSR0EAlMEUUsWb0000FbkZxmTq;sid=qMrDMfYx-LbDJaU4oN5dlpQUrxnF4rtSt rg=?FileName=EUNA14223FRC_001.pdf&SKU=EUNA14223FRC_PDF&CatalogueNumber=EU-NA-14-223-FR-C)

De Pauw E, Ghaffari A, Ghasemi V (2004) Agroclimatic Zones Map of Iran. Explanatory notes. Technical Report, ICARDA, Syria and DARI, Iran, p 45. (Link https://dl.dropboxusercontent.com/u/33140444/Publications/ACZ_Map_Iran_explanatory%20notes.pdf)

FAO (1978–1981) Report on the Agro-ecological Zones Project. vol.1, Methodology and results for Africa vol.2 Results for Southwest Asia vol.3, Methodology and results for South and Central America; vol.4, Results for Southeast Asia. [FAO] World Soil Resources Report, 48/1, 4. Food and Agriculture Organization of the United Nations, Rome, Italy

Fischer G, Nachtergaele FO, Prieler S, Teixeira E, Tóth G, van Velthuizen H, Verelst L, Wiberg D (2012) Global Agro-Ecological Zones (GAEZ v3.0): Model Documentation. IIASA, Laxenburg, Austria and FAO,Rome, Italy. (Link http://www.fao.org/fileadmin/user_upload/gaez/docs/GAEZ_Model_Documentation.pdf)

Hutchinson MF (1995) Interpolating mean rainfall using thin plate smoothing splines. Int J Geogr Info Syst 9: 385–403.

Hutchinson MF (2007) ANUSPLIN version 4.37. User Guide. Center for Resource and Environmental Studies, Australian National University, Canberra. (Link http://fennerschool.anu.edu.au/files/anusplin437.pdf).

Smit B, Waltner-Toews D, Rapport D, Wall E, Wichert D, Gwyn E, Wandel J (1998) Agro-ecosystem health: Analysis and assessment. University of Guelph, Faculty of Environmental, ScienceGuelph, Ontario, Canada, p 114

UNESCO (1979) Map of the world distribution of arid regions. Map at scale 1:25,000,000 with explanatory note. United Nations Educational, Scientific and Cultural Organization, Paris, p 54. ISBN 92-3-101484-6 Link http://unesdoc.unesco.org/images/0003/000326/032661eo.pdf

USDA (2014) Keys to Soil Taxonomy. 12th ed, p 372 (Link http://www.nrcs.usda.gov/wps/portal/nrcs/detail/soils/survey/class/?cid=nrcs142p2_053580)

Author Biographies

Dr. Eddy De Pauw After his studies in geology and soil mapping at Ghent State University, worked for the United Nations in Africa between 1975 and 1991. Afterwards he worked as a coordinator of a research network on Acid Soils in Africa for IBSRAM. From 1996 to 2011, he was responsible for cartographic production at ICARDA, in Aleppo, Syria. Since 2011, he operates as international consultant and provides map products and technical advice related to natural resources and climate change.

Dr. Vahidreza Ghasemi Dehkordi is a senior researcher at Soil and Water Research Institute (SWRI), Iran. His current research interests focus on geomatics, geostatistics and spatial modeling, using similarity analysis to evaluate soil and environment characteristics and potential impact of climate change on soil and environment, agro-climatic zone (ACZ) and agro-ecological zone (AEZ).

Dr. Abdolali Ghaffari is a senior scientist and now serving as Director of Agriculture and Natural Resources Research and Education Center of West Azerbaijan Province in Iran. His main research interests are agro-ecological zone, crop modelling and impact of climate change on land suitability and crop yields. He contributed to the development of Dyland Agricultural Research Institute (DARI) of Iran as Director General of the institute for several years.

Soil Fertility

10

Najafali Karimian and Mohammad Mehdi Tehrani

Abstract

Analysis of the trends of cultivated area, agricultural production, and water resources of Iran shows that the focus of agricultural development should be on intensification to cope with the rising demands for food of the growing population. While this may lead to more pressure on some of our soil resources, thereby threatening sustainable production and food security, it reflects the key role of soil fertility and balanced soil fertilization as one of the main concerns of agricultural sector. However, recent soil fertility evaluation of more than 20,000 soil samples has revealed that more than 50% of cultivated soils suffer from the deficiency of one or more nutrients. Furthermore, low soil organic matter content is the main limitation factor of sustainable soil fertility and production. With the introduction of chemical fertilizers some 50 years ago, their use gained popularity such that the amount of chemical fertilizers consumed by Iranian farmers exceeded three million tons in 2015. Fertilizer recommendation in Iran is mainly based on the *sufficiency range* approach. Fertilizer recommendation trend shows a transition from the general and regional recommendation toward the soil test recommendation along with more attention to research.

Keywords

Soil fertility • Plant nutrition • Fertilizers • Iran • Soil productivity

10.1 Introduction

Of the more than 90 chemical elements found naturally in the earth crust, only 16–21 are considered essential to plant growth depending on the criteria used (Brown et al. 1987; Mengel and Kirkby 2001). These elements called as "Essential Elements" or "Essential Nutrients" are carbon (C), hydrogen (H), oxygen (O), Nitrogen (N), phosphorus (P), potassium (K), sulfur (S), calcium (Ca), magnesium (Mg), iron (Fe), zinc (Zn), copper (Cu), manganese (Mn), boron (B), molybdenum (Mo), chlorine (Cl) and nickel (Ni). Some other nutrients such as sodium (Na), cobalt (Co), vanadium (V), selenium (Se), silicon (Si) and aluminium (Al) are considered as "Beneficial Nutrients". The first three, i.e., C, H, and O are naturally taken up from air and water rather than from the soil; therefore, they are not usually discussed in the soil fertility textbooks. Soil is the main supplier of the remaining 18. A soil with the right concentrations and correct ratios of the nutrients is called a "Fertile Soil." A fertile soil is not necessarily a productive one. To have a productive soil (a soil capable of producing enough plant material to justify its cultivation), it is vital not only to have all the essential elements in the right concentrations and correct ratios but other plant growth factors should also be in proper amounts and proportions as well. More than 50 factors (e.g., light, moisture, temperature, etc.) have been found influential in plant growth and development. Maintaining the right concentrations and proportions of the above-mentioned

N. Karimian
Soil Science Department, Shiraz University, Shiraz, Iran

M.M. Tehrani (✉)
Soil and Water Research Institute, Meshkin Dasht, Iran
e-mail: mtehrani2000@yahoo.com

M.H. Roozitalab et al. (eds.), *The Soils of Iran*, World Soils Book Series,
https://doi.org/10.1007/978-3-319-69048-3_10

essential elements is just one of the many factors affecting the plant growth. Moreover, the growth medium has to be free from growth inhibitors such as noxious gasses, toxic compounds, harmful microorganisms.

The fact that most of the lands of Iran are located in arid and semiarid climate condition, when coupled to the low annual precipitation of most parts of the country, leads one to believe that the soils derived from such formations must be rich in plant nutrients, if only the total concentration of nutrients is considered. But, the total amount of a certain nutrient present in the soil does not determine the amount absorbed by the growing plant. In fact, most of the amount of each nutrient present in the soil is usually in a form not useable by plants because it might be water-insoluble, bound to inorganic or organic soil solids, or constitute the bodies of soil microorganisms. The term "Available Nutrient" is used to indicate that fraction of a nutrient which is actually taken up by plants. This fraction might be very small for some nutrients. Potassium (K) is a good example. Some soils contain more than 10,000 mg K kg^{-1} soil. Yet, only less than 1% of it might be in water-soluble and exchangeable forms, the forms readily taken up by plants. More than 78% of soil air consists of nitrogen gas (N_2), but it is totally useless for the plant unless some soil-borne microorganisms, i.e., non-symbiotic or symbiotic N_2-fixing bacteria convert it to ammonium (NH_4^+) or nitrate (NO_3^-).

Iranian farmers, like other farmers of the world, traditionally used animal manures to achieve the soil fertility level required for growing the plants. A 1514 A. D. book entitled "Crop Production Guide," perhaps the most ancient Farsi book written about the fertility management of soils of Iran, describes and compares the beneficial effects of pigeon, donkey, horse, mule, sheep, goat, and pig manures; and wheat and faba bean crop residues on plants (Moshiri 1977). Apparently, chemical fertilizers were unknown to Iranian farmers before 1945, the year when a small factory near Karaj, Iran, started with production of 5000 MT (metric ton) of single superphosphate, thermo phosphate, bone powder, and potassium nitrate per year (Anonymous 2004). No data are available about the import of chemical fertilizers into the country before 1950, when the import of 100 tons each of ammonium sulfate and nitrate fertilizers plus 50 metric tons (MT) of triple superphosphate was officially reported (Fahimifar 1990). The amount of chemical fertilizers consumed by Iranian farmers, which is estimated to have been about 100 MT of triple superphosphate in 1945, exceeded two million MT (more than 1 million P + approximately 1 million N fertilizers) in 1992. This is a clear indication of the acceptance of chemical fertilizers by Iranian farmers. The demand has increased sharply since then and FAO forecasts that a total of 6.2 million MT of chemical fertilizers would be needed in Iran by 2018 (Anonymous 2004). Nitrogen fertilizers constitute more than 50% of the demand followed by P (3.6 and 1.1 million MT, respectively). It is estimated that 41% of the chemical fertilizers produced in or imported to the country is consumed for growing wheat, 42% goes to other crops, 15% is used in orchards, and the remaining 2% is devoted to other purposes.

The kind and amount of fertilizers used in the early years were based on farmers' observations and was not matched with the actual plant needs. This was so until 1957 when the first scientific experiments were initiated in Khuzestan province and then extended to six experiment stations nationwide by the newly formed "Soil Fertility Division" of Ministry of Agriculture in 1961. The division received help from Food and Agriculture Organization (FAO) of the United Nations (Gerami 1982). The fertilizer trials were conducted in the farmers' fields based on the FAO's experience gained in India (Anonymous 1961, 1978). The number of experiment stations was later increased. Presently, there are more than 156 soil testing laboratories (32 government and 124 privately owned) operating in the country which analyze soil, plant, and water samples and give advice to the farmers (Anonymous 2010).

The status of nutrient elements in cultivated soils of the country is reviewed here based on the analysis of more than 200,000 test results of the country extensive research projects (Tehrani 2013), and the general over view of individual nutrient elements and their distributions in various provinces of the country is specified (Table 10.1).

Survey of the general condition and distribution of plant nutrients in different provinces showed that the amount of organic carbon in 55.7% of the country's soils was less than 1%, indicating the unstable fertility in these soils. Provincial distribution of soils containing less than 1% is shown even in some provinces such as Qazvin, Semnan, Zanjan, Khuzestan, Isfahan, Yazd, Bushehr, and Khorasan reached more than 80% of collected samples.

The amount of available phosphorus in 70.2% of the soils is less than 15 mg per kg, and in most provinces, except Gilan and Mazandaran, more than 70% of soils face phosphorus deficiency. The amount of available potassium in 33.6% of the soils is less than 200 mg per kg; and in Gilan, Mazandaran, Khuzestan, and Yazd provinces, it is more than 40%.

The amount of available iron in 45.5% of the soils is less than 5 mg per kg. In Markazi, East Azerbaijan, Hamedan, and Khorasan provinces, the percent of soils with similar content of Fe is more than 60%; and in Gilan, Golestan, and Mazandaran, it reaches 10% and less. The amount of available zinc in 55% of the soils is less than 0.75 mg per kg, while more than 60% of soils in West Azerbaijan, Kermanshah, Lorestan, Hamedan, and Khorasan provinces have less than this amount.

Table 10.1 General and provincial distribution of nutrient status in cultivated soils (Tehrani et al. 2010)

Province	OC < 1%	No. of samples	P < 15 mg/kg	No. of samples	K < 200 mg/kg	No. of samples	Fe < 5 mg/kg	No. of samples
Tehran	70.7	167	71.0	183	11.6	181	57.2	180
Semnan	90.2	92	84.6	91	14.3	91	76.9	13
Qazvin	90.8	184	72.2	180	6.7	180	18.3	104
Qom	66.7	3	–	–	–	–	0.0	25
Markazi	92.4	458	63.0	459	14.7	463	68.0	125
Golestan	23.0	2730	78.6	2740	29.4	2747	10.2	294
Gilan	13.0	1162	49.7	1218	75.1	1219	6.7	1009
Mazandaran	8.9	1904	43.1	1909	41.5	1913	5.0	1045
East Azerbaijan	65.8	278	74.0	285	17.5	285	73.6	258
West Azerbaijan	45.0	1869	74.8	1890	10.9	1899	56.8	1794
Ardabil	10.7	75	93.3	75	6.7	75	9.3	97
Zanjan	89.8	719	77.7	725	5.5	725	54.2	24
Kurdestan	65.3	101	51.0	102	8.8	102	52.9	102
Ilam	45.6	1035	79.9	1150	31.2	1146	44.1	68
Kermanshah	36.8	592	66.5	597	12.3	595	47.9	597
Lorestan	15.2	33	63.6	33	3.0	33	56.7	60
Hamedan	61.8	1321	64.9	1320	12.0	1320	64.9	1320
Khuzestan	83.3	5939	88.0	5939	46.8	5941	47.1	255
Isfahan	81.9	481	78.4	486	30.7	499	16.8	131
Yazd	93.6	202	52.1	192	50.2	213	100.	38
Fars	67.9	418	56.8	410	18.8	420	25.2	218
Bushehr	96.5	255	93.9	213	–	237	–	–
Kerman	79.0	138	62.7	142	27.0	141	43.1	65
Sistan and Baluchestan	–	–	–	–	–	–	0.0	16
Khorasan	85.5	233	74.9	346	28.7	359	73.1	264
Sum	55.7	20,389	70.2	20,685	33.6	20,874	45.5	8102

(continued)

The amount of available manganese in 20.7% of all the country's cultivated soils is less than 4 mg per kg, while more than 60% of soils in Khuzestan province and more than 30% in Golestan, East, and West Azerbaijan provinces have the same problem. The available copper is less than 0.75 mg per kg in 13.27% of the whole country's soils reaching more than 20% of soils in East Azerbaijan and Khuzestan and more than 30% of soils in Markazi and Khorasan provinces. The results indicated that more than 50% of the soils suffer from one or more nutrient deficiency, which is more severe in some provinces revealing the imbalance between nutrient elements in soils. Low-organic carbon content and, as a result, nitrogen deficiency are the main limiting factors in sustainable fertility and productivity.

10.2 Estimation of the Country's Fertilizer Requirement

Amount of required plants nutrient elements of nitrogen, phosphorus, and potassium was estimated in 2010 based on the results of the aforementioned researches (Tehrani et al. 2010). Estimation of required fertilizer was 4 million tons including 2.9, 0.7, and 0.4 million tons of nitrogen, phos-

Table 10.1 (continued)

Province	Zn < 0.75 mg/kg	No. of samples	Mn < 4 mg/kg	No. of samples	Cu < 0.75 mg/kg	No. of samples
Tehran	44.7	179	13.7	131	5.3	95
Semnan	61.5	13	0.0	10	0.0	9
Qazvin	55.8	104	5.6	90	0.0	87
Qom	52.0	25	27.3	11	0.0	8
Markazi	28.0	125	11.2	116	32.3	99
Golestan	51.7	294	32.3	294	0.4	272
Gilan	32.6	1023	6.6	1009	6.5	1009
Mazandaran	34.6	1047	9.2	1047	1.7	1047
East Azerbaijan	40.7	258	30.8	234	22.2	234
West Azerbaijan	70.1	1795	36.3	1725	16.9	1726
Ardabil	56.7	97	2.3	88	3.5	85
Zanjan	60.9	23	0.0	15	29.4	17
Kurdestan	44.1	102	1.1	87	6.9	87
Ilam	41.8	67	10.5	19	22.6	31
Kermanshah	62.3	597	20.9	584	13.4	584
Lorestan	63.3	60	34.5	58	14.3	49
Hamedan	73.2	1321	17.9	1321	19.5	1320
Khuzestan	54.5	255	67.2	229	23.4	231
Isfahan	31.3	131	4.4	114	4.4	114
Yazd	44.7	38	13.6	22	0.0	20
Fars	52.3	218	5.0	199	8.0	201
Bushehr	–	–	–	–	–	–
Kerman	38.5	65	11.4	35	14.7	34
Sistan and Baluchestan	56.3	16	8.3	12	91.0	11
Khorasan	75.8	264	14.0	236	41.7	240
Sum	55.1	8117	20.7	7686	13.3	7610

Table 10.2 Estimated N, P, and K fertilizers use during recent years in Iran (million tons)

Year	Nitrogen fertilizers	Phosphoric fertilizers	Potassium fertilizers	Total
1999	2.7	1.2	0.6	4.7
2005	2.6	1.3	0.5	4.4
2006	2.8	1.2	1.0	5.0
2010	2.9	0.7	0.4	4.0

phorus, and potassium, respectively. Table 10.2 shows the trend of estimated fertilizer use in recent years.

The following sections present the results of fertility evaluation of agricultural soils of Iran together with the experience of the authors showing a better picture of the soil fertility status of the country. The results for N, P, and K will be discussed under separate headings. Those of the remaining nutrients will be presented in a section entitled "Other Nutrients." Warnings will be given about the consequences of overfertilization and the practices which may lead to the pollution of soils (Figs. 10.1 and 10.2).

10.3 Nitrogen Status of Soils

Nitrogen is perhaps the most limiting nutrient element for crop production in Iran. This is because 95% of soil N is present as a part of soil organic matter (OM) (Serna and

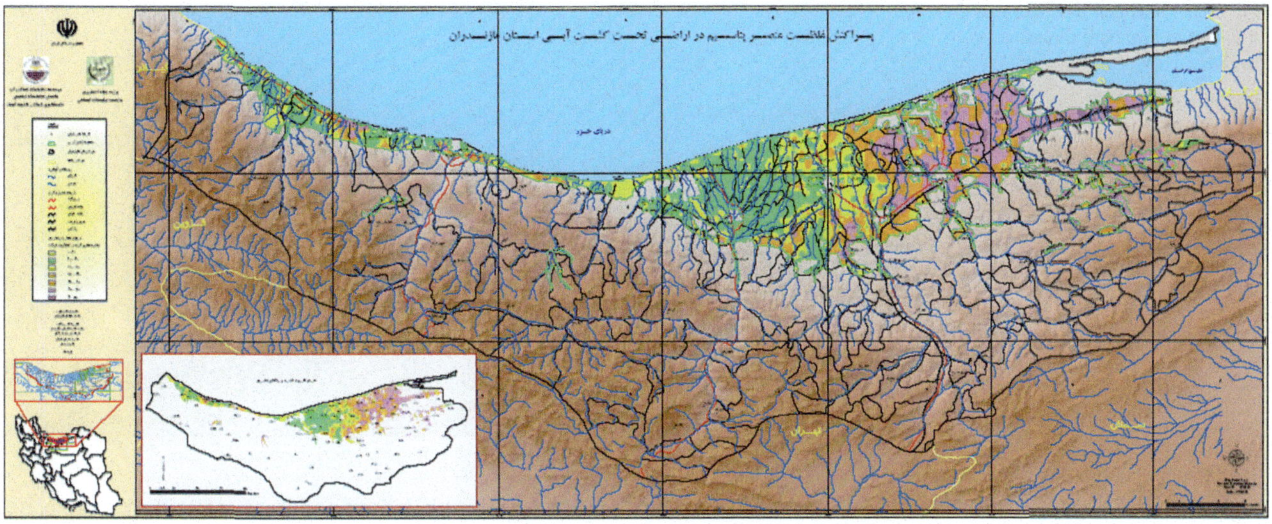

Fig. 10.1 The distribution of potassium in irrigated land of Mazandaran province (Tehrani 2013)

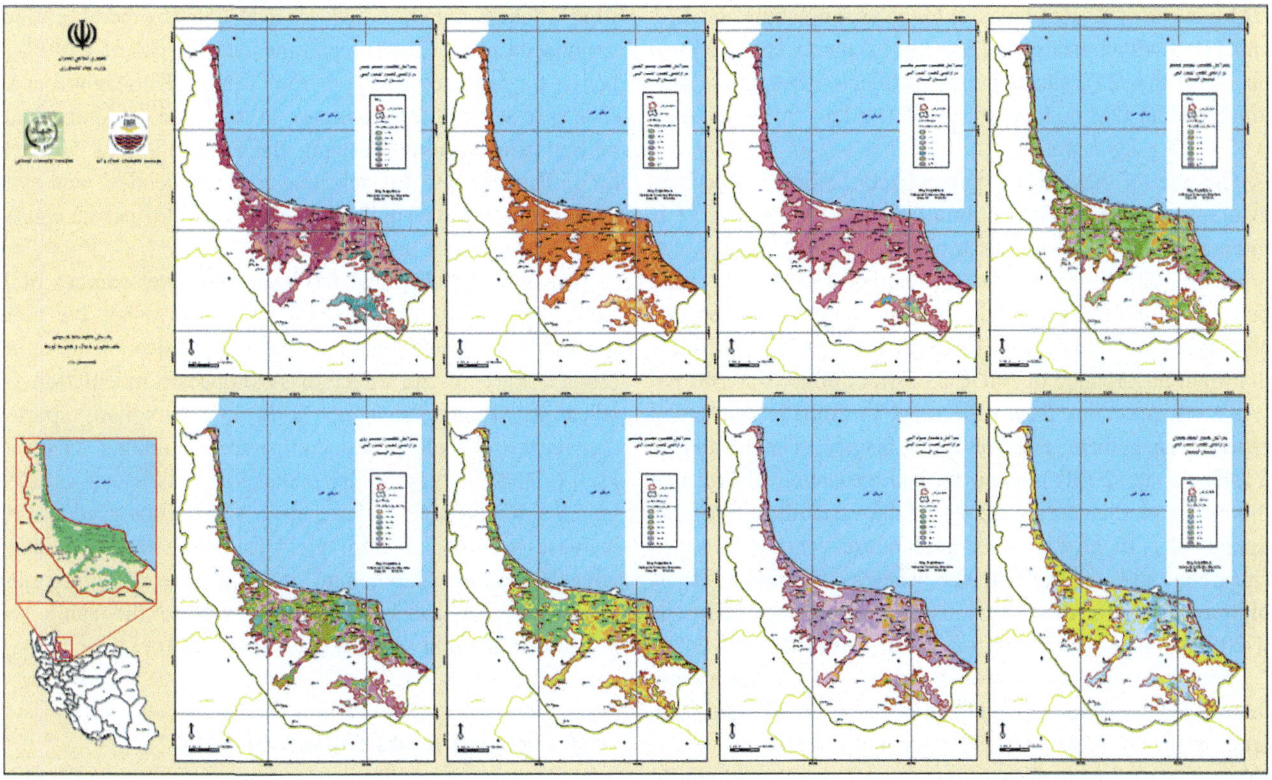

Fig. 10.2 The distribution of nutrients in irrigated land of Gilan province (Tehrani 2013)

Pomares 1992). Generally speaking, the amount of soil OM or its certain fractions is considered to be the index of quality of agricultural soils (Haynes 2005), including their N supplying capacities. Most soils of Iran are said to contain less than 5.0 g OM kg^{-1} (Kalbasi 1996). Mirbagheri et al. (2012) reported the OM content of 30 soil samples from potato fields of Semnan to be 3.0–22.3 with a mean of 9.8 g kg^{-1}. Atarodi and Naderi Khorasgani (2009) reported the OM content of 24 soils from Birjand to range between 3.3 and 17.4 g kg^{-1}. Samavati and Hosseinpour (2011) showed that 30 soil samples under garlic cultivation in Hamedan contained 6.0–31.0 g OM kg^{-1}. Moosavi et al. (2011) observed that 286 soil samples from Hamedan ranged in OM between 0.3 and 43.4 with a mean of 7.7 g kg^{-1}. Hashemi Beni et al. (2010) reported the soil OM content of 205 soil samples from Chaharmahal and Bakhtiari province

Fig. 10.3 Nitrogen fertilizer trial on wheat in Gonbad, Golestan Province

to be 0.5–38.0 g kg^{-1}. Mahmood-Soltani and Samadi (2003) reported the OM content of 39 calcareous soils from Fars province to range from 7.0 to 34 g kg^{-1} with a mean of 19.0 g kg^{-1}. Kavoosi and Malakouti (2006) determined the OM content of 21 rice fields of Gilan to range between 11.8 and 118.9 with a mean of 37.9 g kg^{-1}. Yasrebi et al. (2004a) observed that 25 calcareous soils collected from Fars province had an OM content of 7.8–35.0 g kg^{-1} of which only 0.45–2.03 g kg^{-1} was N.

Organic N has to be mineralized before it becomes available to plants. Only a small part of the organic N is mineralized by microbial processes during any given growing season. Numerous techniques and criteria have been proposed for prediction of the amount of mineralizeable N of soils (Antep 1997). Sharifi et al. (2007) concluded that light absorbance of $NaHCO_3$ extract of soils at 260 nm was a reliable index for a wide range of climatic zones, soil types, management practices, and history. The results of Yasrebi et al. (2004b), however, had previously shown that this index predicted only 22.4% of the variability in total N uptake of corn grown under greenhouse conditions on 25 calcareous soils of Fars province, whereas, this figure was as high as 66.2% for the same soils under the same conditions if 205-nm light absorbance was used instead (Fig. 10.3).

With the evidence presented above, it is not surprising, therefore, to observe beneficial effects of N fertilizers application to crops across the country. The results of selected experiments with different crops across the country are presented below to show the need for N fertilization in soils of Iran. Ali Abbasi et al. (2007) concluded that 240 kg N per ha was required for the desirable yield of rice in Gilan province. In Khuzestan province, the amount needed for corn was 220 kg N per ha (Lack et al. 2007). Noushad et al. (2001) studying the N fertilizer requirement of grain corn in Fars province concluded that 180 kg N per ha produced the highest yield. This was despite the fact that Ghadiri and Majidian (2003) recommended the use of only 92 kg N per ha for corn in a calcareous soil from Fars province. Rezaeinejad and Afyuni (2010) recommended 168 kg N per ha for corn grown in a soil from Isfahan with 0.9 g total N kg^{-1}. Mirlohi et al. (2000) recommended the use of 230 kg N per ha for sorghum on a calcareous soil from Isfahan province with total N of 2.0 g per kg. Gholamhoseini et al. (2008) recommended the use of 270 kg N per ha for rapeseed in a soil from Tehran. Khodagholi et al. (2008) observed 100% increase in the yield of range plants of Semirom region following the application of 50 kg N ha^{-1}. The increase in yield due to P application was even more pronounced. Application of 60 kg P_2O_5 per ha caused a fivefold increase in dry matter.

Not only the chemical fertilizers but other sources of N have also proved to be beneficial in increasing the yield potential of N-deficient soils of Iran. Hajeeboland et al. (2004) studying the effect of Azotobacter inoculation on wheat grown on a calcareous soil from Azerbaijan, reported that its positive effect was comparable to adding 24 mg N kg^{-1} (a 35% increase in dry matter for Azotobacter vs. 27% increase for NH_4NO_3). Raei et al. (2009) observed a 23% increase in soybean grain yield following the inoculation with Rhizobium Japonicum in a soil from Azerbaijan. Application of 138 kg N ha^{-1} as urea to the same soil increased the yield only by 13% (Fig. 10.4).

10.4 Phosphorus Status of Soils

Phosphorus occurs in soils both as inorganic and organic forms. Inorganic forms of P in soils are Ca phosphate (apatites and octa- and di-Ca phosphates), Al and Fe phosphates, and Fe oxide-occluded phosphates. Organic forms are not well known. Less than 50% of soil organic forms of P are known compounds like inositol phosphate and phospholipids. Microbial activities are responsible for transformation of forms into each other (Hedley et al. 1982).

Mahmood-Soltani and Samadi (2003) determined the total P content of 39 calcareous soils from Fars province to range

Fig. 10.4 Balanced fertilization trial on maize in Karaj

Table 10.3 Olsen-P content of calcareous soils of Fars province and their need to DAP fertilizer for wheat crop

P (mg kg^{-1})	No. of soils	% of soils studied	Recommended P fertilizer (kg ha^{-1})
0–5	3	8	200
5–10	6	15	150
10–15	10	26	100
15–20	7	18	50
20–25	4	10	0
>25	9	23	0
Total	39	100	

Adapted from Ghanbari et al. (2000) and Shahrokhnia et al. (1993)

from 254 to 1176 mg kg^{-1}, of which 73% was inorganic and 27% organic. Light-textured soils contained less total P than the heavy-textured ones (mean 690 vs. 482 mg kg^{-1}). Total P content of 20 soil samples from arid and semiarid regions of Isfahan and Shahr-e-Kord was reported to be 302–1135 with a mean value of 715 mg kg^{-1}. Organic P of the soils was 70–238 with a mean of 144 mg kg^{-1} constituting about 11–35% of total P. These results are not much different from those of Turner et al. (2003) who found the total P content of soils from semiarid western USA to range between 220 and 1210 mg kg^{-1} of which 3–36% was organic. Samavati and Hosseinpour (2011) studying 30 soil samples from garlic farms of Hamedan, however, reported that their total P was 931–2686 mg kg^{-1} with a mean of 1501 mg kg^{-1}, probably a consequence of heavy fertilization of garlic with P. Organic P constituted 19% of the total P.

Like N and many other nutrients, not the entire total P is available to plants. Several methods have been used to determine the P availability status of soils. Phosphorus extracted by 0.5 M $NaHCO_3$ buffered at pH 8.5 (Olsen et al. 1954) is perhaps the most widely accepted method for calcareous soils. It is referred to in Iran as Olsen-P, has been tested in different parts of the country, and is believed to be the most reliable method for evaluating the P status of soils. Karimian and Ghanbari (1990) results were perhaps the first published evidence to support this. Karimian and Ghanbari (1990) compared several methods and concluded that 18 mg Olsen-P kg^{-1} could be used as a partitioning point of P-deficient and P-sufficient soils. Shahrokhnia et al. (1993) studied the P status of soils in Fars province by conducting fertility evaluation experiments in the farmers' fields in search of a guideline for P fertilization of wheat. Based on their results, 20 mg Olsen-P kg^{-1} was suggested as the practical critical level for partitioning the soils into those in need of P fertilizer and those that are not.

About 33% of the soils of Fars province, as reported by Ghanbari et al. (2000) are therefore P-sufficient and need no fertilizer P for wheat (Table 10.3). Samadi (2006) determined the Olsen-P of Urmia soils to range between 5 and 27 mg kg^{-1}. Hosseinpour and Shariatmadari (2007) compared several P extractants on soils of Hamedan province, using alfalfa as the test plant. Their results show Olsen-P to significantly correlate with plant response. Atarodi and Naderi Khorasgani (2009) working with 24 soil samples from Birjand determined the Olsen-P to range between 3.6 and 59 mg kg^{-1}. They determined the critical level of P for sorghum to be 17 mg kg^{-1}. Hajabbasi et al. (2008), studying the effect of cultivation on the soil properties of selected pastures of Isfahan province, reported their Olsen-P to be 12–38 mg kg^{-1}. Although cultivation changed some of the soil properties including soil OM, no mention was made of the change, if any, of Olsen-P.

Kavoosi and Malakouti (2006) observed that in 21 rice fields of Gilan province which ranged in Olsen-P of 2–56 with a mean of 24 mg P kg^{-1}, 9 fields had Olsen-P of greater than

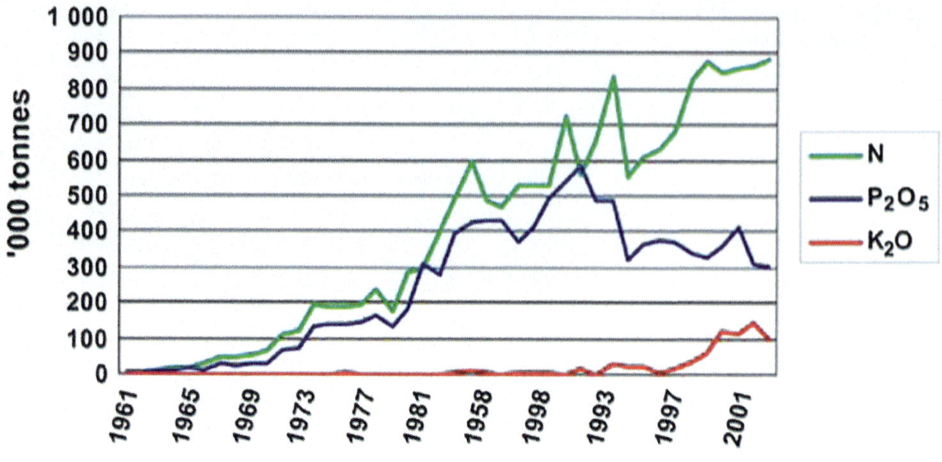

Fig. 10.5 Fertilizers consumed in Iran (1 ton = 1 Mg) during 1961–2001 (Anonymous 2004)

20 mg kg^{-1} and that the relation between rice yield and soil P was negative. Mirbagheri et al. (2012) working with 30 soils from potato fields of Semnan, reported their Olsen-P to be 17.7–48.4 with a mean of 40.1 mg kg^{-1} which they believed to be several times higher than the recommended level. They reported the mean P concentration of potato tubers to be 2.2 g kg^{-1}, whereas the desirable level is believed to be 1.4–1.7 g kg^{-1}. They, therefore, recommended that potato farmers must avoid P fertilization for at least 2–3 years. The recovery of P fertilizers by the first crop is usually low, seldom exceeding 10–15% and, unlike N, the element is not volatilized, essentially not leached, and the portion not used by the crop remains where it was applied and accumulates in the root zone.

The concern about overuse of P fertilizers was stated by Karimian (1994) which was repeated later on in a review article (Karimian 1998). He strongly recommended the use of soil test results before any fertilizer is applied to soil because, he reasoned, "fertilizers are plant medicine and no medicine should be consumed without a prescription." Overuse of P fertilizer has long been associated with induced deficiency of other nutrients, particularly of Zn (Karimian 1995).

Karami and Ebrahimi (2000) surveying the farmers of Fars province concluded that they were using excess P fertilizers, probably because of the beneficial effect they were getting from the N which accompanies P in diammonium phosphate, a major P fertilizer widely used in Iran. Recommendations were, therefore, made to the authorities to lift the ban on production and import of triple superphosphate, in effect during those years, supposedly to cut the transportation expenses of fertilizers, so that enough of it would be available in the market and the farmers can choose the exact kind of fertilizer they need. Triple superphosphate is a P fertilizer with no N but with essentially the same P content as diammonium phosphate. The campaign on prevention of P fertilizers overuse was, apparently, effective as it could be seen in a decline of P consumption after 1990 despite the increase in N and K fertilizer consumption during the same period (Fig. 10.5).

Soluble P of chemical fertilizers is converted to less soluble forms soon after it is added to the soils. The solid phases responsible for this conversion are thought to be oxides and hydrous oxides of Al and Fe in acid soils and calcium carbonate in calcareous soils. Adhami et al. (2012) used 20 calcareous soil samples from Fars province to study the transformation of P under field capacity and waterlogged moisture regimes. They monitored different forms of P in soils at 80 and 160 days after addition of 300 mg P kg^{-1} as KH_2PO_4, a highly soluble form of P. The $NaHCO_3$-extractable P decreased under both moisture regimes as the time of contact increased. The decrease was higher under waterlogged than field capacity moisture. Oxides of Al and Fe were found more important in the transformation than previously thought for calcareous soils, especially under waterlogged condition. Care must be taken when interpreting such results because the reactions of P in calcareous soils might differ depending on the application rate and the incubation conditions.

10.5 Potassium Status of Soils

Total K in soils ranges between 0.4 and 30 g kg^{-1} and occurs in four major forms of structural, non-exchangeable, exchangeable, and soluble (Sparks 1987). An equilibrium exists between these forms. Potassium reserves in soils of arid and semiarid zones, like Iran, are high, and relatively little research data are available about the fertilizer needs of these soils regarding K. Some, however, argue that intensive agriculture and cultivation of nutrient-demanding varieties of plants would soon deplete the agricultural soils from K and this necessitates the studies of K even in soils of arid and semiarid regions. Jafari and Baghernejad (2007) observed that 40 years of sugarcane cultivation increased the expanding type clay content of Khuzestan soils and also their K concentration. They, however, point to the fact that even these soils are non-responsive to K fertilizers.

A large number of procedures and techniques including different reagents have been suggested for assessing the K status of soils. Equilibration with 1 M neutral ammonium acetate (NH_4OAc) and determination of K concentration in the extract is perhaps the most common procedure used in Iran. The index, referred to in this chapter as NH_4OAc-K, has been disapproved by some Iranian researchers. Kavoosi and Kalbasi (2000) compared 15 K extractants on 23 soil samples from Gilan province using rice as the test plant and reported that NH_4OAc-K was not superior to others in predicting the response of rice to K fertilizer. Their results show a range of 60–240 with a mean of 105 and a critical level of 112 mg kg^{-1} for NH_4OAc-K of their studied soils. The literature review made by Kavoosi and Kalbasi (2000) indicates the disagreements between researchers in different parts of the world about the best procedure for prediction of K availability for rice. They found out that some researchers believe in NH_4OAc-K, some use 0.01 M $CaCl_2$, others use procedures that determine the non-exchangeable K of soil, and a few have found no extractants suitable. Hosseinpour (2004) announced that NH_4OAc-K was not a good index for estimation of available K of garlic in soils of Hamedan.

Sharifi and Kalbasi (2001), who compared 14 extractants to study the K status of 26 soils from Isfahan, using corn as the test plants, were not able to establish a critical level of K because they believed the soils were K-sufficient. They, however, suggested extraction with 2 M NaCl as an easy, economical, and practical procedure. Kavoosi and Malakouti (2006) determined the NH_4OAc-K concentration of 21 rice fields of Gilan province to be 77–254 with a mean of 140 mg K kg^{-1} and suggested that 110 mg K kg^{-1} could be used as the critical level. Davatgar et al. (2006) studied the K status of 109 soil samples from rice fields of Gilan province and presented maps for the distribution of available K throughout the province based on their NH_4OAc-K which ranged between 45 and 340 with a mean of 156 mg K kg^{-1}. They ranked the K availability status of 68.8% of their studied fields as "low to medium."

10.6 Status of Micronutrients in Soils

Calcium (Ca), Mg, and S are usually abundant in soils of Iran because the parent material of soils in most parts of the country is calcareous and/or gypsiferous (Moazallahi and Farpoor 2012). The majority of soils in Iran are calcareous with variable amounts of calcite and dolomite. Polluted air resulting from consumption of S-containing fossil fuels must not be overlooked as a source of S. Although the solubility of these minerals is low, enough concentration of these nutrients are often available to plants in most cases and crops. Malakouti et al. (1999) reported that spraying with $CaCl_2$ solution enhanced the fruit firmness of Red Delicious apple at harvest and after 4.5 months storage at 2 °C. Ghaffari Nejad and Sabbah (2009) observed that date palm benefited from trunk injection of $Ca(NO_3)_2$ solution. Mobility of Ca in plant tissues is believed to be low and its concentration might drop below the sufficiency level during the drought and high temperature periods. High levels of available K might also have depressing effect on Ca and Mg uptake. Sulfur (S) supplement is seldom used for its direct nutritional effects. It is usually used as the element accompanying N, P, K, etc., fertilizers; as S-containing pesticides; and as acidifying agent in calcareous soils. Kalbasi et al. (1988) used it to increase the availability of soil Fe, Zn, and Mn for corn, sorghum, and soybean.

Calcareous soils have high pH values leading to impaired availability of nutrients like Cu, Fe, Mn, and Zn. Gypsiferous soils have lower pH than calcareous soils, and therefore, it is expected that crops growing on them have less problem with the availability of the just mentioned nutrients because the lower pH increases the solubility of carbonates, phosphates, oxides, and hydroxides of the nutrients. The presence of gypsum in calcareous soils complicates the situation. On the one hand, gypsum increases the solubility of Cu, Fe, Mn, and Zn, but on the other hand, the increased solubility of Ca might interfere with their nutrition.

Yasrebi et al. (1994) reported the total Zn concentration of 20 soils from Fars province to range between 32.4 and 66.7 mg kg^{-1}. Darjeh et al. (1991) compared five Zn extractants and concluded that EDTA-ammonium carbonate and DTPA may be used for partitioning soils into Zn deficient and non-deficient. ReyhaniTabar et al. (2006) reported total Zn of 20 soil samples from Tehran province to range between 70 and 170 mg kg^{-1} of which only about 0.1% was in soluble and exchangeable form. Soluble Zn fertilizers are rapidly converted to less soluble forms. Yasrebi et al. (1994) showed that about 60% of $ZnSO_4$ fertilizer was converted to carbonate form, which is not available to the first crop but may serve as a reserve for the subsequent crops (Karimian and Yasrebi 2003).

Gholamalizadeh-Ahangar et al. (1995) compared several extracting solutions for prediction of available Mn of soils of Fars province which had high capacity for retention of this plant nutrient (Karimian and Gholamalizadeh-Ahangar 1998). Ghaffari Nejad and Karimian (1998) used 22 soil samples from Fars province to compare five Mn-extractants using soybean as the test plant and concluded that DTPA with 13 mg Mn kg^{-1} as the critical level was suitable for assessing Mn status of soils. The same extractant has been suggested for Cu (Lindsay and Norvell 1978). Although this extractant was originally suggested also for Fe, no reliable data are available in the Iranian soil fertility literature to prove that this soil test method is also suitable for assessing the Fe status of calcareous soils of Iran. Roomizadeh and Karimian (1996) compared several extractants including DTPA for Fe using soybean as the test plant, but found that none of them were reliable. They concluded that antagonistic effects between Fe and Mn may be the reason.

Pirzadeh et al. (2012) studied the Zn status of 136 rice fields of Isfahan, Fars, and Khuzestan provinces and found out that available Zn in 16% of the fields in Isfahan was below the sufficiency level, whereas this figure was 66% in Fars and 75% in Khuzestan. They used 2 mg DTPA-Zn kg^{-1} soil as the sufficiency level for their classification. This was despite the fact that mean total Zn of the fields were 47.2 (18.7 to 67.5), 35.9 (17.5 to 116.5), and 38.3 (14.1 to 49.6) mg kg^{-1} for Isfahan, Fars, and Khuzestan, respectively. Kheirabadi et al. (2012) determined the DTPA-Zn of 11 soil samples from Isfahan to range between 0.42 and 1.38 mg kg^{-1}.

Boron (B), being an essential nutrient, is vital to plant growth; therefore, information about its status is important in soil fertility. The range between B levels of soil solution causing deficiency and plant toxicity is small. Its deficiency is found most often in humid regions, whereas its toxicity is expected in soils of arid and semiarid regions that are irrigated with high B waters (Mezumen and Keren 1981).

Chlorine (Cl) and sodium (Na) deficiencies are not of concern in arid and semiarid regions. On the contrary, their accumulation in the soils are usually encountered in such regions, which is dealt with as salinity problem in case of sodium chloride (NaCl) accumulation, as sodicity in soils with high exchangeable sodium percentage (ESP), and in irrigation waters as high sodium adsorption ratio (SAR). These terms are defined in Eqs. [10.1] and [10.2].

$$\mathrm{ESP} = 100(\mathrm{exch.\ Na})/(\mathrm{CEC}), \qquad (10.1)$$

$$\mathrm{SAR} = (\mathrm{Na}^{+})/\left[(\mathrm{Ca}^{2+} + \mathrm{Mg}^{2+})/2\right]^{0.5}, \qquad (10.2)$$

where exch. Na is soil exchangeable sodium, cmol kg^{-1}; CEC is soil cation exchange capacity, cmol kg^{-1}; and Na^{+}, Ca^{2+}, and Mg^{2+} are concentrations of soluble Na, Ca, and Mg in water, cmol L^{-1}, respectively.

A review of salinity studies in Iran by Momeni (2010) concludes that, of approximately 165 million ha total area of the country, about 6.8 million ha suffer from different degrees of salinity out of which 4.3 million ha are those with salinity but no other major restriction. The remaining 2.5 million ha have salinity problem plus other restrictions like soil type, topography, erosion, and groundwater problems. Only 8.4% of the 6.8 million ha are faced with groundwater problem in addition to salinity. Excess NaCl is usually the cause of salinity followed by excess Na_2SO_4. In some areas of the country, excess B is also present in addition to salinity.

Molybdenum (Mo) is usually not deficient in soils with high pH. On the contrary, high concentrations of Mo in soil which is often observed on such soils may lead to accumulation of Mo in pasture plants which, although not harmful to plant itself, may induce molybdenosis in ruminants grazing on them. It is a disease of ruminants caused by grazing on high Mo plants especially in cases where Cu intake is low. It is characterized by persistent diarrhea. Rasooli et al. (2010) determined the Mo and Cu contents of soil and pasture plant samples from four different regions of Khuzestan province and reported the mean Mo content of 48 soil samples to be 18.00–36.57 and that of 48 pasture plant samples between 11.47 and 12.97 mg kg^{-1}. They concluded that the Mo content of pastures was dangerously high. A Mo concentration of below 3 mg kg^{-1} on the dry matter basis of pasture is considered safe, whereas concentrations greater than 10 mg kg^{-1} is dangerous unless enough Cu is supplied.

Little has been reported about the status of Co, V, and Si in soils and crops of Iran. Sharafi et al. (2013) reported the beneficial effect of Co on beans in a greenhouse experiment. Fallah et al. (2011) reported that applied Si increased the yield of rice by 10%.

10.7 Obstacles to Soil Productivity

As mentioned earlier, a fertile soil is not necessarily a productive one because many factors other than nutrient concentration have to be at the right amount and proportion. Some of the soil conditions which might interfere with the productivity of soils in Iran are discussed in brief, here. Water shortage, high calcium carbonate, salinity, and pollution are perhaps the most important factors contributing to the low productivity of soils in Iran.

10.7.1 Water Shortage

Iran is located in a region with semiarid to arid climate. Modares and Sarhadi (2011) studied the annual time series of 137 synoptic stations in Iran covering the period of 1952–2003. Their data show that annual rainfall ranges between 1800 mm in the north to less than 100 mm in the central region of the country, with a coefficient of variation between 20% in the north to 75% in the south. Except for a relatively very small region in the north and smaller regions elsewhere, map of the mean annual rainfall shows that most parts of the country receive less than 300 mm of annual rainfall. This low precipitation coupled with high annual evapotranspiration of more than 3600 mm leaves no choice to the farmers other than irrigation. Due to scarcity of surface water, most irrigated areas of the country are dependent on groundwater, which, due to overuse, has been largely depleted in recent years (see Chap. 12). The balance of groundwater has become negative for the major agricultural areas of the country like Fars province (Hojjati and Boustani 2010).

10.7.2 Salinity

A soil is considered saline if it contains enough soluble salts to interfere with normal plant growth of the agronomic crops. Electrical conductivity of higher than 4 dSm^{-1} of soil saturation extract is usually defined as the boundary separating saline from non-saline soils. Excess boron and higher solubility of Cd are also a problem associated with saline soils. Higher concentrations of Cl increase the plant uptake of Cd presumably because of formation of $CdCl^{+}$ complex (Khoshgoftarmanesh et al. 2004).

10.7.3 Pollution

Moosavi et al. (2011) studied heavy metals content of 286 soil samples of Hamedan province and using a geostatistical method, presented maps showing the distribution of arsenic (As), chromium (Cr), antimony (Sb), and copper (Cu).

Amini et al. (2007) studied 255 soil samples from seven regions of Isfahan province and determined the mean total Cd, Cu, Pb, and Zn to be 1.73, 16.74, 25.6, and 4.58 mg kg^{-1}, respectively. No guideline is available for the maximum permissible concentration of these heavy metals in Iran; but, using those of Switzerland, they concluded that 95% of the studied soils exceeded the limit for Cd. Using the mass balance modeling; they estimated the accumulation rate of Cd to range between 3 and 18 g ha^{-1} y^{-1}. The calculated values for Pb were 10–260 g ha^{-1} y^{-1}. In addition to atmospheric fallout, P fertilizer was believed to be the source of Cd, whereas Pb was thought to originate from animal manures. Ingestion of Pb deposited on the forage by fallout was thought to be the source for manure Pb. These metals are usually immobile in soils and stay within the surface plow layer but preferential flow and organic acids produced in the rhizosphere might help them to penetrate to deeper zones.

Gharaie et al. (2002) demonstrated that the capacity of calcareous soils of Iran for Pb retention was very high. Rajaie et al. (2006) observed that a major portion of added Cd to a calcareous soil from Fars province was converted to carbonate form. Khodakarami et al. (2012) studied the distribution of Cr, Co, and Ni in soils of Hamedan province and reported their mean concentrations to be 63.1, 17.6, and 88.9 mg kg^{-1}, respectively. Using the geostatistical approach, they concluded that the distribution was related to the geological formations of the region and not to the agricultural land use. Jahangiri (2011) studied the relationship between chemical forms of As and soil properties of Kurdestan province. Soils with more than 40 mg As kg^{-1} are supposed to be toxic to plants. They, however, concluded that in soils with higher amounts of clay, as is converted to arsenate form, is less available to plants.

10.8 Conclusion

Beyond soil test as a basis for fertilizer recommendations, other factors such as the climate (drought, temperature), soil characteristics (salinity), plant (uptake efficiency), and agronomic management (rotation) should be included in the fertilizer recommendations system. Fortunately, the first steps for integration of these factors have been taken by development of an "*automated fertilizer recommendation program,*" which needs to be improved and updated by more comprehensive investigations. Fertilizer consumption estimation mostly uses *crop-based* models due to more availability of soil and plant analysis data. The consumption trend shows higher fertilizer requirement in the future. However, it is needed to distinguish between the total fertilizer requirement and its consumption on the farm. This means that proper use of fertilizers should be among the main governmental policies. A *national soil testing program* should be developed in a way to include all stakeholders (private soil test lab, producers, etc.) in fertilizer program. With the provision of soil test infrastructure and facilitation of farmers' access to soil lab and appropriate fertilizers, balanced fertilization and, consequently, sustainable soil fertility, food security, and environmental protection could be achieved.

References

Adhami E, Ronaghi A, Karimian N, Molavi R (2012) Transformation of phosphorus in highly calcareous soils under field capacity and waterlogged conditions. Soil Res 50:249–255

Ali Abbasi HR, Esfahani M, Rabiei B, Kavousi M (2007) Effect of nitrogenfertilizing management on rice (Cv. Khazar) yield and its components in a paddy soil of Gilan province. J Sci Technol Agric Natur Resour 10(4):293–307 (in Farsi with English abstract)

Amini M, Afyuni M, Khademi H (2007) Modeling cadmium and lead balances in agricultural lands of Isfahan region, central Iran. J Sci Technol Agric Natur Resour 10(4):77–90 (In Farsi with English abstract)

Anonymous (1961) Results of using chemical fertilizers in Iran. Technical Report no. 1. Soil fertility division ministry of agriculture, Tehran, Iran (In Farsi)

Anonymous (1978) Summary of research results from studies since the establishment till 1978. Bull. No. 545. Institute of pedology and soil fertility, Ministry of agriculture and rural development, Tehran, Iran (in Farsi)

Anonymous (2004) Fertilizer use by crops in Islamic Republic of Iran. http://www.fao.org

Anonymous (2010) Soil and water research institute introduced. http://www.swri.ir (in Farsi)

Antep S (1997) Evaluation of some chemical methods of available soil nitrogen based on ^{15}nitrogen technique. Commun Soil Sci Plant Anal 28:537–550

Atarodi B, Naderi Khorasgani M (2009) Determination of suitable extractant and critical level of phosphorus for sorghum (Sorghum bicolor L. Var. Speedfeed) in calcareous soils of Birjand region, southern Khorasan, Iran. J Sci Technol Agric Natur Resour 13 (50):133–146 (in Farsi with English abstract)

Brown H, Welch RM, Cary E (1987) Nickel: A micronutrient essential for higher plants. Plant Physiol 85:801–803

Darjeh Z, Karimian N, Maftoun M, Abtahi A, Razmi K (1991) Correlation of five Zn extractants with plant responses on highly calcareous soils of Doroodzan Dam area Iran. Iran Agirc Res 10:29–45

Davatgar N, Kavoosi M, Alinia MH, Paykan M (2006) Study of potassium status and effect of physical and chemical properties of soil on it in paddy soils of Gilan province. J Sci Technol Agric Natur Resour 9(4):71–89 (in Farsi with English abstract)

Fahimifar J (1990) World market of chemical fertilizers. World goods market series no. 11. Institute for business research, Tehran, Iran (in Farsi)

Fallah A, Osko T, Khosravi V, Mohammadian M, Rostami M (2011) Reduction of chemical pesticides by using of silicate fertilizers in paddy fields. In: Proceedings of the 5th international conference on silicon in agriculture, p. 46. Beijing, China, Sep 13–18 2011

Gerami B (1982) Summary of the results of soil and water research activities. Agriculture and natural resources organization of ministry of agriculture, Tehran, Iran (in Farsi)

Ghadiri H, Majidian M (2003) Effect of different nitrogen fertilizer levels and moisture stress during milky and dough stages on grain yield, yield components and water use efficiency of corn (*Zea mays* L.). J Sci Technol Agric Nat Resour 7(2):103–113 (in Farsi with English abstract)

Ghaffari Nejad SA, Karimian N (1998) Correlation of five manganese extractants with soil properties and soybean responses in calcareous soils of Fars province. J Sci Technol Agric Nat Resour 2(4):65–76 (in Farsi with English abstract)

Ghaffari Nejad SA, Sabbah A (2009) Effect of trunk injection of calcium salts on dieback of Mazafati dates cultivar. J Soil Res (Soil Water Sci) 23(2):163–171 (in Farsi with English abstract)

Ghanbari A, Maftoun M, Karimian N (2000) Greenhouse and laboratory evaluation of several extractants for determination of available phosphorus for corn in some calcareous soils of Fars province. J Sci Technol Agric Nat Resour 3(4):41–56

Gharaie HA, Maftoun M, Karimian N (2002) Lead adsorption characteristics of selected calcareous soils of Iran and their relationship with soil properties. In: 17th world congress of soil science, pp. 1–8 Thailand. Paper no. 1961, 14–21 August 2002

Gholamalizadeh-Ahangar A, Karimian N, Abtahi A (1995) Growth and manganese uptake by soybean in highly calcareous soils as affected by native and applied manganese and predicted by nine different extractants. Commun Soil Sci Plant Anal 29:1441–1451

Gholamhoseini M, Aghaalikhani M, Malakouti MJ (2008) Effect of natural Zeolite and nitrogen rates on canola forage quality and quantity. J Sci Technol Agric Nat Resour 12(45):537–548 (in Farsi with English abstract)

Haynes RJ (2005) Labile organic matter fractions as central components of the quality of agricultural soils: an overview. Adv. Agron. 85:221–268

Hajabbasi MA, Besalatpour A, Melali AR (2008) Impacts of converting rangelands to cultivated land on physical and chemical properties of soils in west and southwest of Isfahan. J Sci Technol Agric Nat Resour 11(42):525–534 (in Farsi with English abstract)

Hajeeboland R, Asgharzadeh N, Mehrfar Z (2004) Ecological study of Azotobacter in two pasture lands of the north-west Iran and its inoculation effect on growth and mineral nutrition of wheat (*Triticum aestivum* L. cv. Omid) plants. J Sci Technol Agric Natur Resour 8(2):75–90 (in Farsi with English abstract)

Hashemi Beni O, Salehi MH, Beigi Harchegani H (2010) Estimation of soil organic matter by loss-on-ignition procedure in four major plains of Chaharmahal and Bakhtiari province. J Sci Technol Agric Natur Resour 13(50):77–89 (in Farsi with English abstract)

Hedley MJ, Stewart JWB, Chauhan BS (1982) Changes in inorganic soil phosphorus fractions induced by cultivation practices and by laboratory incubations. Soil Sci Soc Am J 46:970–976

Hojjati MH, Boustani F (2010) An assessment of groundwater crisis in Iran, case study: fars province. World Acad Sci Eng Technol 46:475–479

Hosseinpour AR (2004) Application of kinetic models in describing non-exchangeable potassium release in some soils of Hamedan. J Sci Technol Agric Nat Resour 8(3):85–94 (in Farsi with English abstract)

Hosseinpour AR, Shariatmadari H (2007) Phosphorus soil testing: 1-Evaluating of a few extractants in determining alfalfa available phosphorus. J Sci Technol Agric Nat Resour 10(4):107–119 (in Farsi with English abstract)

Jafari S, Baghernejad M (2007) Effects of wetting and drying, and cultivation systems on potassium fixation in some Khuzestan soils. J Sci Technol Agric Nat Resour 11(4):75–90 (in Farsi with English abstract)

Jahangiri S, Souri B, Badakhshan H (2011) Relationship between different physico-chemical properties of some calcareous soils of Ghorveh watershed with soil arsenic. J Soil Res (Soil Water Sci) 25 (4A):337–348 (in Farsi with English abstract)

Kalbasi M (1996) Organic matter status of soils of Iran. Abst. 5th Cong. Soil Sci. Soc. Iran. Karaj, Iran. P 7 (in Farsi)

Kalbasi M, Filsoof F, Rezai Y (1988) Effect of sulfur treatments on yield and uptake of Fe, Zn, and Mn by conr, sorghum, and soybean. J Plant Nutr 11:1353–1360

Karami E, Ebrahimi HR (2000) Overfertilization with phosphorus in Iran: A sustainability problem. J Ext Sys 16:100–120

Karimian N (1994) Consequences of overfertilization with phosphorus fertilizers. Abstr 4th Soil Sci Congr Iran, pp 20–21 (in Farsi)

Karimian N (1995) Effect of nitrogen and phosphorus on zinc nutrition of corn in a calcareous soil. J Plant Nutr 18:2261–2271

Karimian N (1998) Consequences of over-fertilization with phosphorus fertilizers. Soil Water 12(4):1–14 (in Farsi with English abstract)

Karimian N, Ghanbari A (1990) Evaluation of different extractants for prediction of plant response to applied P fertilizers in highly calcareous soils. 10th World Fertilizer Congress of CIEC Abstract p. 25

Karimian N, Gholamalizadeh-Ahangar A (1998) Manganese retention by selected calcareous soils as related to soil properties. Commun Soil Sci Plant Anal 29:1061–1070

Karimian N, Yasrebi J (2003) Effect of growing wheat, spinach, and lettuce on chemical forms of zinc in calcareous soils. Yemeni J Sci 5(1):13–21

Kavoosi M, Kalbasi M (2000) Comparison of soil potassium extracting methods to determine suitable extractants for sepeedroud rice variety in some Gilan rice paddy fields. J Sci Technol Agric Natur Resour 3(4):57–70 (in Farsi with English abstract)

Kavoosi M, Malakouti MJ (2006) Determination of potassium critical level with ammonium acetate extractant in Gilan rice fields. J Sci Technol Agric Natur Resour 10(3):113–123 (in Farsi with English abstract)

Kheirabadi H, Khoshgoftarmanesh AH, Khanmohamadi Z (2012) The effects of some soil properties on Zn availability for corn in certain calcareous soils in Isfahan province. J Sci Technol Agric Natur Resour 16(62):1–10 (in Farsi with English abstract)

Khodagholi M, Eskandari Z, Saeidfar M, Chavoshi S (2008) Effect of nitrogen and phosphorus fertilization on range species production in Semirom region. J Sci Technol Agric Natur Resour 12(44): 333–345

Khodakarami L, Soffianian A, Mirghafari N, Afyuni M, Golshahi A (2012) Concentration zoning of chromium, cobalt and nickel in the soils of three sub-basin of the Hamedan province using GIS technology and the geostatistics. J Sci Technol Agric Natur Resour 15(58):243–254 (in Farsi with English abstract)

Khoshgoftarmanesh AH, Shariatmadari H, Karimian N (2004) Effects of saline irrigation water and Zn application on soil Cd solubility and Cd concentration in wheat. J Sci Technol Agric Natur Resour 7 (4):53–60 (in Farsi with English abstract)

Lack SH, Naderi A, Saidat SA, Ayenehband A, Nour–Mohammadi GH, Moosavi SH (2007) The effects of different levels of irrigation, nitrogen and plant population on yield, yield components and dry matter remobilization of corn at climatic conditions of Khuzestan. J Sci Technol Agric Natur Resour 11(42A):1–14 (in Farsi with English abstract)

Lindsay WL, Norvell WA (1978) Development of soil test for zinc, iron, manganese, and copper. Soil Sci Soc Am J 42:421–428

Mahmood-Soltani S, Samadi A (2003) Determination of phosphorus in some calcareous soils in Fars province and their relationships with physico-chemical properties. J Sci Technol Agric Natur Resour 7 (3):119–128 (in Farsi with English abstract)

Malakouti MJ, Tabatabaei SJ, Shahabil A, Fallahi E (1999) Effect of calcium chloride on apple fruit quality of trees grown in calcareous soils. J Plant Nutr 22:1451–1456

Mengel K, Kirkby EA (2001) Principles of plant nutrition, 5th edn. Kluwer Academic Publishers, Dordrecht, The Netherlands

Mezumen U, Keren R (1981) Boron adsorption by soils using a phenomenological adsorption equation. Soil Sci Soc Am J 45:722–726

Mirbagheri E, Abbaspour A, Rohani A, Ghorbani H (2012) Evaluation of phosphorus status of some potato fields of Majan Semnan province. J Soil Res (Soil Water) 26(3A):235–243

Mirlohi A, Bozorgvar N, Bassiri M (2000) Effect of nitrogen rate on growth, forage yield and silage quality of three sorghum hybrids. J Sci Technol Agric Natur Resour 4(2):105–116 (in Farsi with English abstract)

Moazallahi M, Farpoor MH (2012) Soil genesis and clay mineralogy along the xeric-aridic climotopo sequence in south central Iran. J Agric Sci Technol 14:683–696

Modares R, Sarhadi A (2011) Statistically-based regionalization of rainfall climates of Iran. Glob Plan Change 75:67–75

Momeni A (2010) Geographic distribution and salinity status of soils of Iran. J Soil Res (Soil Water) 24(3):203–215

Moosavi E, Sofianian A, Mirghaffari N, Khodakarami L (2011) Spatial distribution of some heavy metals in soils of Hamedan province. J Soil Res (Soil Water) 25(4):323–336

Moshiri M (1977) Crop production guide: a GhasemibnYousefe Haravi book, 2nd edn. Amir Kabir Publication, Tehran, Iran (in Farsi)

Noushad H, Ronaghi A, Karimian N (2001) Improving N use efficiency in grain corn using soil nitrate test and leaf chlorophyll content. J Sci Technol Agric Natur Resour 5(3):65–78 (in Farsi with English abstract)

Olsen SR, Cole CV, Watanabe FS, Dean LA (1954) Estimation of available phosphorus in soils by extraction with sodium bicarbonate. USDA Circ. 939. U. S. Gov. Print. Office, Washington, DC

Pirzadeh M, Afyuni M, Khoshgoftarmanesh AH (2012) Status of zinc and cadmium in paddy soils and rice in Isfahan, Fars, and Khuzestan provinces and their effect on food security. J Sci Technol Agric Natur Resour 16(60):81–93 (in Farsi with English abstract)

Raei Y, Sedghi M, SeiedSharifi R (2009) Effects of rhizobial inoculation, urea application and weed on growth and seed filling rate in soybean. J Sci Technol. Agric Natur Resour 12(43A):81–91 (in Farsi with English abstract)

Rajaie M, Karimian N, Maftoun M, Yasrebi J, Assad MT (2006) Chemical forms of cadmium in two calcareous soil textural classes as affected by application of cadmium-enriched compost and incubation time. Geoderma 136:533–541

Rasooli A, Nouri M, Razi-Jalali M (2010) Influence of antagonistic minerals in soil and pastures on the blood and liver copper in goats in Khuzestan province, Iran. J Vet Res Shiraz Univ 11(1, Ser. no. 30):46–50

ReyhaniTabar A, Karimian N, Muazardalan M, Savaghebi GR, Ghannadha MR (2006) Zinc fractions of selected calcareous soils of Tehran province and their relationship with soil characteristics. J Sci Technol Agric Natur Resour 10(3A):125–136 (in Farsi with English abstract)

Rezaeinejad Y, Afyuni M (2010) Effect of organic matter on soil chemical properties and corn yield and elemental uptake. J Sci Technol Agric Natur Resour 4(4):19–20 (in Farsi with English abstract)

Roomizadeh S, Karimian N (1996) Manganese-iron relationship in soybean grown in calcareous soils. J Plant Nutr 19:397–406

Samadi A (2006) Phosphorus sorption characteristics in relation to soil properties in some calcareous soils of Western Azerbaijan province. J Agric Sci Technol 8:251–264 (in Farsi with English abstract)

Samavati M, Hosseinpur AR (2011) Phosphorus fractions and availability in some calcareous soils in Hamedan province. J Sci Technol Agric Natur Resour 15(55):127–138 (in Farsi with English abstract)

Serna MD, Pomares F (1992) Evaluation of chemical indices of soil organic nitrogen availability in calcareous soils. Soil Sci Soc Am J 56:1486–1491

Shahrokhnia A, Rajea A, Keshavarz H (1993) The results of using phosphorus fertilizer according to the soil test in wheat fields of Fars province. In: Proceedings of wheat research symposium, Shiraz, Iran. September 1993 (in Farsi)

Sharifi M, Kalbasi M (2001) Selection of suitable extractant to extract available potassium for corn in soils of central region of Isfahan Province. J Sci Technol Agric Natur Resour 5(1):77–92 (in Farsi with English abstract)

Sharafi M, RanjbarFordoui A, BeigiHarchegani H, Iranipour R (2013) Effect of cobalt on growth characteristics of bean. J Soil Res (Soil Water Sci) 27(1A):85–96 (in Farsi with English abstract)

Sharifi M, Zebarth BJ, Burton DL, Grant CA, Cooper JM (2007) Evaluation of some indices of potentially mineralizable nitrogen in soil. Soil Sci Soc Am J 71:1233–1239

Sparks DL (1987) Potassium dynamics in soils. Adv Soil Sci 6:1–63

Tehrani MM (2013) Micro nutrient distribution in irrigated soils of Iran. Research Bulletin, Soil and Water Research Institute, Iran

Tehrani MM, Balali MR, Moshiri F, Daryashenas AM (2010) Fertilizer recommendation and forecast in Iran: challenges and strategies. Iran J Soil Res 2, Iran

Turner BL, Cade-Menun BJ, Westermann DT (2003) Organic phosphorus composition and potential bioavailability in semi-arid arable soils of the western United States. Soil Sci Soc Am J 67:1168–1179

Yasrebi J, Karimian N, Maftoun M, Abtahi A, Ronaghi A, Assad MT (2004a) Distribution of nitrogen forms in cultivated calcareous soils of Fars province and their relationships with soil characteristics. J Sci Technol Agric Natur Resour 7(4):39–52 (in Farsi with English abstract)

Yasrebi J, Karimian N, Maftoun M, Abtahi A, Ronaghi A, Assad MT (2004b) Laboratory and greenhouse evaluation of ultraviolet light absorption methods of estimating nitrogen-supplying capacity of calcareous soils. Commun Soil Plant Anal 35:219–232

Yasrebi J, Karimian N, Maftoun M, Abtahi A, Sameni AM (1994) Distribution of zinc forms in highly calcareous soils as influenced by soil physical and chemical properties and application of zinc sulfate. Commun Soil Sci Plant Anal 25:2133–2145

Author Biographies

Dr. Najafali Karimian, (1943–2015) was an emeritus and distinguished professor at Soil Science Department of Shiraz University. His major field was soil chemistry and fertility. He received M.Sc. in soil chemistry in 1970 from Texas A&M University and Ph.D. in soil science in 1977 from North Carolina State University, Raleigh, North Carolina, USA. Dr. Karimian was Vice President of Soil Science Society of Iran from 1992 to 2009 and contributed significantly to the advancement of soil science and soil fertility in Iran. He published many research papers in international refereed journals and was the lead author of this chapter on soil fertility. He died in March 2015 in Shiraz, Iran.

Dr. Mohammad Mehdi Tehrani is a senior researcher at Soil and Water Research Institute of Iran. His current research interests focus on soil fertility indices, plant nutrition, soil chemistry, fertilizer recommendations, and the role of biostimulants in agriculture. He has wide scientific experiences in preparing soil fertility map and modeling of soil nutrient status in various region of Iran.

Soil Biology and Biotechnology

11

Hossein Besharati, Nasser Aliasgharzad, Kazem Khavazi, and Hadi Asadi Rahmani

Abstract

In this chapter, we attempt to present the status of soil biology, the corresponding research activities, and the biological properties of the soils of Iran. After a brief history, biological characteristics associated with soil characteristics, climate, and vegetation are presented. Then, the biological properties of the dominant soils of the country are illustrated, followed by research findings on exploiting soil biological potentials in agriculture, especially production and commercialization of technological knowledge of biofertilizers. Also, the list and characteristics of useful soil microorganisms that exist in the country and the ones which are available at the "Microbial Bank" are presented. Finally, the priority research areas of the country in the field of soil biology and biotechnology are presented.

Keywords

Biofertilizer • Inoculation • Mycorrhiza • PGPR • S-Biooxidation • Soil Biotechnology

11.1 A Brief History of Soil Biology in Iran

Soil biology research began in Iran with the establishment of soil science programs and the subsequent launching of soil biology programs in the local universities. The department of soil science that offered an undergraduate program was established in 1960 at the College of Agriculture, University of Tehran. This was followed by the launching of similar soil science degree programs in Shiraz and Tabriz Universities in 1962. In subsequent years, other national universities also offered new degree programs in soil science. The first graduate program in soil science was inaugurated at the University of Tehran in 1966. Other universities gradually started offering similar programs. It took some time, however, before the University of Tehran approved the proposal for establishing an independent division of soil biology and biotechnology within the department of soil science. Graduate students were admitted to the MSc and Ph. D. programs in this division in 2003.

Research in the field of soil biology and biotechnology in Iran is of a more recent history. The first MSc Thesis in soil biology entitled "*Bio-oxidation* of soil sulfur" (Golchin 1987) was written in 1987. The first Ph.D. dissertation entitled "Frequency and distribution of *Arbuscular Mycorrhizal fungi* in Iranian soils" (Aliasgharzad 1991) was submitted in 2000 at the University of Tehran. The 2000s during which enrollment in the departments of soil biology took a growing trend witnessed a diversified range of research topics in soil biology that included molecular and genetic microorganisms, soil enzymes, and bioremediation (Saleh-Rastin 2005; Asadi et al. 2011).

Soil and Water Research Institute (SWRI) has also conducted a significant part of the research on soil biology and biotechnology. In 1980, a joint committee was mandated to study the feasibility of producing soybean inoculant and much needed biofertilizer in Iran. During the early years in the 1990s, SWRI established and equipped a soil biology

H. Besharati (✉) · K. Khavazi · H. Asadi Rahmani
Department of Soil Biology, Soil and Water Research Institute, Agricultural Research, Education and Extension Organization, Karaj, Iran
e-mail: Hbesharati@areeo.ac.irBesharati1350@yahoo.com

N. Aliasgharzad
Department of Soil Science and Engineering, University of Tabriz, Tabriz, Iran

M.H. Roozitalab et al. (eds.), *The Soils of Iran*, World Soils Book Series,
https://doi.org/10.1007/978-3-319-69048-3_11

and microbiology research laboratory which came into operation in late 1995, followed by the founding of Biological Fertilizers Manufacturing Lab in 1999 (Saleh-Rastin 2005).

Once the infrastructure was established, research proposal began to be drafted in varied areas such as sulfur-oxidizing bacteria, phosphate-solubilizing bacteria, mycorrhizal fungi, and plant growth stimulators (*Azotobacter*, *Pseudomonas*, etc.). The qualitative and quantitative achievements in these research projects ultimately led to the development of biofertilizers production technologies, some of which have already been marketed to the private sector for mass production. The 12th National Soil Sciences Congress witnessed a rapid rise in the number of papers on soil biology in such a way that they accounted for 14% of the total number of papers (Asadi et al. 2011). The Journal of Soil Biology (in Persian) is the first academic periodical published in Iran that has been the forum for presenting research findings on soil biology. Along these activities, "the standard procedures for biofertilizer evaluation and the soil biology strategic program" have been developed by Iranian scholars that have had valuable contributions to the field of soil biology in Iran.

11.2 Relationships Between Soil Biological Characteristics and Environment

The organisms living in the soil cover a wide spectrum, and their ecological requirements are equally diverse. The relative numbers, diversity, and composition of soil-living organisms are functions of soil characteristics (moisture content, aeration, nutrients, soil pH, etc.), climatic characteristics, and vegetative cover. In other words, soil biological conditions that are part of its dynamic characteristics indicate its adaptability to the requirements of soil organisms.

Iran has mostly an arid to semiarid climate with an average annual precipitation of only around one-third of the global average and an annual evaporation rate three times of the world average (see Chaps. 3 and 9). This is a major cause of the low soil organic content of the soils in Iran. The organic carbon content in over 60% of the cultivated areas is less than 1% and in more than 84%, is less than 1.5% (Shahbazi and Besharati 2014). Most soil-living bacteria nutritionally belong to the heterotrophic species, which depend for their survival on soil organic matter. The autotrophic species, which produce organic matter in soil, account for a small portion of the soil-living organisms and, thus, produce only small amounts of organic matter in soil (Saleh-Rastin 2005). This is why soil organic matter plays a significant role in enhancing bacterial populations and activities through its contributions to improved physical (enhanced aeration, water holding capacity, etc.) and chemical (releasing nutrients, adsorption and storage of elements, etc.) characteristics of soils. Hence, shortage of soil organic matter accounts for a major portion of the poor quality of many farm soils in Iran. This is confirmed by the comparison of results from independent studies across the nation.

The average total population of bacteria, phosphate-solubilizing bacteria, and fungi (which are heterotroph and rely on soil organic matter), in 50 samples in both rice paddies and other farms in the north of Iran were 5×10^7, 7.4×10^5, and 3.2×10^4 cells per gram of soil, respectively (Fallah et al. 2004). The organic matter content of the studied soils was around 7%. Another study showed that the organic content of farm soils in Fars Province was 0.6%, and their average bacterial count was 5.2×10^5 cells per gram of soil (Anonymous 2012). A similar study conducted in crop farms in Tehran Province found that the average organic content of soil was 0.5%, and its *Azotobacter* content (heterotrophic relying on soil organic matter) was as low as 1.5×10^3 cells per gram of soil (Khosravi 1997). Characteristics of soil and climate not only affect soil microorganism counts but also have significant effects on their diversity and composition. Compared to other parts of the country, population and activities of microorganisms in soils of the Caspian Sea region are higher due to humid and sub-humid climatic conditions (see Chap. 7 and Banai et al. 2005). Heterotrophic microorganisms including all types of heterotrophic bacteria and fungi have greater populations in these regions such that the average total populations of bacteria and fungi have been reported to be 5×10^7 and 2×10^5 cells per gram of soil (Fallah et al. 2004). This is why, compared to wet areas, bacterial and fungal populations are far lower in areas whose soils remain dry for most of the year and contain a low soil organic matter content. *Azospirillum* populations in soils in Karaj, Pakdasht, Savochbolagh, and Hashtgerd are reportedly to be around 1×10^4 cells per gram of soil (Esbati et al. 2014). Maximum and minimum mycorrhizal spores in farmlands under bean cultivation in the Provinces of Markazi, Lorestan, and Zanjan have been reported to be around 2.8 and 0.35 spores per gram of soil (Soltani et al. 2014).

Iran is characterized by a climate diversity, which naturally entails diversity in soil type (see Chap. 7). This soil and climate diversity plays an important role in the diversity and relative composition of soil-living organisms. Due to the high temperatures in central parts of Iran and the low soil moisture and organic matter content in these areas, the total populations of microorganisms are low, but the microbes surviving these conditions are also actively present in soil. For example, sporiferous bacteria, or those resistant to high salinity and temperature, as prevailing conditions in Iranian soils, are the prevailing bacteria.

The type of vegetative cover and farm management are also the factors involved in the numbers and relative compositions of soil microorganisms. A survey carried out in four provinces revealed that soils previously subjected to treatment with elemental sulfur contained abundant sulfur-oxidizing bacteria while the untreated soils lacked these bacteria or their populations were negligible (Besharati and Fallah 2008). Symbiotic rhizobia associated with leguminous plants exhibit heterotrophic nutrition habits in the absence of their host plants using soil organic matter as their source of carbon and energy and soil mineral matter as their nitrogen source. Rhizobium populations significantly decrease in soils in which no leguminous plants have been grown for long periods, while those under legume plantations will be rich in rhizobia.

Alfalfa is a plant indigenous to Iran; the symbiotic bacteria associated with this plant are usually abundant in Iranian soils, especially in those under alfalfa. The average number of symbiotic bacteria associated with alfalfa in perennial farms in Hamedan Province was found to be 300 times greater than that in the neighboring uncultivated land. The average bacterial populations in 1-, 2-, 3-, 4-, 5-, 6-, 7-, and 15-year-old alfalfa-cultivated farms as well as in neighboring fallow plots were measured as 6.92×10^4, 3.02×10^4, 1×10^4, 1.26×10^4, 1×10^4, 4.47×10^3, 1.8×10^2, and 2.14×10^2 cells per gram of soil, respectively (Khavazi et al. 2003). This is why a symbiotic bacterial inoculant is required for cultivating soybean which is not indigenous to Iran as its symbiotic bacteria are extremely low in numbers in Iranian soils. Bacterial counts in 55 soil samples from northern Iran revealed that the highest population (around 1 million cells per gram of soil) of symbiotic bacteria associated with soybean was observed in farms which had been regularly under alternate soybean cultivation, and the lowest (nearly 10 cells per gram of soil) was detected in farms in which the alternate soybean cultivation had been suspended, and the relevant inoculants had not been used in a timely manner (Khavazi et al. 2001).

Generally speaking, plants may affect soil microbial populations through the type and quantity of their root secretions, alteration of the prevalent conditions in the rhizosphere, introduction of organic matter into soil as plant remains, and relationships (symbiosis and cooperation) with soil microorganisms. Populations of mycorrhizal fungi will definitely rise in the farms where sorghum, alfalfa, corn, and potato are grown as their favorable hosts compared to farms where non-mycorrhizal crops such as canola are cultivated. Crop management and land use are also expected to affect soil enzymes whose origins are mainly soil-living organisms, especially soil-dwelling microorganisms (Safari-Sanjani and Sharifi 2006; Matinizadeh et al. 2012).

The effect of land use on soil enzymes was investigated in a plain in western Iran to observe the highest activities by dehydrogenase, cellulose, unease, acidic and alkaline phosphomonoesterase, and organic carbon in rangelands. It was further observed that conversion of natural rangeland to cultivated crop farms or to apple orchards led to the greatest decrease in the above indices such that the organic content and soil enzyme activities exhibited sharp declines (Ebrahimzadeh et al. 2013). In another study, the tea plantations in northern Iran were evaluated for their mycorrhizal fungi. Results revealed no indication of symbiosis between the mycorrhiza and the tea roots despite the abundant mycorrhizal fungal spores in the farm soils studied. This finding was further confirmed by investigation of the patterns of phospholipidic fatty acids (Aliasgharzad et al. 2011). Application of 0.5, 1, and 2 t/ha of sulfur in farms under corn cultivation has led to significant reduction in activities of both acidic and alkaline phosphatase enzymes as compared to the control while the reduced activity was greater with acidic phosphatase than with its alkaline counterpart (Anonymous 2012). Recent investigations of the patterns of phospholipidic fatty acids in salty soils near Urmia Salt Lake and its protected islands revealed that the mycorrhizal fungi have significantly decreased in number while the non-useful gram-positive and gram-negative bacteria have been on the rise. It has been, however, found that, despite the recent droughts and the loss of microbial diversity in rangelands, orchards, and farms around the lake, microbial diversity in the lake has been much retained. These should be considered as alarming symptoms of an imminent crisis in other parts of the country that need to be duly managed in time before they become irreversible (Aliasgharzad et al. 2011).

Investigation of the likely relationships existing between soil biological characteristics and the global warming challenge forms a major topic for research in soil biology. Greenhouse gases have left great impacts on global warming. This is while farm soils have served as a major source of carbon dioxide released into the atmosphere. Clearly, the roles played by soil biology investigations and management of soil biological characteristics in reducing carbon dioxide emissions and enhancing carbon sequestration are of significant importance. One way to compensate for the damages incurred is to intensify the process of carbon precipitation in soils through glomeral fungi to increase glomalin production in them. A major portion of the organic carbon received by the fungi precipitates as glycoprotein in hyphae and spore walls and subsequently enters the soil. Given the rather long half-life of glomalin in soil, it can store considerable quantities of carbon in soil. Examination of rangelands with different types of soil and vegetative cover revealed severe declines in *Azotobacter* populations in soils lacking any vegetative cover. *Azotobacter* populations have been found to be the highest in the rhizosphere of the wheat family but the lowest in the rhizosphere of the pea family. Moreover, *Azotobacter* populations have been reported to have a significantly positive relationship with soil organic matter content but a significantly negative relationship with soil pH (Hajeeboland et al. 2004).

11.3 Biological Characteristics of Dominant Soils in Iran

Soil biological properties constitute its dynamic characteristics which rapidly change with changes in time, place, and other soil characteristics. These changes and the associated dynamism are so great that they lead to significant differences in the biological characteristics of different parts of one single soil. Due to the scant information and data on the biological characteristics of soils across the country, some selected characteristics of 40 soil samples collected from different parts of Iran representing prevailing soil subgroups (see Chap. 7) and families have been analyzed and measured. Table 11.1 shows the biological characteristics of the major soil families across Iran.

11.4 Research Findings and Exploiting Soil Biological Potentials in Agriculture

11.4.1 Azotobacter

The first research on *Azotobacter* in Iran dates back to the mid-1980s when a graduate dissertation at the University of Tehran investigated *Azotobacter* (Khosravi 1997). This was later followed by a large number of studies at other universities and research institutes addressing evaluation of *Azotobacter* effects on dry-farmed wheat (Khosravi and Mahmoudi 2013), corn (Biyari et al. 2011), medicinal plants (Hosseinpour et al. 2011), and apple tree (Khosravi et al. 2009). Field studies of *Azotobacter* have been principally carried out at the SWRI. In one such research program, experiments were conducted on 362 soil samples collected from wheat fields in different provinces (East Azerbaijan, West Azerbaijan, Kurdestan, Golestan, Fars, and Khorasan), from which 217 strains of nitrogen-fixing *Azotobacter* were isolated, of which 102 were confirmed as belonging to *Azotobacter chroococcum*. Finally, application of those *Azotobacter* inoculants led to enhancements in grain yield in most cases (Khosravi et al. 2015). Also, related researches on plant growth-promoting properties as well as identification and molecular cloning of the gene encoding the enzyme ACC-deaminase in strains native to farms around Tabriz University have been conducted at Tabriz University (Farajzadeh et al. 2012).

Results of the research on *Azotobacter* over the past years indicates that its use has led to improved plant growth and that it led to better results in combination with application of different organic materials (Khosravi and Mahmoudi 2013). However, cases have also been reported in which application of *Azotobacter* did not produce the expected effects mainly because *Azotobacter* is a heterotrophic bacterium that depends on organic substances which are deficient in most soils in Iran (Shahbazi and Besharati 2014). On the other hand, *Azotobacter* is inferior to Rhizobial bacteria in terms of its nitrogen-fixing capacity so that, in most cases, it plays only a small role in supplying nitrogen to plants. This is why certain strains that enjoy appropriate growth promoters exhibit considerable effects on plant growth. Finally, *Azotobacter* naturally occurs in most farms, rangelands, and forest soils in Iran, and, thus, they serve their function in plant growth without the need for man's intervention. This is evidenced by its population ranging between 10^3 and 10^5 cells per gram of soil in most soils in Iran (Khosravi 1997; Hajeeboland et al. 2004) (Fig. 11.1).

11.4.2 Beans

In a series of experiments performed during the period from 1995 to 2005, over 80 strains of symbiotic rhizobia associated with beans were isolated and identified in several provinces. Greenhouse and field experiments revealed that efficient rhizobium strains enjoy considerable capacity for nitrogen fixation which makes them capable of supplying 70% of the nitrogen requirement by the bean plant. This finding was further confirmed in studies using the nitrogen-15(^{15}N) isotope. In some field experiments, inoculation of the bean plant with superior rhizobium strains led to an average enhancement of 30% in bean grain yield compared to the control (Asadi et al. 2005).

There is also a considerable volume of published literature on the biological fixation of nitrogen in bean plants cultivated in different parts of Iran (Khalaj et al. 2013; Afshari et al. 2006). It may be concluded from all the research findings across the nation that appropriate rhizobia inoculated into bean varieties are capable of providing for a considerable portion of the nitrogen requirements of the plant and can, thus, be effective in improving plant performance. Presently, around 60% of the symbiotic rhizobia associated with bean are conserved at the Culture Collection of Soil Microorganisms (CCSM) at SWRI. The technical know-how emerging from this vast number of studies has been marketed to the private sector for the mass production of biofertilizers bearing symbiotic rhizobia associated with bean (Asadi et al. 2005) (Fig. 11.2).

11.4.3 Faba Bean

A few studies on nitrogen fixation by faba bean have been carried out at the local universities. These limited studies

Table 11.1 Some biological properties of the dominant soil families in Iran

Location of survey	Soil classification (family)	Acidic phosphatase (μg P-NF $g^{-1}h^{-1}$)	Alkaline phosphatase (μg P-NF $g^{-1}h^{-1}$)	Total bacteria (10^6 cfu g^{-1} Soil)	Urease activity (μg NH_4 $g^{-1}2\ h^{-1}$)	Total fungi (10^3Prop. g^{-1} Soil)	Sulfur-oxidizing Bacteria (10^3 cfu g^{-1} Soil)		Soil respiration (mgCO_2 g^{-1} Soil $24h^{-1}$)	Microbial biomass (mg C g^{-1} Soil)
							Heterotroph	Autotroph		
Fereidown, Isfahan	Fine, mixed, mesic, Calcic Haploxerepts	351.9	852.3	6.2	25.5	1.0	150	0	14.52	277.57
Central Region of Isfahan (Kouhpayeh Sagzi)	Fine-loamy, carbonatic, thermic, Fluventic Haplocambids	348.07	665.4	0.68	22.1	5	1.7	0	14.61	289.7
Zarrin-shahr, Talkhuncheh, Mahyar and Charmhin	Loamy-skeletal, gypsic, thermic, shallow, Typic Haplogypsids	164.4	275.0	0.46	14.5	<1	8.8	0	7.27	3.38
Mianeh Region	Fine-loamy, mixed, mesic, Calcic Haploxerepts	309.6	648.0	1.7	6.5	50	64	0	14.7	191.81
Hashtrood Region	Coarse-loamy, mixed, calcareous, mesic, Typic Xerofluvents	258.5	725.0	18	8.4	110	10	0	14.51	285.98
Shahin Dedge Region	Coarse-loamy, mixed, mesic, Oxyaquic Xeropsamments	142.3	335.1	8.4	2.1	90	8.7	0	14.54	727.85
Boukan Region	Fine, mixed, mesic, Calcic Haploxerepts	213.4	634.6	9.9	3.2	20	3.7	0	14.7	481.93
Plain Lands, Poshtekouh and Palang, Boushehr	Fine-loamy, carbonatic, hyperthermic, Ustic Haplocambids	305.6	707.6	11	3.7	30	14.8	0	11.73	637
Plain Lands, Poshtekouh and Palang, Boushehr	Coarse-loamy, carbonatic, hyperthermic, Ustic Torripsamments	212.5	573	35	2	40	6.3	0	9.98	802.24
Plain Lands, Poshtekouh and Palang, Boushehr	Fine-loamy, carbonatic, hyperthermic, Ustic Torriorthents	522	778.8	1.1	23.5	<1	0.15	0	8.79	83.14
Southwest of Tehran (Shahriyar-Western Cave)	Fine-loamy, mixed, thermic, Typic Haplpocalcids	925.6	2568.4	11	10.9	17	14.5	0	11.5	578.39
Southwest of Tehran (Shahriyar-Western Cave)	Fine-loamy, mixed, thermic, Fluventic Haplocambids	796.1	1563.0	8.2	21.1	1	15.5	0	11.67	21.18

(continued)

Table 11.1 (continued)

Location of survey	Soil classification (family)	Acidic phosphatase (μg P-NF $g^{-1}h^{-1}$)	Alkaline phosphatase (μg P-NF $g^{-1}h^{-1}$)	Total bacteria (10^6 cfu g^{-1} Soil)	Urease activity (μg NH_4 $g^{-1}2$ h^{-1})	Total fungi (10^3Prop. g^{-1} Soil)	Sulfur-oxidizing Bacteria (10^3 cfu g^{-1} Soil)		Soil respiration ($mgCO_2$ g^{-1} Soil $24h^{-1}$)	Microbial biomass (mg C g^{-1} Soil)
							Heterotroph	Autotroph		
Damavand Region	Loamy-skeletal, mixed, mesic, Typic Xerofluvents	188.45	313.5	0.9	7.6	1	0.1	0	5.42	238.61
Dashte Falard Lands	Fine, carbonatic, thermic, Typic Calcixerolls	137.5	407.1	0.8	13.9	1	2.5	0	7.51	758.91
Esfarayen Plain	Fine-loamy, mixed, mesic, Typic Haplosalids	165.4	373.0	1	41.8	<1	1.45	0	8.95	355.64
Esfarayen Plain	Fine, mixed, mesic, Gypsic Haplosalids	147.06	213.5	1.5	6.5	4	14	0	6.62	295.8
Plain Land, Gonabad	Coarse-loamy over sandy skeletal, thermic, Typic Haplogypsids	150.9	586.5	4.1	63.8	2	13.8	0	11.64	438.02
Plain land, Sabzevar	Fine-loamy, mixed, thermic, Typic Haplosalids	141.3	523.08	44	22.6	<1	0.5	0	9.29	375.79
North of Shadegan Lands	Fine, carbonatic, hyperthermic, Typic Torriorthents	214.7	338.4	46	17.8	30	12.8	0	11.27	67.51
North of Shadegan Lands	Coarse-loamy, carbonatic, hyperthermic, Typic Aquisalids	162.5	271.15	11	27.4	11	12.7	0	11.28	234.33
Omidiyeh	Coarse-loamy, carbonatic, hyperthermic Aridic Ustipsamments	94.26	338.5	0.57	4.1	2	0	0	7.3	285.1
Ramhormoz	Fine-loamy, gypsic, hyperthermic, shallow Typic Haplosalids	69.23	190.4	4.9	21.5	1	11	0	8.42	187.39
Bastam Plain	Fine, carbonatic, mesic, Typic Haplocambids	118.26	205.8	4	48.8	<1	0.4	0	9.32	409.85
Neyriz, Tang-e-Hana and Chah-e-Sorkh plains	Fine, gypsic, thermic, Gypsic Haplosalids	420.2	771.1	5.5	8.5	2	4.3	0	11.79	612.6
Neyriz, Tang-e-Hana and Chah-e-Sorkh plains	Coarse-loamy over clayey, gypsic, thermic, Leptic Haplogypsids	374.02	961.5	6.7	4.7	12	20	0	13.77	958.05

(continued)

Table 11.1 (continued)

Location of survey	Soil classification (family)	Acidic phosphatase (μg P-NF $g^{-1}h^{-1}$)	Alkaline phosphatase (μg P-NF $g^{-1}h^{-1}$)	Total bacteria (10^6 cfu g^{-1} Soil)	Urease activity (μg NH_4 $g^{-1}2$ h^{-1})	Total fungi (10^3Prop. g^{-1} Soil)	Sulfur-oxidizing Bacteria (10^3 cfu g^{-1} Soil)		Soil respiration ($mgCO_2$ g^{-1} Soil $24h^{-1}$)	Microbial biomass (mg C g^{-1} Soil)
							Heterotroph	Autotroph		
Neyriz, Tang-e-Hana and Chah-e-Sorkh plains	Fine, mixed, thermic, Typic Haplosalids	158.86	425.0	5.1	8.9	3	1.65	0	9.35	525.42
Larestan, Darz and Sayeban Plains	Coarse-loamy, carbonatic, hyperthermic, Aridic Ustorthents	254.6	482.7	4.3	29.4	3	19.7	0	13.86	732.68
Jiroft and Mosafer-abad	Loamy-skeletal over fragmental, mixed, hyperthermic, Typic Haplocalcids	49.03	273.1	7.1	30.4	2	0.1	0	7.57	316.03
Jiroft and Mosafer-abad	Fine-loamy, mixed, hyperthermic, Typic Haplocambids	175.1	690.4	8.6	41.0	2	2.6	0	14.02	488.89
Jiroft and Mosafer-abad	Loamy-skeletal over fragmental, mixed, hyperthermic, Typic Haplocalcids	53.8	113.5	7.5	9.1	6	1.15	0	9	81.86
Roudbar-e-Kahnouj	Sandy-skeletal, mixed, hyperthermic, Typic Torriorthents	68.26	150.0	0.1	34.9	<1	0	0	9.08	967.77
Songhor and Koliai Region	Fine, mixed, mesic, Calcic Haploxerepts	201	498.0	4.1	37.2	1	6.6	0	9.62	187.78
DamHabib Ishan	Fine-loamy, mixed, thermic, Calcic Haploxerepts	46.74	121.2	1.6	42.9	5	7.7	0	11.93	276.28
Southwest of the Gonbad city (Lands of Peyman Cooperative Production Co.)	Coarse-loamy, mixed, thermic, Typic Aquisalid	528.83	890.4	3.8	6.8	1	2.9	0	13.94	711.69
Khorram–Abad and Boroujerd Plains	Fine, mixed, mesic, Typic Haploxerept	231.7	1121.2	41	0	32	390	0	12.35	190.26
Lands, West of Mazandaran Province	Fine, mixed, calcareous, thermic, Typic Fluvaquents	89.43	955.8	31	13.1	14	6150	0	9.97	93.00

(continued)

Table 11.1 (continued)

Location of survey	Soil classification (family)	Acidic phosphatase (μg P-NF $g^{-1}h^{-1}$)	Alkaline phosphatase (μg P-NF $g^{-1}h^{-1}$)	Total bacteria (10^6 cfu g^{-1} Soil)	Urease activity (μg NH_4 $g^{-1}2\ h^{-1}$)	Total fungi (10^3Prop. g^{-1} Soil)	Sulfur-oxidizing Bacteria (10^3 cfu g^{-1} Soil)		Soil respiration ($mgCO_2$ g^{-1} Soil $24h^{-1}$)	Microbial biomass (mg C g^{-1} Soil)
							Heterotroph	Autotroph		
Lands, West of Mazandaran Province	Fine, mixed, nonacid, thermic, Mollic Fluvaquents	205.77	1026.9	40	6.5	14	440	0	18.05	522.02
Jask Region	Coarse-loamy, mixed, hyperthermic, Typic Aquisalids	87.5	253.8	15	0.0	80	6.8	0	15.44	350.32
Taherlouon Region	Fine, mixed, mesic, Calcic Haploxerepts	159.7	248.0	5.1	8.9	18	98	0	11.57	305.33
Damagh, Churmagh and Saravak Regions	Fine, mixed, mesic, Calcic Haploxerepts	99.04	253.9	2.9	2.2	320	3.2	0	11.03	108.61

started in the first half of the 2000s in which 168 rhizobium nodules were collected at the flowering stage of faba bean growth from different fields in some provinces, and their symbiotic bacteria were isolated. In the plant infection test, 110 isolates were confirmed as symbiotic rhizobia associated with faba bean, and 44 isolates were considered for greenhouse experiments to determine their nitrogen-fixing efficiency and impact on faba bean growth. Based on the results obtained, three to five *Rhizobium leguminosarum bv. viciae* were selected for each province to make a total number of 19 bacterial isolates. The faba bean inoculant (a mixture of perlite and bacterial suspension) was applied on seeds to be planted in agricultural research farms in four provinces (Golestan, Mazandaran, Khuzestan, and Lorestan) over a period of two years. Crop yield showed increase of 17–69% due to inoculation. The inoculant used on farmers' fields led to an increase of 30% in pod yield and 65% in nitrogen uptake by the vegetative parts. Currently, 45 symbiotic bacteria associated with faba bean are conserved at the SWRI, and the technical know-how on mass production of biofertilizers is ready for marketing to the private sector (Khosravi et al. 2005) (Fig. 11.3).

11.4.4 Mycorrhiza

Given the environmental (saline, thermal, and drought) stresses in most Iranian soils, and considering the significance of mycorrhiza in enhancing the resistance of their host plants against these stresses and the roles fungi play in improving water and nutrient uptake and, thereby, in the quality and quantity of farm products, numerous studies have been devoted in the universities and research institutes across the nation to the identification of mycorrhizal fungi and the functions they serve in enhancing the growth and performance of farm crops and horticultural plants.

Research on arbuscular mycorrhizal fungi dates back to 1986,[1] which mainly concentrated on collecting fungi, their purification, identification, study of their diversity, and their pot culture. These studies led to the identification of the dominant mycorrhizal fungi in Iranian soils. Some of these have been inscribed at the international mycorrhiza Web site (www.mycorrhizas.org). Research dealing with the effects of these fungi on improving plant growth started at the University of Tehran in 1991 as a research dissertation (Aliasgharzad 1991). This work was later supplemented by a project that aimed to identify and purify the mycorrhizal fungi in the soils around the Urmia Salt Lake. The findings

[1]Ferdowsi University of Mashhad (Dr. Hormozdyar Kiyanmehr), Shahid Chamran University of Ahvaz (Dr. Vaheh Minasiyan), and University of Urumia (Dr. Hamid Mehravaran).

Fig. 11.1 Effects of *Azotobacter* on wheat growth in comparison with the control infield experiments in Khorasan Province (Khosravi et al. 2015)

Fig. 11.2 Root nodules of bean (left) and a locally produced biofertilizer specific for beans (right) (Asgharzadeh et al. 2007)

were confirmed by international expert bodies in the field which led to the international certification documents for four fungal varieties now available at the University of Tabriz (Aliasgharzad 1991). A number of studies have also been conducted on the different aspects of mycorrhiza including their nutritional effects as well as their contribution to plant resistance to salinity, drought, and heavy metals. Concentrated and concerted research on these useful fungi, however, was launched in 1996 at the Department of Soil Biology at SWRI (Rejali et al. 2004a).

The fungi indigenous to soils in crops such as wheat fields in different provinces have been isolated and identified based on their spore characteristics in many studies (e.g., Balali-Aliabadi et al. 2001; Sadravi 2005; Rejali et al. 2004a). Applications of these fungi under environmental stresses such as drought (Rejali et al. 2003), salinity (Daei et al. 2009; Tavasoli and Aliasgharzad 2009; Rejali et al. 2003), soil compaction (Miransari et al. 2009), and heavy metals pollution (Alizadeh-Oskoui et al. 2009) have led to great improvements in the host plants growth and resistance to adverse environmental conditions. The performance of these fungi in the phytoremediation of heavy metals contaminated soils have been investigated (Zarei et al. 2011).

Numerous studies have also been conducted on the roles mycorrhiza play in supplying plant nutrients such as phosphorus (e.g., Rejali et al. 2003, 2004b). It has been reported that application of phosphate fertilizers can be reduced greatly at no cost to plant performance if the right mycorrhizal fungi are properly used. In a different line of research, the effects of indigenous fungi on enhanced growth and

Fig. 11.3 No root nodule in non-inoculated control (upper left), the root nodules in faba bean (upper right), and a field test of faba bean (bottom) (Khosravi et al. 2005)

performance of medicinal plants revealed that the symbiotic association between these plants and mycorrhizal fungi leads to improved growth through enhanced mineral uptake which, in turn, leads to increased essence production by medicinal plants (Aliabadi-Farahani and Valadabadi 2010).

Distribution of mycorrhizal fungi in different climatic and soil conditions, their nonspecific relationship to host plants, their different mechanisms for affecting host plants (such as improving plant nutrient uptake), their diversity and the wide spectrum of host plants, their synergistic relations with other useful soil-living microorganisms, and their stability in the inoculants (spore) are among the many advantages of using these useful fungi in both agriculture and natural resources. It is due to these considerations and their advantages that research is being continued to develop their applications for farm, horticultural, and range plants as well as forest trees. One disadvantage or barrier to their reproduction, however, is that they can only be reproduced in the presence of their host plants. For the same reason, their reproduction is associated with serious difficulties as it is cost-intensive and time-consuming. Despite all these difficulties, however, the technique that has much facilitated their reproduction is in vitro culture. In 2000, SWRI purchased the in vitro culture technology of fungi from India. Once an expert team from SWRI was trained by the Indian experts from TERI, the production of mycorrhizal inoculants was initiated using the in vitro culture method at the institute (Rejali et al. 2003). There is now a well-developed technology available at the Department of soil biology at the SWRI for culturing *Rhizophagus intraradices* in vitro culture. Moreover, the mycorrhizal fungi *Funneliformis mosseae*, *Rhizophagus intraradices*, *Claroideoglomus etunicatum*, *G. versiform*, *Rhizophagus clarus*, *Rhizophagus fasciculatus*, *and Funneliformis caledonius* are available at the culture collection of the institute. Their culture and production of their inoculants are accomplished by the private sector, and their products are being marketed to both agricultural researchers and producers (Fig. 11.4).

11.4.5 Lentils

Few studies have been conducted on nitrogen fixation of lentil plants (Afshari et al. 2007; Gorji-Anari et al. 2007). However, the greatest portion of research in this area has been carried out at SWRI. In one series of such research lasting from 2003 to 2005, the nitrogen-fixing efficiency of indigenous isolates of symbiotic rhizobia associated with lentil plant (*R. leguminosarum bv. viciae*) was investigated, and 76 isolates were ultimately purified. The capacity of the

Fig. 11.4 In vitro cultures of mycorrhizal fungi (Rejali et al. 2003)

Fig. 11.5 Sampling and investigation of nodulation of lentil bushes in Ardabil Province (Afshari et al. 2007)

isolates to form nodules on lentil plantlets revealed that 45 from the 76 purified isolates were capable of nodulating the host roots. Investigation of the capability of the nodulating indigenous isolates to fix molecular nitrogen revealed that over 60% (51% effective and 11% very effective) of the isolates exhibited a rather high nitrogen-fixing capability. Some of the isolates led to 19–25% increases in the dry weight of the vegetative parts, 5–12% enhancement in the nitrogen content of the vegetative parts, and an enhanced yield in the range of 28–36% relative to the nitrogen and control treatments (Afshari et al. 2007). The future research in this area shall focus on obtaining isolates indigenous to the lentil fields, enrichment of the collection of indigenous isolates, evaluation of the efficiency of the isolates thus obtained under farm conditions, and, finally, identification of symbiotic isolates for each variety of lentil as well as the specific climatic and soil conditions in places where lentil is cultivated (Fig. 11.5).

11.4.6 Sulfur-Oxidizing Bacteria (*Thiobacillus*)

Calcareous soils constitute a major portion of agricultural land in the country. The high pH and abundance of Ca cations in these soils cause most nutrients (especially, phosphorus, iron, and zinc) to be fixed and unavailable to

Table 11.2 Effects of applying one ton of biofertilizers per hectare together with *Thiobacillus* inoculant on increased performance of soya, canola, wheat, and corn in different provinces (Anonymous 2012)

Location of farm experiment	Increased grain yield of soybean relative to control (%)	Increased yield of canola grains relative to control (%)	Location of farm experiment	Increased performance of wheat grains relative to control (%)	Increased performance of corn grains relative to control (%)
Alborz Province	18	–	Khorasan Razavi	10.93	11
Khorasan Razavi	21	–	East Azerbaijan	19.89	–
Mazandaran (Dasht Naz)	26	48.45	Fars (Eghlid)	11.3	–
Mazandaran (Baye Kala 1)	29	34.87	Fars (Zarghan)	40.22	5.45
Mazandaran (Baye Kala 2)	–	66.57	Semnan	39	–
Golestan (SWRI)	41	–	Kermanshah	3.38	8.63
Golestan (Gorgan)	26	1.5	Jiroft	–	16.65
Ardabil	16	4.67	Qazvin	–	2.26
Isfahan	6	–	Isfahan	25.63	5.26
Qom	–	42.6	Kerman	–	4.8
Khuzestan	–	13.93	Khuzestan	12.46	0.02
Mean	23	30.4	Mean	20.36	11

plants. As Iran produces over two million tons of elemental sulfur in its petroleum and natural gas refineries, there is a great potential for exploiting the abundant sulfur thus produced as an acid-generating substance toward improved plant nutrition. Exploiting this potential toward beneficial uses, however, can only be realized in the presence of *Thiobacillus* to oxidize sulfur in soil. This has encouraged much research on the bio-oxidation of sulfur in Iran. The research in this area in Iran dates back to late 1980s (Golchin 1987) with an effort to isolate *Thiobacillus* bacteria in soils and their utilization toward enhanced sulfur bio-oxidation in saline and alkaline soils. This was followed by other studies on the isolation and efficiency of sulfur-oxidizing bacteria (Yakhchali 1987). SWRI first started research in this field in 1996. Sulfur-oxidizing bacteria were isolated from many soil samples, purified, and characterized while their capacity for oxidizing elemental sulfur was evaluated, and the effective and efficient isolates were selected (Besharati et al. 1998; Kariminia et al. 2000). From the mid-1980s onward, research focused on the effects of applying *Thiobacillus* and sulfur on plant growth characteristics and absorption of soil micronutrients. The findings of most studies confirmed the positive effects of applying *Thiobacillus* and sulfur on both soil and plant characteristics. Considering the difficulty of long-term conservation of these bacteria as a result of both their nature and their activities, a number of studies were designed and implemented to develop different formulations that could be used for the long-term conservation of the bacteria and their marketing as biofertilizers. For this purpose, different kinds of materials that were both inexpensive and locally available were combined to develop materials that could serve as vehicles for the microorganisms. Investigations were also carried out to determine the stability of bacteria in these vehicle materials, and finally, the best and most appropriate formulations were selected (Besharati et al. 2003, 2007).

Some of the studies concentrated on the application of the biofertilizers on farms in different provinces across the country with a variety of crops including corn, soybean, canola, wheat, and cotton. The results obtained are summarized in Table 11.2 (Anonymous 2012).

The effective and efficient strains of these bacteria are currently conserved at the microbe culture collection of the SWRI, and the technologies developed for production of the *Thiobacillus* biofertilizer are being marketed to the private sector for mass production. Arrangements have been made for the 2015–16 farming season to deliver 300,000 packs (each 1 kg) of these biofertilizers to farmers (Figs. 11.6 and 11.7).

Isolation and identification of *Thiobacillus* bacteria indigenous to soils from different parts of Iran, evaluation of the capacity of the bacteria to oxidize elemental sulfur,

Fig. 11.6 Corn-sulfur project sites in Fars (left) and Kermanshah (right) Provinces (Anonymous 2012)

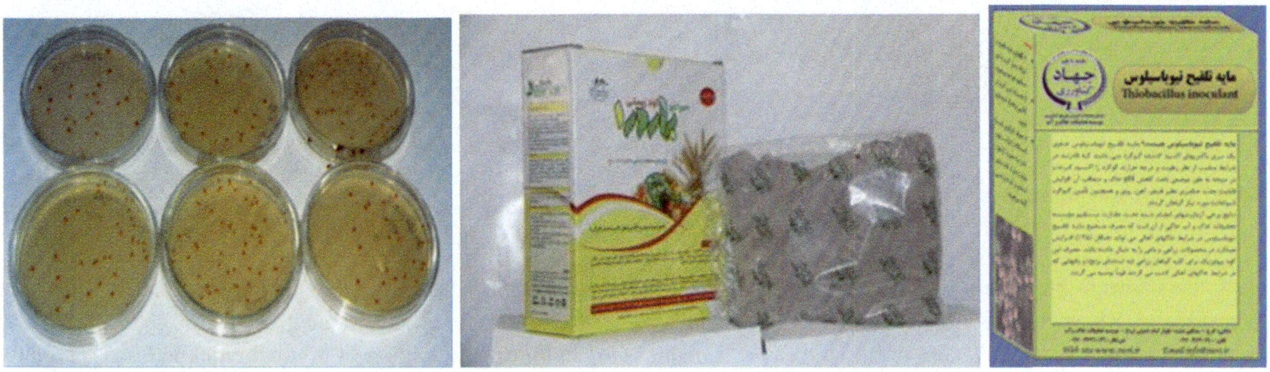

Fig. 11.7 *Thiobacillus* bacterial colonies on the Postgate-specific culture medium (left) (Besharati et al. 1998) and *Thiobacillus* inoculant manufactured by the local private sector (middle and right)

identification of the best inoculants-sulfur mixes, determination of the quantities and application stages of sulfur in soils with different buffering capacities, methods of facilitating their application for different crops, and investigations of the genetic manipulation of indigenous isolates and the consequent effects on other useful soil-living microorganisms are among the most important areas of future research.

11.4.7 Soybean

Soybean is one of the grain legumes that despite considerable cultivation areas is not native to Iran. Therefore, insufficient population of its symbiotic bacteria in soils is not unexpected. So that, in soybean cultivation, inoculation of soybean seeds with symbiotic bacteria is always recommended. In the 1995s, SWRI decided to produce the inoculants domestically and provide it for soybean farmers. Therefore, a series of targeted researches in this direction were carried out. In 1997, a large number of root nodule and soil samples were collected from soybean-cultivated lands in the north of the country. The population of *B. japonicum* bacteria in soil samples ranged from 1.07 to 5.86 logarithmic units per gram of soil samples (Khavazi et al. 2001). After achieving effective symbiotic isolates, *Rhizobium* biofertilizers were prepared, and their impact on growth and yield of soybean were studied. Considering the history of cultivation and diversity of climate, some experimental plots which had no soybean cultivation were considered, and field tests with four SWRI produced inoculants (IR-A, B, C, and D), two Russian inoculants (R-626 and R-634), an Indian inoculant (IND), an Italian inoculant (ITA), 120 kg N ha^{-1}, and control were carried out (Table 11.3).

In more than 60 ha soybean-cultivated fields of Lorestan Province, the results indicated that in fields where Italian inoculant (ITA) and SWRI produced inoculants (IR-D) were

Table 11.3 The soybean grain yield (kg/ha) related to different inoculants application at different field experiment sites

Experiment sites	R-626	R-634	IND	ITA	IR-A	IR-B	IR-C	IR-D	Fert	Control
Karaj	2100d	2625c	1825de	2710bc	1550ef	2920ab	3125a	3050a	1450f	1000g
Dezful	2890bc	2556bc	2920b	2890bc	2840bc	3720a	2780bc	2780bc	2500c	1445d
Jiroft	1275e	1400c	1280de	1570a	1585a	1520b	1350cd	1300de	1272e	825f
Gonbad	1770d	1865abc	1300e	1835bcd	1960a	1750d	1800cd	1935ab	1800cd	1000f
Sari1 (with soybean cultivation history)	3390de	3830cd	3280ef	3380de	3345de	2780f	4950b	6170a	4280c	3350ef
Sari2 (with soybean cultivation history)	2300b	2600ab	2200b	2180b	2800a	2600ab	2360b	2900a	1400c	1050c

Fig. 11.8 Soybean project sites in different provinces of Iran (Khavazi et al. 2001)

used, the soybean grain yield were 2172 and 2307 kg ha^{-1}, respectively (Khavazi et al. 2001) (Fig. 11.8). In another field experiment, the effectiveness of the soybean inoculants produced in the country was compared with the imported inoculants in Alborz, Fars, Mazandaran, Golestan, Lorestan, Ardabil, Chaharmahal and Bakhtiari, Isfahan, and Khorasan agricultural research centers. There was no significant difference among control and other inoculants (native and imported) (Afshari et al. 2012). Due to the acceptable efficiency of the domestic inoculants compared with imported ones and proof of their effectiveness in several trials, eventually, the technical know-how of soybean inoculant production in the country was registered. This knowledge was used for mass production, replacing imported inoculant with domestic ones, in the private sector on several times. Presently, superior strains with proper efficiency are held in SWRI microbial collection. Also, good quality inoculants are produced by private sector under license and supervision of SWRI (Fig. 11.14).

11.4.8 Chickpea

Among grain legume crops, chickpea had the highest cultivated area in the country. The main chickpea-cultivated lands are related to the West and North West regions of the country. In these areas, the plant is mostly cultivated under dryland conditions. Due to the prevailing conditions in the rainfed areas, the application of fertilizers is less than irrigated lands; therefore, the supply of nitrogen through biological nitrogen fixation in dryland chickpea is of particular importance. Researches on biological nitrogen fixation in chickpea began in 1996 at the Soil and Water Research Institute. From chickpea-cultivated lands of East Azerbaijan, West Azerbaijan, Kurdestan, Hamedan, Kermanshah, Fars, Lorestan, Ardabil, Khorasan, and Tehran Provinces, 400 samples of chickpea root nodules were collected, and about 200 isolates attributable to *Mesorhizobium ciceri* were isolated. Using plant infection test indicated that only 130 isolates could create nodule in chickpea plant roots. 76 isolates were tested to determine their symbiotic effectiveness (SE). From the point of chickpea shoot dry weight and nitrogen uptake in shoot, only about 40 and 10% of isolates were significantly different from control, respectively. To evaluate the biological nitrogen fixation efficiency of isolates, 16 bacteria were used in field trials. 2 strains from ICARDA and one strain from ICRISAT were obtained, and the rest 13 strains were isolated from chickpea-cultivated lands in different areas, of which 2 isolates were from West Azerbaijan (SWE6 and SWRI4), 3 isolates from Lorestan (SWRI1, SWRI11, and SWRI13), 2 isolates of Azerbaijan (SWRI5 and SWRI10), 2isolates of Kurdestan (SWRI7 and SWRI12), 2 isolates of Kermanshah (SWRI8 and SWRI14), and 2 isolates from Khorasan (SWRI9 and SWRI16). Field trials were conducted in 9 sites in Zanjan, Kurdestan, Kermanshah, Ilam, Khorasan, Azerbaijan (Tabriz), Azerbaijan

Fig. 11.9 Chickpea root nodes from chickpea-cultivated lands (left), chickpea biofertilizer which produced by private sector (middle), and field test of chickpea (right) (Asgharzadeh et al. 2007)

(Maragheh), and Markazi Provinces for one to three years (one year in some areas, in regions 2 and in some places 3 years).

According to the results of grain yield, in Lorestan and Ilam sites, none of the bacteria could increase grain yield relative to control, whereas in Khorasan and Azerbaijan (both in Maragheh and Tabriz), all bacteria increased grain yield compared to the control. In Zanjan, Kurdestan, Kermanshah, and Markazi, 5, 8, 6, and 8 bacteria increased grain yield, respectively. SWRI7 and SWRY11 bacteria in no region increased yield, while SWRI9 and SWRI5 as the best bacteria increased it in 8 sites (Asgharzadeh et al. 2007). After field assessment, genetic diversity of *Mesorhizobium species, indig*enous to Iran soils, was conducted (Asgharzadeh et al. 2013). Now, approximately 70 *Mesorhizobium strains* are stored in the Soil and Water Research Institute microbial collection. It should be noted that in addition to the research conducted at the Soil and Water Research Institute, many researches have been conducted in universities in terms of MSc. and Ph.D. thesis on biological nitrogen fixation in chickpea plants (e.g., Alimadadi et al. 2010; Tavasoli et al. 2011). Technical know-how derived from researches done in this regard for the mass production of biofertilizer containing *Mesorhizobium* was rented to private sector (Fig. 11.9). Preparing inoculants that have proper efficacy in soil and climatic conditions of the chickpea-cultivated areas, and also in the pea varieties that are cultivated in the country, are the research priorities in this area.

11.4.9 Alfalfa

Iran is the origin of alfalfa, and more than 600,000 ha of its agricultural land go under this crop. Therefore, its symbiotic bacteria exist naturally in soils of the country. Biological nitrogen fixation in alfalfa depends on plant properties (variety, cultivar, etc.), symbiotic bacteria (effectiveness, adaptability to environmental conditions, etc.), soil properties (pH, salinity, etc.), climate (temperature, rainfall, etc.), and farm management. Biological research on this crop has included identification of symbiotic bacteria with high N-fixing potential, determining the effect of environmental conditions (mainly drought) on biological nitrogen fixation, selecting appropriate cultivars as well as the right combination of alfalfa cultivars and bacterial strains, and also determining the role of farm management (especially fertilizer use), to improve biological nitrogen fixation potential. Examples of such research are those on selection of appropriate plant varieties and the right combination of varieties and strains (Panahpour 2008; Gholipoor et al. 2009). Other researches have attempted to obtain strains with high efficiency; the Symbiotic Effectiveness (SE) of 31 isolates from alfalfa-cultivated lands of Kerman Province was evaluated. SE of the isolates varied from 146% to 13.5%. 27, 23, 16, and 36% of isolates were highly effective, effective, partially effective, and non-effective, respectively (Ebrahimi and Akhgar 2014). The SE of 195 isolates from Hamedan Province indicated that 8.6, 65.8, and 25.6% of them had SE 50–75, 76–100, and more than 100%, respectively (Khavazi et al. 2003). 50 isolates obtained from perennial alfalfa-cultivated lands in the Provinces of Zanjan and Tehran has been assessed. The highest and lowest SE were 316.7 and 46.1%, respectively. 72, 16, 10, and 2% of bacteria were highly effective, effective, partially effective, and non-effective, respectively (Ghasem et al. 2010, 2012).

Due to drought, salinity, and pollution of heavy metals stresses in some alfalfa-cultivated lands in the country, researches have also been done to improve biological nitrogen fixation in these conditions. 80 bacteria indigenous to Kerman soils were evaluated for drought tolerance. Alfalfa inoculation with drought resistant strains increased nitrogenase activity, shoot dry weight, and proline content compared with sensitive strains (Abolhasani et al. 2010). Alfalfa cultivars Hamadani, Qareyonjeh, and Qarghaloq which were inoculated with salt-tolerant bacteria had no yield reduction under salt stress condition, while salt-sensitive bacteria did not have this capability (Fazaeli and Besharati 2012). These three varieties of alfalfa for resistance to salinity were the same, while Yazdi was more resistant to salinity compared with the Hamedani one. 42% of the bacteria isolated from saline soils could grow at 700 mM salinity (NaCl) (Ghasem et al. 2010). In general, the use of native rhizobia with high efficiency and tolerant to salinity, and also use of salinity resistant varieties, can be considered as a mechanism to cope with salinity stress in alfalfa-cultivated lands (Ghasem et al. 2010).

Fertilizer use management, in particular the type and amount of nitrogen fertilizer, on legume-cultivated land such as alfalfa, will have a great impact on biological nitrogen fixation. The evaluation of nutrient role in nitrogen fixation of alfalfa plants showed that good nutrient supply and soil fertility has a very important role in plant response to inoculation. With optimal nutrient supply, plant response to inoculation at 6 soils from 16.7, 3.6, 4.6, 4.1, 1.9, and 1% increased to 32, 4.6, 12.4, 37, 54.2, and 21.2%, respectively. Despite the occurrence of indigenous bacteria in the soil, inoculation increased alfalfa dry weight at least by 4.6% and a maximum of 54.2% (Khavazi et al. 2003).

The conclusion of researches in the biological nitrogen fixation in alfalfa suggests that selecting proper alfalfa varieties suitable to soil and climate conditions, inoculating alfalfa with symbiotic bacteria with high efficiency and resistance to stress (salinity, drought, pollution, etc.), determining the right combination of the strains and varieties, proper nutrients supply, and farm management (especially fertilizer usage) are the most important issues that should be considered to improve the biological nitrogen fixation potential in alfalfa-cultivated land in the country.

11.4.10 Soil Enzymes

Increasing attention to the issue of sustainable agriculture and soil quality conservation led to various criteria used in the assessment of soil quality. Among soil biological indices, soil enzymes react to changes and reflect them faster than other soil indicators, so they are useful indicators for assessing soil changes and considered as the most important criteria of soil quality. Soil enzymes activity and stability depend upon soil characteristics, type, and quantity of materials added to the soil, vegetative cover, and field management. Many researches have been done on these issues in the country.

The urease activity of 20 different soils in Isfahan region ranged from 5 to 79 μg N g^{-1} soil 2 h. Urease enzyme activity had the highest correlation with soil organic carbon, but no correlation observed with soil pH, SAR, CCE, CEC, and total soil fungi and bacteria (Noorbakhsh et al. 2001). L-asparaginase enzyme activity correlated with soil properties such as organic carbon and total N content, but no correlation observed with soil pH, CCE, CEC, and soil particle size distribution (Noorbakhsh et al. 2006). Application of sewage sludge in 10 calcareous soils of Chaharmahal and Bakhtiari indicated that alkaline phosphatase activity correlated with total phosphorus, organic phosphorus, and iron oxides bonded phosphorous (Raeisi and Hosseinpur 2014).

Soil organic matter and clay minerals by adsorbing enzymes are effective in their maintenance and stability. Soil organic carbon is the most important factor controlling the enzymes activity, and it is a reliable indicator to estimate soil enzyme activities. In addition to soil properties, vegetation or type of land use also affects the activity of enzymes.

Study of cellulase activity in pastures, broadleaf forests, conifer forests, dry land, and irrigated fields of Hamedan Province revealed that the enzyme activity in different land use ranged from 0.17 in dryland areas to 0.38 μg Glucose g^{-1} soil in broadleaf forest (Safari-Sanjani and Sharifi 2006). Acid phosphatase activity in hazelnut habitats in Ardabil, Gilan, and Alborz Provinces in summer was 2.5 times more than spring while alkaline phosphatase activity was 2 times or less than it in three mentioned habitats (Moraghebi et al. 2012). During spring and fall seasons, the activity of acid phosphatase, alkaline phosphatase, and dehydrogenase in juniper habitats of Khorasan, Golestan, Ardabil and Chaharmahal, and Bakhtiari under shelter were more than in open air (Matinizadeh et al. 2010). The activity of acid phosphatase, alkaline phosphatase, urease, and dehydrogenase in non-grazed juniper forest habitats condition was more than in grazed condition (Matinizadeh et al. 2012).

Field management such as fertilizers application and residue management are the factors affecting enzyme activity in soils. In corn-cultivated calcareous soils, chicken manure was more effective than urea in the activity of urease, acid phosphatase, alkaline phosphatase, and sucrase enzymes, because of microbial activity stimulation (Fereidooni Naghani et al. 2009).

Application of sewage sludge on basil-cultivated lands increased acid and alkaline phosphatase activity, and chemical fertilizers usage along with sewage sludge intensified enzymes activity (Dehghan et al. 2012). Application of phosphate fertilizer and saline water in clover reduced acid and alkaline phosphatase activity (Ghollarata et al. 2009).

Fig. 11.10 Fla-Wheat (a biofertilizer that contains PGPRs) produced by a local private sector (left) and its know-how certificate (right)

Evaluation of the role of farm management (barley residue application, burning residue, tillage, and use of nitrogen fertilizer) showed that residue burning and tillage reduced the activity of alkaline phosphatase and urease. No-tillage and residue maintenance were the most effective management to maintain and increase enzymes activity (Hosseini et al. 2012). Water logging stress in mandarin orchards in soils of east Mazandaran increased catalase activity (Asadi Kangarshahi et al. 2013).

11.4.11 Plant Growth-Promoting Rhizobacteria

Plant growth promoting rhizobacteria are symbiotic and/or non-symbiotic heterogeneous group of bacteria in the plants rhizosphere, which are capable of increasing plant growth directly or indirectly through one or more specific mechanisms. A wide variety of soil bacteria can be plant growth-promoting rhizobacteria, but PGPR bacteria mainly belonged to *Pseudomonas, Bacillus, Azospirillum, Azotobacter, and Flavobacterum* genera. Given the wide range of PGPR bacteria, diversity of target plants, their distribution in different climate and soil conditions, and plural mechanisms of their impact, many studies have been done about them over the world. Also, in Iran, many research efforts have been conducted across the nation as institutionally funded research projects or graduate student theses and dissertations whose objectives were to investigate potential uses of plant growth-promoting rhizobacteria on different plants and issues. Including germination, growth, and yield of corn (Nezarat and Gholami 2011), growth rate, growth, yield, and yield component of wheat (Zabihi et al. 2009), catalase, superoxide dismutase, and glutathione peroxidase activities, growth, growth parameters, and levels of micronutrients in salinity in canola (Soltani et al. 2014), nitrogen uptake and wheat grain yield (Mohammadi et al. 2010), forage and grain sorghum (Keshavarz et al. 2011), the quantitative and qualitative characteristics of sesame (Jahan et al. 2013), on nitrogen fixation and growth of alfalfa plant (Ebrahimi and Akhgar 2014), yield and quality of potatoes (Behbood et al. 2012), and yield and essence of saffron (Rasouli et al. 2015) that in most above mentioned cases, positive effect on the measured indices have been reported.

In a study that was conducted at the Soil and Water Research Institute, wheat rhizosphere soil samples from different provinces were collected, *Pseudomonas, Azospirillum* and *Flavobacterium* were isolated, and then, their plant growth-promoting activities were evaluated. The bacteria with significant growth-promoting ability in greenhouse and then in wheat fields in the Provinces of Fars, Kermanshah, Mazandaran, Semnan, Khorasan, and Khuzestan were evaluated. Isolates used in most provinces increased growth and yield of wheat crop. Finally, the more effective bacteria were evaluated in wheat-cultivated farms in 28 provinces, and it was found that, in 25 provinces, wheat grain yield increased by 15% compared with non-inoculated control. The effective bacteria have been formulated as biofertilizer, and the technologies developed for mass production of biofertilizer are being marketed to the private sector (Fig. 11.10).

11.5 Useful Soil-Living Microorganisms and Their Characteristics in Different Regions

Over the past decade, considerable number of microorganisms have been isolated, screened, and evaluated within the framework of research projects as well as graduate theses and dissertations. Some have been already prepared as biofertilizers or marketed to the private sector for mass production. Table 11.4 provides some of the microorganisms and their characteristics presently preserved at the culture collection of SWRI. No doubt, the availability of such a collection is part and parcel of a prerequisite to any

Table 11.4 Characteristics of microorganisms preserved at the culture collection of SWRI

Form of deposit	Place of isolation	Property	No. of isolates in culture collection	Microorganism	Row
Slant, Lyophilized vials	Different provinces	Ability of symbiotic nitrogen fixation	60	Bean symbiotic rhizobia	1
Slant, Lyophilized vials	Mazandaran, Golestan, Lorestan, Khuzestan	Ability of symbiotic nitrogen fixation	50	Soybean symbiotic rhizobia	2
Slant, Lyophilized vials	Different provinces	Ability of symbiotic nitrogen fixation	72	Chickpea symbiotic rhizobia	3
Slant, Lyophilized vials	Wheat rhizosphere from different provinces	Plant growth-promoting properties	40	*Flavobacterium*	4
Slant	Different provinces	Free-living nitrogen-fixing and plant growth-promoting properties	10	*Azotobacter*	5
Slant	Lorestan, Hamedan, Mazandaran	Ability of symbiotic nitrogen fixation	30	Clover symbiotic rhizobia	6
Slant, Lyophilized vials	Ardabil	Ability of symbiotic nitrogen fixation	120	Lentil symbiotic rhizobia	7
Slant, Lyophilized vials	Gilan, Fars, Mazandaran, Kermanshah	The ability to dissolve insoluble organic and inorganic phosphates	300	Phosphate-solubilizing bacteria	8
Slant, Lyophilized vials	Gilan, Mazandaran, Fars, Kermanshah	The ability to dissolve insoluble organic and inorganic phosphates	350	Phosphate-solubilizing Fungi	9
Slant, Lyophilized vials	Wheat, canola and rice rhizosphere from different provinces	Plant growth-promoting properties	209	Fluorescent pseudomonads	10
Lyophilized vials	Various agricultural residues	Decomposing of cellulose containing organic matter	8	Cellulolytic bacteria	11
Slant, Lyophilized vials	Khuzestan, Bushehr	The ability to use aromatic and aliphatic compounds as substrate	90	Petroleum decomposing bacteria	12
Slant, Lyophilized vials	Hamedan, Zanjan, Tehran	Ability of symbiotic nitrogen fixation	210	Alfalfa symbiotic rhizobia	13
Slant, Lyophilized vials, Carriers	Gilan, West Azerbaijan, Kerman, Kermanshah, Lorestan, Tehran, Alborz	Ability of elemental sulfur oxidation	40	Sulfur-oxidizing bacteria	14
Slant, Lyophilized vials		Making root proliferation and root initiation in tissue culture	4	Agrobacterium	15
Slant, Lyophilized vials	Different provinces	Plant growth-promoting properties	80	Bacillus	16
Spore, Carriers	East Azerbaijan Fars, Khorasan, Golestan	Forming mycorrhiza symbiosis with plants and improve plant growth by different mechanisms	5	Mycorrhizal fungi	17

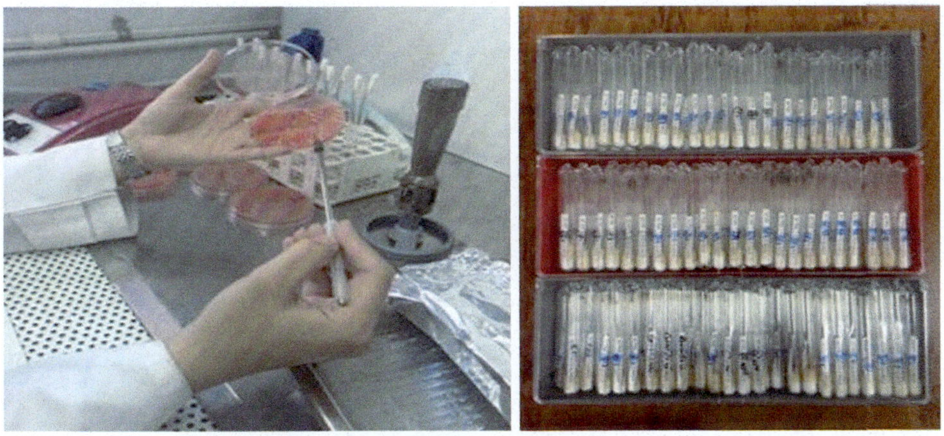

Fig. 11.11 Culturing and isolation of symbiotic rhizobia associated with leguminous plants at SWRI (left) and Lyophilizedvials of microorganisms collected at the Culture Collection of Soil Microorganisms (CCSM) (right)

Fig. 11.12 View from the workshop (left) and storehouse (right) at the biofertilizer producing private firm in Iran

research work that aims at application of biology and biotechnology in the fields of agriculture and natural resources. It is hoped that the enrichment and further development of this culture collection toward a general microbial culture collection and its exchanges with other collections at the international level will lead to improved and enhanced exploitation of soil biological potentials toward the development of agriculture and conservation of natural resources at the national level (Fig. 11.11).

11.6 Capability and Potentials for Biofertilizer Production

The technologies for preparing most of the required *biofertilizers* are locally available. The availability of such indigenous technologies that are the outcome of intensive and targeted research in many academic and research institutions across the nation prepares the grounds for enhanced production and application of these biofertilizers. Up until now, 13 technologies have been patented, some of which have been marketed to the private sector (Besharati and Khavazi 2011).

The private sector is in possession of the equipment and the competence to produce and market *biofertilizers*. The diversity of production plants and highly equipped private firms involved in the production of biofertilizers are the strengths that can be exploited toward further development of soil biology in Iran. Over 30 registered firms are presently active in the field, some of which have gained the required experience and qualifications for producing biofertilizers (Besharati and Khavazi 2011) (Fig. 11.12).

Farms and orchards are potential markets for *biofertilizers*. The considerably vast area under legume and cereal cultivation in Iran provides the opportunity for applying the useful microorganism to beneficial uses. Around 1.8 million ha of land farms both under irrigated and dry land farming system are under legume cultivation; these plants are capable of establishing symbiotic association with nitrogen-fixing rhizobia. On the other hand, over 50% of the total land farm is under cereal cultivation, the most important of which is

Fig. 11.13 Abundant perlite supplies (left), sulfur (middle), and rock phosphate (right) used in biofertilizers production

Fig. 11.14 Samples of biofertilizers produced using the technologies emerging from the research conducted by the private sector in Iran

wheat as a strategic crop that accounts for the largest cultivated area and the greatest share of agricultural production. Research has shown that PGPR bacteria employ a variety of mechanisms to improve wheat growth and performance. This provides the opportunity for investigating, producing, and applying rhizobial biofertilizers as well as those containing PGPR (Besharati and Khavazi 2011).

The raw materials required for producing *biofertilizers* are available in local markets. Availability of raw materials required for producing biofertilizers or for use in pilot research projects creates an opportunity for the production and application of these fertilizers. In most countries, peat is used as a preservative in biofertilizers. Peat supplies in Iran are limited and of very low quality. Instead, there are plenty of perlite, rock phosphate, sulfur, and mineral (vermiculite, bentonite, etc.) supplies of high quality which are capable of being used in different types of biofertilizers (Besharati and Khavazi 2011) (Fig. 11.13).

Legal support for growing production and application of *biofertilizers*. There is a good collection of legislation and acts that support the production and consumption of biofertilizers and organic fertilizers while they also require due attention to and investment in the biological potentials of soil for producing healthy agricultural products and preserving basic production resources. These supports provide useful opportunities for producers and consumers alike. One such legal Item is Paragraph H under Item 143 of the Fifth Five-Year National Development Plan as well as the Act for Enhanced Productivity in Agriculture and Natural Resources (Besharati and Khavazi 2011) (Fig. 11.14).

11.7 Priorities of Soil Biological Research in Iran

The main future research priority areas in soil biology and biotechnology in Iran may be summarized as following:

1. Monitoring the quality of soil biological resources (including variations in populations, activities, diversity, and proportions of soil-living organisms, especially the useful ones), exploitation of research findings toward management of soil biological colonies and communities, creation of a national microbial culture collection, and establishing relations with similar institutions around the world.
2. Promoting public attitude and raising awareness among the public and the beneficiaries on the importance and significance of management and exploitation of soil biological resources, their roles in both healthy food

products and public health, and their contributions to reduced use of agrochemicals and conservation of soil and water resources

3. The requirement to draft and pass necessary regulations and laws and develop standards and procedures on the preservation, enhancement, and exploitation of soil biological resources.
4. Creating the infrastructure required for the production and consumption of soil biological resources, and supervision and monitoring of manufacturing processes aimed at preservation of basic resources, production of healthy food products, maintaining public health, and reducing the consumption of agrochemicals and relaxing plant stresses and conducting research to investigate the facilitated use and application of microorganisms in agricultural production (e.g., covering seeds with vermiculite);
5. Preparing the grounds for the qualitative and quantitative enhancement of soil biological communities (consumption of organic materials in soil, returning plant remains into soil, pollutant control and removal, correction of soil physical characteristics as by drainage, etc.)
6. Paying due heed to biotechnology and microbial manipulations in an attempt to identify newly formed strains of desirable quality and characteristics, production of plant growth stimulating bacteria or soil and water purifying bacteria, production of transgenic bacteria and fungi for improving compost production, and also, employment of new techniques such as identification of phospholipidic fatty acids to determine diversity and frequency of microbial communities in ecosystems.
7. Investigation and monitoring of microbial communities as a result of global warming, droughts, soil salinization, improper application of chemical fertilizers, soil pollution, changes in vegetative cover, etc. Management of soil biota in order to combat phenomena such as global warming, salinization of soil and water resources, etc.

References

Abolhasani ZM, Lakzian A, Gholamhossien PJA, Akhgar A (2010) Effect of water stress on nitrogen fixation of *Sinorhizobium* and compatible solutes accumulation in Alfalfa (*Medicago sativa cv. Bami*). J Water Soil 24(3):407–416

Afshari M, Soudi MR, Nouhi A (2006) Providing Rhizobial inoculant from native strains of Rhizobium on common bean. J Sci Uni Tehran 32(1):1–9

Afshari M, Arzanesh MH, Khalafi H, Alizadeh A (2007) Evaluation of the efficiency of indigenous *Rhizobium legominosarum bv. Viciae* in symbiosis with Lentil in soils of Ardabil province. Final Research Report, No. 85/1309. Soil and Water Research institute. Karaj, Iran

Afshari M, Rejali FM, Asadi-jelodar A, Rezai M, Khavazi K, Asadi H, Hemmati V, Shamshiripour M, Alizadeh N, Esmaili-zad A, Saghafi K (2012) Evaluation of soybean inoculants efficiency on increasing soybean yield and biological nitrogen fixation. Final Research Report, No. 41282. Soil and Water Research Institute. Karaj, Iran

Aliabadi-Farahani H, Valadabadi SAR (2010) The role of arbuscular mycorrhizal fungi on the medicinal plant coriander (*Coriandrum sativum* L.) under drought stress conditions. Iranian J Soil Res 24 (1):69–79

Aliasgharzad N (1991) The abundance and distribution of arbuscular mycorrhizal fungi in saline soils of the Tabriz Plain and their inoculation effects on the improvement of salt tolerance in onion and barley. Ph.D. Dissertation, Soil Science Department Faculty of Agriculture, University of Tehran, Karaj, Iran

Aliasgharzad N, Hajiboland R, Olsson PA (2011) Lack of arbuscular mycorrhizal colonisation in tea (*Camellia sinensis* L.) plants cultivated in Northern Iran. Symbiosis 55(2):91–95

Alimadadi A, Jahansouz MR, Besharati H, Tavakkol-Afshari R (2010) Evaluating the effects of biofertilizers and seed priming on chickpea (*Cicer arietinum* L.) seed quality. Int J Food Agri Environ-JFAE-89–10

Alizadeh-Oskoui P, Aliasgharzad N, Shariatmadari H, Asgharzadeh A, Baghban-Sirous Sh (2009) Effects of two species of arbuscular mycorrhizal fungi in reducing the toxicity of cadmium in tomato plants with different levels of phosphorus. Iranian J Soil Res 23 (2):217–228

Anonymous (2012) Evaluation the effect of sulfur and sulfur—oxidizing bacteria on nutrient uptake, yield of some selected crops and physical, chemical and biological properties of soil. Final Report of Research Plan, Soil and Water Research Institute and National Iranian Gas Company (NIGC), Research and Technology Management (RTM), Tehran, Iran

Asadi Kangarshahi A, Savaghebi GhR, Samar SM, Akhlaghi Amiri N (2013) Effects of waterlogging stress on catalase activity and chlorophyll reading change trend in Satsuma mandarin with different rootstocks in soils of East Mazandaran. Iranian J Water Soil Res 44(3):299–309

Asadi H, Khodshenas MA, Hemati A, Tokasi M, Kalhor M, Mehnatkesh AM (2005) Improving the yield of inoculated common bean using supplementary N-fertilizers. Final Report of Research Plan, No. 89/13. Soil and Water Research Institute. Karaj, Iran

Asadi H, Khavazi K, Asgharzadeh A, Rejali F, Afshari M (2011) Biofertilizers in Iran: opportunity and challenges. 1st Iranian Fertilizer Challenges Congress: Half a Century of the Fertilizer Consumption. Tehran, Iran

Asgharzadeh A, Afshari M, Tavasoli A, Toshih V, Daneshi N, Ganjehie M, Rostami A, Soleimani R, Mahmudi H, Asadian MS, Kalhor M, Belson W, Asadi H, Khavazi K, Hemmati V, Alizadeh N, Shamshiripour M (2007) Decreasing the utilization of nitrogen fertilizers through improvement the biological nitrogen fixation (BNF) in chickpea (*Cicer arietinum*) cropping area. Final report of Research No.86/567. Soil and Water Research Institute. Karaj, Iran

Asgharzadeh A, Rafiee Dolat Abady M, Kargar M, Asadi H (2013) Genetic diversity of chickpea –nodulating *Mesorhizobium* species native to Iranian soils. Iranian J Soil Res 27(1):121–129

Balali-Aliabadi M, Kianmehr H, Mehravaran H, Hajian-Shahri M (2001) The introduction of 6 new species of vesicular-arbuscular mycorrhizal (VAM) fungi from Khorasan (Mashhad). J Sci Khwarizmi Univ 1(2):68–80

Banai MH, Moameni A, Bybordi M, Malakouti MJ (2005) Thesoils of Iran. Soil and Water Research Institute, Karaj, Iran

Behbood M, Golchin A, Besharati H (2012) The Effects of phosphorus and inoculation with plant growth promoting rhizobacter (PGPR), *Pseudomonas fluorescent*, on yield and quality of potato tuber (*Agria* Cultivar). J Water Soil 26(2):260–271

Besharati H, Fallah AR (2008) The study of occurrence of sulfur oxidizing microorganisms and sulfur oxidation potential in agricultural soils of four provinces of Iran. Final Report of Research, No. 1541. Soil and Water Research Institute, Karaj, Iran

Besharati H, Khavazi K (2011) Use of native potential for producing of biological and organic fertilizers and production of safe agricultural products. 1st Iranian Fertilizer Challenges Congress: Half a Century of the Fertilizer Consumption. Tehran, Iran

Besharati H, Saleh-Rastin N, Ardalan M, Yakhchali B (1998) Effects of sulfur application and *Thiobacillus* inoculation on increasing of nutrient availability in soil. Ms. Thesis, University of Tehran, Faculty of Agriculture, Department of Soil Science. Karaj, Iran

Besharati H, Saleh-Rastin N, Malakouti MJ, Alizadeh A (2003) A new carrier for *Thiobacillus* bacteria and their interactive effects with Arbuscular Mycorrhizal fungi on yield and nutrient uptake of wheat crop. Ph.D. Thesis, University of Tarbiat Modarres, Faculty of Agriculture, Department of Soil Science. Tehran, Iran

Besharati H, Atashnama K, Hatami S (2007) Biosuper as a phosphate fertilizer in a calcareous soil with low available phosphorus. Afr J Biotech 6(11):1325–1329

Biyari A, Gholami A, Rahmani HA (2011) Effect of Different Plant Growth Promotion Bacteria (*Azotobacter, Azospirillum*) on Growth Parameters and Yield of Field Maize. J Water Soil 25(1):1–10

Daei G, Ardekani MR, Rejali F, Teimuri S, Miransari M (2009) Alleviation of salinity stress on wheat yield, yield components, and nutrient uptake using arbuscular mycorrhizal fungi under field conditions. J Plant Physiol 166:617–625

Dehghan MH, Bahmanyar MA, Lakzian A, Salek Gilani S (2012) Effect of application of sewage sludge and sewage sludge enriched with chemical fertilizer on the rate of organic carbon, respiration and enzyme activity of soil under Basil cultivation. Iranian J Water Soil 26(3):554–562

Ebrahimi M, Akhgar AR (2014) Effect of co-inoculation of *Sinorhizobium* sp. and plant growth-promoting rhizobacteria on N-fixation and growth of alfalfa. J Soil Manage Sustainable Prod 4(1):183–199

Ebrahimzadeh SA, Aliasgharzad N, Najafi N (2013) Effect of land use changes on soil enzyme activity in Sulduz plain (Naqadeh-West Azarbaijan). J Soil Manage Sustainable Prod. 3(2):133–149. (http://ejsms.gau.ac.ir)

Esbati M, Akhavansepahy A, Asgharzadeh A, Khosrowchahli A (2014) Isolation, identification and evaluation of *Azospirillum* inoculation on tomato growth under greenhouse condition. Iranian J Soil Biol 2 (1):43–54

Fallah AR, Rahimian H, Saleh-Rastin N, Malakouti MJ (2004) Distribution of phosphate solubilizing microorganisms in some soils of Gilan province. Iranian J Soil Water Sci 17(2):162–176

Farajzadeh D, Yakhchali B, Aliasgharzad N, Sokhandan-Bashir N, Farajzadeh M (2012) Plant growth promoting characterization of indigenous *Azotobacter* isolated from soils in Iran. Curr Microbiol 64:397–403

Fazaeli A, Besharati H (2012) Effect of salinity on some growth indices and total protein content of alfalfa genotypes inoculated with *Sinorhizobium meliloti* strains under greenhouse conditions. Iranian J Sci Technol Greenhouse Cult 3(9):25–38

Fereidooni Naghani M, Raiesi F, Fallah S (2009) The effect of nitrogen source and level on the trend of enzymatic activity in a calcareous soil cultivated with forage maize. Iranian J Water Soil 23(4):127–136

Ghasem F, Poustini K, Besharati H, Mohammadi V (2010) Evaluation of *Sinorhizobium meliloti* isolates and biological nitrogen fixation and growth in Alfalfa (*Medicago sativa*) cultivars under salt stress. Ph.D. Thesis, University of Tehran, College of Agriculture and Natural Resources, Faculty of Agricultural Science and Engineering, Department of Agronomy. Karaj, Iran

Ghasem F, Pustini K, Besharati H, Mohammadi VA, Goettfert M (2012) Pre-inoculation of Sinorhizobium meliloti with luteolin, methyl jasmunate and genistein affecting alfalfa (medicago sativa) growth, nodulation and nitrogen fixation under salt stress. J Agric Sci Technol 6(14):1255–1264

Gholipoor H, Nabizadeh A, Nasrolahzadeh A (2009) A scrutiny on the dimension of nitrogen fixation in annual medics (*M. rigidula and M. truncatula*) by native rhizobium strains in greenhouse conditions. J Res Crop Sci 2(2):45–54

Ghollarata M, Raiesi F, Nadian H (2009) Investigation the effects of salinity and soil phosphorus on respiration, microbial biomass and phosphatase activity in the rhizosphere of berseem clover (*Trifolium alexanderinum* L.). Iranian J Soil Res 22(2):217–223

Golchin A (1987) Biological oxidation of sulfur in saline and alkaline soils and enhancing oxidation level by inoculation with authotrophic bacteria. MSc. Thesis, Soil Science Department Faculty of Agriculture, Tehran University, Karaj, Iran

Gorji-Anari M, Refahi HGh, Alikhani HA (2007) Effects of manure and biofertilizer (*Rhizobium*) application on lentil production. Iranian J Agric Sci 38(3):305–311

Hajeeboland R, Asgharzadeh N, Mehrfar Z (2004) Ecological study of *Azotobacter* in two pasture lands of the north-west Iran and its inoculation effect on growth and mineral nutrition of wheat (*Triticum aestivum* L. cv. *Omid*) Plants. J Sci Technol Agric Nat Resour Water Soil Sci 8(2):75–90

Hosseini MS, Haghnia Gh, Lakzian A, Emami H (2012) Short-term effects of barley residue management on urease and alkaline phosphatase activities. Iranian J Water Soil 26(3):545–553

Hosseinpour M, Pirzad A, Habibi H, Fotokian MH (2011) Effect of biological nitrogen fertilizer (*Azotobacter*) and plant density on yield, yield components and essential oil of Anise. Iranian J Sustain Agric Prod Sci 2(21):69–79

Jahan M, Aryaee M, Amiri MB, Ehyae HR (2013) The effect of plant growth promoting rhizobacteria (PGPR) on quantitative and qualitative characteristics of *Sesamum indicum* L. with application of cover crops of *Lathyrus* sp. and Persian clover (*Trifolium resopinatum* L.). Iranian J Agroecol 5(1):1–15

Kariminia A, Saleh-Rastin N, Yakhchali B (2000) Identification the strains of *Thiobacillus* indigenous to Iranian soils and assess their impact on reducing soil pH. Ms. Thesis, Faculty of Agriculture, Soil Science Department, Tarbiat Modarres University, Tehran, Iran

Keshavarz AR, Chaichi M, Alipour Jahangiri A, Ansari Jovini M, Moghaddam H, Ehteshami SM, Khavazi K (2011) Effects of foliar application of plant growth promoting *Rhizobacteria* on forage and grain yield of forage sorghum (var. Speedfeed). Iranian J Field Crops Sci 42(3):575–584

Khalaj MA, Moshiri F, Asadi Rahmani H (2013) Evaluation of the N_2-fixing ability of Rhizobia strains in common bean cultivated region of Qazvin. J Water Soil 27(1):54–60

Khavazi K, Rejali F, Arzanesh MH, Khademi Z, Ghalebi S, Daryashenas AM, Charati-araee A, Sepahvand M, Asadi H, Besharati H, Fallah AR, Rezaee H, Ghaderi J, Afshari M, Khosravi H (2001) A comparison between soybean inoculants produced by SWRI and some imported kinds. Final Report of Research, No. 1112. Soil and Water Research institute. Karaj, Iran

Khavazi K, Rahimian H, Malakouti MJ, Saleh-Rastin N (2003) Evaluation of the soil nutrient status and *Sinorhizobium meliloti* population densities, symbiotic effectiveness and their symbiotic nitrogen fixing potentials in the alfalfa fields of Hamadan province. Ph.D. Thesis, Faculty of Agriculture, Soil Science Department, Tarbiat Modarres University, Tehran, Iran

Khosravi H (1997) The occurrence, distribution and some physiological characterizations of *Azotobacter chroococcum* in crop fields areas. Ms. Thesis. Soil Sci. Department, Faculty of Agriculture, Tehran University

Khosravi H, Mahmoudi H (2013) Evaluation of effects of *Azotobacter* inoculation and manure on growth of rainfed wheat. J Soil Manage Sustain Prod 3(2):205–219

Khosravi H, Mirzashahi K, Kalhor M, Ramazanpour MR, Mirrasouli E, Rejali F, Sadjadi H, Karimiun MH, Kheiravaran M, Hajiabadi H (2005) Achieving the know how for production of fababean *Rhizobium* inoculant. Final Report of National Research Design, No. 84/631. Soil and Water Research Institute. Karaj, Iran

Khosravi H, Samar SM, Fallahi E, Davoodi H, Shahabian M (2009) Inoculation of 'Golden Delicious' apple trees on M9 rootstock with *Azotobacter* improves nutrient uptake and growth indices. J Plant Nutr 32:946–953

Khosravi H, Tavasoli AR, Sedri MH, Ziaeain AH, Zabihi HR, Montazeri EZ (2015) Effects of inoculation of native *Azotobacter* on yield and growth indices of wheat in Iran. J Soil Biol 2(2):149–157

Matinizadeh M, Khoshnevis M, Teimouri M, Ghasemi MH (2010) Impact of canopy and seasoning on activity of soil enzymes in some Juniper habitats of Iran. Iranian J For Poplar Res 18(1):152–162

Matinizadeh M, Khoshnevis M, Teimouri M, Shirvany A (2012) The effect of preservation on soil enzymes activities in *Juniperus excelsa* plantation of Chahartagh. Iranian J Water Soil Resour Conserv 1 (4):45–52

Miransari M, Bahrami HA, Rejali F, Malakouti MJ (2009) Effects of soil compaction and arbuscular mycorrhiza on corn (*Zea mays* L.) nutrient uptake. Soil Tillage Res 103:282–290

Mohammadi R, Olamaee M, Ghorbani NR, Chakeralhossaini MR (2010) Effects of urea fertilizer, organic matter and plant growth promoting Rhizobacteria on N uptake and yield of wheat (*Triticum aestivum C.V Alvand*). J Plant Prod 17(2):77–92

Moraghebi F, Matinizadeh M, Khanjani Shiraz B, Teimoori M, Afdide F (2012) Comparison of acid phosphatase and alkaline phosphates activity in soil of three natural and planted habitats of hazel in two seasons. Iranian J Forest Poplar Res 20(2):233–242

Nezarat S, Gholami A (2011) The effects of plant growth promoting rhizobacteria (*Azospirillum, Pseudomonas*) on growth and yield of corn (*Zea mays* L.). Iranian Agron J (Pajouhesh & Sazandegi) 91:44–51

Noorbakhsh F, Hajrasuliha S, Emtiazy G (2001) Factors affecting, urease enzyme activity in some soils in Isfahan province. J Sci Technol Agric Nat Resour Water Soil Sci 5(3):95–106

Noorbakhsh F, Shariatmadari H, Rezai A (2006) Using path analysis to interpret relationships of L-asparaginase and urease enzyme with some soil properties. Sci J Agric 29(2):117–126

Panahpour H (2008) Effects of *Sinorhizobium* strains inoculation on seedling growth and establishment of alfalfa (*Medicago sativa* L.) in arid and semi-arid regions of Iran. Iranian J Range Desert Res 15 (1):129–138

Raeisi T, Hosseinpur A (2014) Evaluation of alkaline phosphatase activity and availability of various P fractions for bean (*Phaseolus vulgaris*) in some calcareous soils amended with municipal sewage sludge. J Sci Technol Greenhouse Cult 5(18):39–50

Rasouli Z, Farahani SM, Besharati H (2015) Saffron (*Crocus sativus* L.) yield as affected by different fertilizing systems. Iranian J Med Aromat Plants 31(2):204–219

Rejali F, Tavasoli A, Hemati V, Esmaelizadeh A, Saghafi K (2003) In-vitro production of arbuscular mycorrhizal inoculant and the study of its effects on increasing plant growth and yield under salinity and drought stresses. Final Report of Research, No. 86.289. Soil and Water Research Institute. Karaj, Iran

Rejali F, Hemati V, Shamshiripour M, Saghafi K, Mohammadi Z, Derakhshandeh F, Kheiravaran M, Alizadeh A, Nejati A, Daee G (2004a) Identification of indigenous arbuscular mycorrhizal fungi and determination of their ability to colonize dry land wheat. Final Report of Research, No. 89.741. Soil and Water Research Institute. Karaj, Iran

Rejali F, Hemati V, Shamshiripour M, Saghafi K (2004b) The effects of different species of arbuscular mycorrhizal fungi in absorption of macro and micro nutrients in wheat. Final Report of Research, No. 88.1019. Soil and Water Research Institute. Karaj, Iran

Sadravi M (2005) Arbuscular mycorrhizal fungi of wheat fields in the Golestan province. Iranian J Bot 7(2):129–136

Safari-Sanjani AA, Sharifi Z (2006) The outcome of some soil properties on cellulase activity in a number of Hamadan soils. Iranian J Agric Sci 37(7):654–652

Saleh-Rastin N (2005) Sustainable management from the perspective of soil biology. In: Khavazi K et al (ed) Necessity for production biofertilizers in Iran. Soil and Water Research Institute, Karaj, Iran

Shahbazi K, Besharati H (2014) Overview of agricultural soil fertility status of Iran. Iranian J Land Manage 1(1):1–15

Soltani TAA, Abbaszadeh Dahaji P, Rejali F, Akhgar AR (2014) Effect of soil physical and chemical properties on common bean root colonization, spore number and infection potential of arbuscular mycorrhizae. Iranian J Soil Biol 2(1):65–78

Tavasoli AR, Aliasgharzad N (2009) Effect of arbuscular mycorrhizal fungi on nutrient uptake and onion yield in a saline soil at field conditions. Iranian J Water Soil Sci 19(1):145–158

Tavasoli A, Aliasgharzad N, Salehi GR, Mardi M, Asgharzadeh A, Akbarivala S (2011) Effects of co-inoculation with arbuscular mycorrhizal fungi and rhizobia on fungal occupancy in chickpea root and nodule determined by Real-Time PCR. Curr Microbiol 63:107–114

Yakhchali B (1987) Reproduction and maintenance of sulfur oxidizing bacteria. Final Report of Research. Research and Scientific Services Center, Ministry of Petroleum, Tehran, Iran

Zabihi HR, Savagebi GR, Khavazi K, Ganjali A (2009) Response of wheat growth and yield to application of plant growth promoting rhizobacteria at various levels of phosphorus fertilization. Iranian J Field Crops Res 7(1):41–51

Zarei M, Saleh-Rastin N, Savaghebi Gh (2011) Effectiveness of arbuscular mycorrhizal fungi in phytoremediation of Zinc polluted soils using maize (*Zea mays* L.). J Sci Technol Agric Nat Res Water Soil Sci 15(55):151–168

Author Biographies

Dr. Hossein Besharati is Professor of soil biology at the Soil and Water Research Institute in the Biology Department and member of the Iranian Soil Science Society. His main research interests are sulfur oxidation and soil microbiology, especially sulfur-oxidizing bacteria. He is one of the laureates of the 28th Khwarizmi International Award.

Dr. Naser Aliasgharzad is Professor of soil biology and biotechnology at the Department of Soil Science and Engineering, University of Tabriz, Iran. He is also resident coordinator of SAFE NETWORK in Iran. His research interests focus on the impact of arbuscular mycorrhizal symbiosis on improvement of crop production, fungal glomalin detection by using monoclonal antibody, and assessment of soil microbial communities by analyzing lipid biomarkers.

Dr. Kazem Khavazi is Professor of soil microbiology and researcher at Department of Soil Biology, Soil and Water Research Institute, Karaj, Iran. His research interests are biological nitrogen fixation, especially forage legumes, and plant growth-promoting rhizobacteria.

Dr. Hadi Asadi Rahmani is an Associate Professor of soil microbiology and researcher at Department of Soil Biology, Soil and Water Research Institute, Karaj, Iran. His research interests are legume-rhizobium symbiosis and plant–microbe interactions.

Human-Induced Land Degradation

12

Abbas Farshad, Ebrahim Pazira, and Ali Akbar Noroozi

Abstract

This chapter begins with generalities, including definitions and data availability. After a glance at environmental conditions, including physiography, climate, soils and water supply, this chapter continues with agricultural systems and their constituting features, followed then by an account of how the systems have been affecting soil quality. While a number of human-induced degradations, including erosion, salinization, compaction, flooding and ponding, and aridification, are discussed, the GLSOD maps (scales of 1:1000,000 and 1:5000,000) and the maps of landslide, erosion features and intensity (PSIAC-based) of Iran are presented, including a brief description of the mapping methodology. This chapter is concluded by short discussions about some of the strategies, such as land reform, dam construction, control on land use conversion, afforestation, and choosing suitable irrigation method, that were and still will have to be followed to mitigate the degradation. Iran, being an oil producing country, chose for a dualistic development policy including agricultural and industrial sectors. In order to mitigate degradation, equilibrium between the socio-economic and the physical environment is inevitably reachable through a thorough planning. Such a planning can never be copied from elsewhere, or planned based on dreams, but must be fully adapted to the local social systems.

Keywords

Causes of degradation • GLASOD map of Iran • Mitigation strategies

12.1 Generalities

The term 'degradation' is used to specify the status of environment or one of its constituents, in terms of quality deterioration. Degradation of environment, degradation of vegetation cover, or of soil and/ or of land are some of the commonly used terms. The term is often associated with 'improper use'. The fitness of soils for '*land-use*' cannot be assessed in isolation from other aspects of the environment (FAO 1976). Improper use or mismanagement implies disregarding compatibility between the *land* and the *use* (Sokouti and Farshad 1999; Farshad 2011).

Soil degradation occurs where human activities (either directly or indirectly) reduce the capacity of the soil to support life. Soil degradation reduces crop production and, hence, has become a great concern of the involved institutions worldwide. Soil degradation is a consequent of a variety of practices such as intensive farming, irrigation, aridification, tillage and pollution. Acidification, salinization,

A. Farshad (✉)
Earth System Analysis Department (ESA) of ITC,
Twente University, Enschede, The Netherlands
e-mail: farshad@itc.nlabbasfarshad@gmail.com

E. Pazira
Department Land and Water Resources Engineering
and Head of Soil Science, Islamic Azad University, Tehran, Iran

A.A. Noroozi
Soil Conservation and Watershed Management Research Institute
(SCWMRI), Tehran, Iran

M.H. Roozitalab et al. (eds.), *The Soils of Iran*, World Soils Book Series,
https://doi.org/10.1007/978-3-319-69048-3_12

nutrient depletion, chemical contamination and erosion are different forms of soil degradation.

12.1.1 Definitions

Land comprises the physical environment, including climate, relief, soils, hydrology and vegetation, to the extent that these influence potential for land use that is based on the functional dimension of land for different human purposes or economic activities (FAO 1976).

Land degradation is a process, which would naturally occur too, but the natural tendency to a relatively slow process of deterioration is accelerated by human activities. Land degradation involves the natural ecosystem and the human social system (Shrestha 2011).

The term desertification is used to imply the land degradation in (semi-) arid and dry sub-humid areas.

Next, to the above definitions, the term soil degradation is used when speaking of declining soil qualities that is commonly caused by improper use of soil by man (Bergsma et al. 1996; Oldeman et al. 1990). It includes physical (e.g. compaction, sealing), chemical and/or biological deterioration (e.g. soil fertility decline, salinization/sodification, acidification).

The term 'aridification' refers to the human-induced changes of soil moisture regime towards increasing soil aridity, caused by disturbances in the soil/topography/vegetation/climate system (Oldeman 1988).

12.1.2 Data Availability

Despite the fact that so many people have been publishing on this subject, a nationwide database on soil degradation is still missing. However, the information at atlas-level (e.g. GLASOD and SCWMRI) can be used for general purposes.

12.2 Environmental Conditions

12.2.1 Physiography

A complex of mountain chains (Alborz and Zagros ranges) encloses a series of interior basins, varying in altitudes from 40 m below sea level to 5678 m above sea level (see Chap. 4).

12.2.2 Climatic Condition

As a whole, the climatic condition is arid to semi-arid, but the big topographic lobs and their effect on forming air masses cause a reasonable variation in seasonal climatic condition. The average annual precipitation is about 250 mm, of which more than half is directly evaporated.

Notably, in most parts of the country, annual evaporation is more than 10 times of the annual precipitation (see Chap. 3).

12.2.3 Soils

Here, while it is recommended to refer to Chaps. 6 and 7, a glance is shed on the common inherent soils quality, with emphasis on the role of the soil forming factors (ClORPT) in producing unfavourable conditions that trigger degradation processes (See also Banaei et al. 2005).

12.2.3.1 Climate (Cl)

Except for the Caspian Sea coast with humid climatic conditions, the inherited nutrients in the soils of the (semi-) arid regions (about 97% of the country surface area, see Chaps. 3 and 9) cannot be properly used by plants simply because of shortage in moisture (Banaei 1997). Because in a large part of the country, evapotranspiration exceeds precipitation, upward movement of soluble salts, gypsum and lime through capillary rise leads to their concentration in the topsoil to the extent that some soils become toxic to plants. Saline/ alkali soils and gypsiferous/ calcareous soils are good examples to be mentioned here.

12.2.3.2 Vegetation/ Soil Fauna (O)

Except for the Alborz Mountain range and the Caspian Sea region, natural vegetation is scanty and of Xerophytic types (see Chap. 5). This explains why soil organic carbon content is low; varying from 0.30 to 2.5%. According to *SWRI*, 63.2% of the soils have organic carbon by less than 1%. However, both Xerollic /Mollic and Typic subgroups of Inceptisols (Xerosols and Yermosols, according to the FAO Soil Classification System) have been mapped in the (semi-) arid regions.

The constructive action of the macro soil animals such as earthworms, mice, ants, beetles, in terms of improving soil physical properties is limited to areas with a reasonable vegetation cover. Their action in arid soils is rather destructive.

12.2.3.3 Topography (R)

Iran has a diverse topography, varying altitude from 40 m below sea level to 5671 m (Mount Damavand). The well-known effect of relief on soil formation in terms of climate, removal/deposition and soil drainage holds fully true for the soils of Iran. Entisols occur on unstable surfaces, whereas Inceptisols and, to a lesser extent, Alfisols occur on the more stable surfaces. In terms of soil drainage condition, next to the dominant soil moisture regimes, i.e. Aridic, Xeric and Ustic, the Aquic moisture regime occurs locally in low-lying areas. This explains the presence of the hydromorphic soils (see Chap. 7). Besides, the effect of slope gradients and forms in combination with lithology (composition, texture and bedding), climate (infrequent but heavy rainfall) and the scanty vegetation cover on erosion and landslide is remarkable in a large part of the country.

12.2.3.4 Parent Material/ Lithology (P)

Although all types of parent materials, derived from sedimentary, metamorphic and igneous rocks, occur in Iran, sedimentary rocks such as limestone, marls (including the gypsiferous and saline types),sandstones and dolomites are the dominant rocks (see Chap. 4). This means that many soils are calcareous, gypsifereous, and in some places inherently saline and alkaline (e.g. soils derived from the saline/gypsifereous marls of the Qom red formation).

Evaluation of soil fertility of agricultural lands based on 50,000 soil samples analysed in *SWRI* laboratories of 30 provinces between 2002 and 2012 period shows that over 87% of farming lands have more than 5% equivalent calcium carbonate. About 83% of soil acidity are 7.5–8.5, and 97% of the soils are alkaline.

12.2.3.5 Time (T)

Theoretically, soil formation in the (semi-) arid regions is active only for short periods when the soil is moist (Buringh 1970). It is, however, not seldom to find rather well-developed Argillic horizons in the arid regions, implying a wetter climate in the past. Normally, Argillic horizon, with clear clay skins (between 50 and 100 cm depth), occurs in regions with (wet) xeric moisture regime. Presence of calcic horizon within or below the argillic horizon, Typic and Calcic Haploxeralfs (in some cases gypsifereous) is common in many places (Farshad 1978).

The strength of soils to resist degradation (erosion, mass wasting), particularly in sloping areas, is for a great part controlled by the occurrence and the sequence of the horizons, which are rarely constant over long periods. This means that land degradation is quite dynamic.

With a bit of exaggeration, one may say a map that is prepared today may not be valid tomorrow; though the principle of 'the past and present are keys to the future' would remain valid.

12.2.3.6 Man

Man, as a farmer and stockbreeder, in Iran, has been influencing soil formation since over eight millennia (Fisher 1968). Unfortunately, this has often been in a negative way, that is, accelerating salinization/ alkalinization, or through deforestation, overgrazing and over-irrigation (Sect. 12.4).

12.2.4 Water Supply and Irrigation

Water resources are considered as the most important natural factor that can limit expansion and development of agriculture. 'Everything is alive, thanks to water' (a Persian proverb). The annual volume of precipitation was estimated between 400 and 413 Bm^3 and additional 12 Bm^3 from the border rivers' flow; only about 130 Bm^3 is potentially available for use. Although in recent years, the average annual volume of precipitation has decreased by % 20 due to frequent drought and climate change. The difference is lost to evaporation from bare soil surfaces, evapotranspiration by natural vegetation and from outflow of the rivers into the Persian Gulf, the Caspian Sea and to the neighbouring countries. Roughly, agriculture sector consumes about 90% of the total water used. In the absence of favourable climatic conditions, the shortage of the water supply has since centuries (back in Sassanian times: ± 210– ± 600 AD) been compensated by tapping subsurface water—the *ghanat*.

- Ghanat (also written as *qanat*), the unique Persian underground irrigation tunnels used to function well, for centuries. Many ghanats are nowadays out of function, partly due to the drop of groundwater table, resulted from uncontrolled excavation of deep wells, and partly because of the lack of proper management, after the 1962 land reform (Sect. 12.6). In many arid areas, ghanat, though with much less water, is still an important source, since surface water is frequently absent or inadequate, and deep wells are costly and in places provide only brackish or highly saline water.
- The river water has also been distributed and conveyed to different altitudes by means of channels angling off from the river.
- With the introduction of the modern techniques in water supply and irrigation, because of the population growth, and industrialization policy, many deep wells and huge dam constructions were added to the above sources.

12.3 Agricultural Systems

Iranian economic stability depends on two sources, oil and agriculture (Roozitalab 1993), with the remark that agriculture in 2013 constituted only about 13% of the GDP whereas the share of industry and services have been increased (pers. comm. Siadat). The strength of the former depends on global circumstances, whereas the latter is primarily a matter of internal organization of the exploitation of territorial resources and can be permanent. In terms of 'major kind of land use', rainfed cultivation is practised in a relatively large part of northern regions, whereas (semi-) irrigated cultivation is practised in (semi-) arid regions. Semi-arid regions are where the dry farming, at least in one season, is possible (Hall et al. 1979).

Agricultural activities are mainly concentrated in the inter-mountain valleys, although the steep slopes of the mountains are increasingly used too. For centuries, production activities in villages were organized in a hierarchically ordered structure, the '*Boneh*', a unit composed of several households (Safinezhad 1989). In areas where the *Boneh* was common, the peasant did not work an assigned plot of land alone; rather, it was typical for the arable land of a village to be organized into units farmed cooperatively by teams of shareholder farmers. The entire population of the village was divided into bonehs. Nearly, all the labour power in a village was trained and channelled towards production inside the traditional institutions.

Despite the fact that the agricultural systems and the socio-economic conditions vary from one region to another, it is still possible to give a general view on some of their features such as landholdings, land-tenure: absentee owners, landless peasants and the cropping patterns (Table 12.1). An attempt is made to cover a brief background to the agricultural systems which should be of help to analyse their effects on land degradation (see under Sect. 12.4), and also in discussing the strategies to mitigate the degradation (see under Sect. 12.6).

12.3.1 Landownership

The landownership in Iran has periodically changed during this century (Fisher 1968). Available statistics of before 1962 indicate that about 56% of the cultivated land was owned by 1% of the population (big landlords). Small holders and peasant owners are estimated to have owned between 10 and 12% of the total cultivated land (Lahsaeizadeh 1993).

In a lengthy analysis of events, Lahsaeizadeh (1993) tackles the issue of the land distribution in Iran. For implementing the land reform, as one of the strategies to mitigate degradation, central headquarters of land reformation and devolution were established in the capital city. At the same time, Seven–Member Committees of Land Devolution (SMCLD) were installed in different cities. By 1980, 150,000 hectares barren lands and 35,000 hectares uncultivated, cultivated and confiscated lands were distributed among the landless and poor peasants. About 60,000 hectares of barren lands was turned over to the Organization of Productive Services Expansion for establishment of rural productive cooperatives. Finally, 850,000 hectares of disputed land was temporarily handed over to the peasants.

Lahsaeizadeh (1993) concludes that there is no doubt that the proceeding of SMCLD increases the agricultural lands of the country. Land reforms carried out in isolation from a more general programme of development, resulted in a

Table 12.1 Generalized features of the agricultural systems

The major agricultural systems: dry farming, irrigated-farming and to a lesser extent nomadic-farming, with the following generalized rural/agricultural features(a picture of over 60 years ago)
• Landownership: Land was owned by: a. The state (public domain), b. Village communities (tribal holdings), c. Royal family (Saltanati), d. Owghaf—(or vaghf-) holdings (religious endowment; 99 years lease-based), e. Large and smaller landowners (all as absentees). Seldom actual farmers owned the land. In almost all types of absentee ownerships, the system of share cropping[a] was practised, often not to the benefit of the farmer
• Farm-size: Not expressed in metric system, but in qualitative terms (*Joft*[b], *Jarib, khish, Ghandom*). Each share cropping farmer had '*Nasagh*[c]' rights on both poor and suitable lands. These turned out to be a complicated problem in the execution of the 1962 land reform
• Land preparation: Done in a traditional way using draft animals and very simple equipments, regardless of any conservation measure
• Soil management: Traditional way of promoting fertility through crop rotation and fallowing the land, and/or through manuring
• Livestock/animal husbandry: This activity was often at a subsistence level. The nomadic pastoralists, concentrated mainly in the mountain ranges (Zagros and Alborz), were very much dependent on the natural pastures
• Off-farm activities (role of women): Women, next to their normal occupations in housekeeping, continued to cover a part of the family expenses through handicrafts (e.g. wool carpets)

[a] In this system, the five essential shares are land, water supply, seeds, land preparation costs and human labour. Often land and water were supplied by the landlord, with right on 2/5 of the yield

[b] Joft (described here as an example) was a surface area which could be cultivated by a pair of oxen and requires a known amount of wheat or barley; roughly, it is about 2–2.5 hectares

[c] Communal right: each member of the sharecroppers could have nasagh in different parts of a village, both in poor and in good lands

redistribution of poverty in rural areas. Since the SMCLD distributes only barren and natural pasture, the commercial and mechanized agricultural lands continue their operations. The country on the one hand faced more inequality in the rural areas, and on the other hand, an environmental disequilibrium because of converting pastures into agricultural lands.

12.3.2 Farm-Size

Changes in landownership and the advances inland preparation led to larger farm-size. Most cultivators work and live on plots of land, usually less than 10 hectares in extent, only part of which is carrying crops at any one time.

12.3.3 Land Preparation

Moving from the traditional devices to the advanced ones took place rather quickly (see Chap. 13). Next to several machineries' making factories, several plants were emerging throughout the country to produce fertilizers and other chemical products.

12.3.4 Application of Chemicals

Application of fertilizers and pesticides became soon a common practice, in the last 50 years. The traditional way of promoting soil fertility, that is, through crop rotation and fallowing the land for every other or every 2 years is less practised. However, because there is far more land than the required water for irrigation, parts of lands, which cannot be irrigated, will lie on fallow for one season or more. Statistics indicate that about 3–4 m ha of the land was under fallow in 2014.

12.3.5 Livestock/Animal Husbandry

This is another component of some agricultural systems. This activity, which was often at a subsistence level, attracted the attention of some investors who started with well-organized animal husbandries. On the other hand, the nomadic pastoralists, still common in a great part of the Zagros Mountain Ranges, are far more dependent on the natural pastures of the high altitudes (an argument to support the idea of nationalizing pastures).

12.3.6 Rural Crafts (Rural Non-Farm Activities)

With the changes in landownership and the advances in technology, country went more industrialized; many carpets and weaving factories, and also agribusiness plants emerged, which had an impact on farmers revenue, so seriously that many of them were urged to migrate to larger cities: a migration with all its direct and indirect consequences.

12.3.7 Yield

According to the agricultural statistics in 2016, wheat production amounted to about 14 million tons and that of barley to about 3 million tons. Differences in yield between dry and irrigation farming remain considerable. On an average, in 2016, wheat yield was about 4 metric ton per hectare in irrigated lands and more than one ton under dry farming. Fisher (1968) referring to the First National Census of Agriculture (issued in 1960) reports the yield of dry-farmed wheat in central Iran between 290 and 360 kg, and that of irrigated land between 850 and 1900 kg. Comparing these figures may say something about 'intensification', for a part being triggered by advances in technology.

12.4 How Agricultural Systems Affect the Soil Quality

A agriculture is a kind of man's interference in nature. In other words, there will be little or no degradation under native conditions and before man's influence (Farshad 1997). No doubt that the man's influence increases with intensification, which may become imperative when extensive agriculture does not meet the needs. For this reason, the terms 'sustainable agriculture' and more recently 'conservation agriculture' (see Chap. 13) were introduced, and received ample attention internationally. The effect of agricultural system on the quality of soil (Farshad 1990, Banaei 2005) can be summarized as follows:

12.4.1 Water Erosion

Water erosion is especially serious in the sloping lands. According to an unpublished FAO report, erosion causes an enormous soil loss/year. Different statistics on soil loss/year are reported: between 2 and 2.5 billion metric tons (BMT) and between 3 and 4 BMT, which is not officially confirmed. According to the sediment survey data from 209 stations scattered throughout the country (SCWMRI), the rate of sedimentation and erosion is, respectively, about 350 million and 1 BMT per year. 1 BMT means a soil loss of 5–6 ton/ha (see under 12.5). The most sedimentation occurs in Hamoon, Jazmoorian, Minab and Southern Baluchestan, Marun and Zohre river basins in the south of the Iran. Momeni (2003) has accepted the officially published figures indicated that the soil loss in Iran is almost twice as much compared to the average soil loss in the world. These areas are influenced mainly in the following ways:

- Over-irrigation. This is quite common in irrigated fields, which cover around 8.5 M ha, about 6.5 Mha of which is traditionally irrigated (Niknami 2009). Application of too much water to the field has caused both drainage problems and loss of surface soil due to large flow rates.
- Deforestation. Despite the forest conservation and the controls exerted, cutting and uprooting of natural vegetation goes on. In comparison with the forested Alborz mountains, the (already) degraded forests of the Zagros mountains have received less attention. During the past few decades, the total forested surface area has decreased from 18 million hectares to about 12.4 million hectares (Banaei et al. 2005).
- Overgrazing the natural pastures. Some 650,000 km^2 of the country's surface area is, in one way or another, misused through deforestation and/or overgrazing. During the last 15–20 years, this has imbalanced the very sensitive natural ratio—vegetation cover/ environment—resulting in more frequent flooding and erosion.
- Improper cultivation. Tillage of the sloping areas, where soils are shallow and quite susceptible to erosion and mass movement (landslide, etc.).
- Fallowing. Leaving the sloping land without vegetation cover, that is, to leave the land bare for one year or two without any crop cover. The silting up of the modern reservoirs is a good indication in understanding the severity of the loss of soils.
- Land use change, e.g. by conversion of rangelands to rainfed agriculture (Farshad 1997). This does not only result in water erosion, but also accelerates wind erosion and indirectly plays a role in soil salinization. Conversion of dry land to irrigated land showed to have caused compaction (Farshad et al. 2005).

12.4.2 Wind Erosion

Wind erosion is associated with the flatter areas, mainly in the central plateau, the areas surrounding the '*Kavirs* = deserts', with an approximate surface area of 450,000 km^2. The area under sand dunes is estimated 12 million hectares, half of which are active or semi-active. According to SCWMRI, the density of air blown deposits in a 40-year period shows a growing trend. Comparing the interpretation results of the Landsat Satellite Images of 1972 and that of 2009 shows an increase of 27 0km^2 in the sand dunes' surface area. Wind erosion, although for the great part, is natural; it has been and still is accelerated by man through:

- Overgrazing: The peasants of the areas surrounding the deserts are poorer than those of the rest of the country. This means that their survival is very much dependant on other activities rather than cultivation, among which livestock activity can be named. The poor vegetation of these areas is badly overgrazed. This in combination with the natural seasonal strong winds and storms not only accelerates the wind erosion but also diminishes the soil organic content, which consequently leads to deterioration of the soil structure (see Chap. 13).
- Cutting vegetation for household heating purposes (e.g. charcoal making), and in some cases for medicinal uses (e.g. *Glycyrrhiza Glabra*; a factory was established in Marvdasht district, Fars province, to produce the plant extract, that is used in pharmaceutics).
- Wind erosion also occurs in poorly managed rainfed (dry farming) areas. This, though obvious, has been investigated in different parts of the country, for instance in *Orazan* sub-basin, near *Taleghan* (Vafakhah and Mohseni Saravi 2011). The results revealed that optimizing land use leads to a decrease of about 10–22% in soil erosion rate.

12.4.3 Salinity/Sodicity

Country-wise, extensive areas are covered by Salorthids and/or saline/sodic phases of other soil orders such as Aridisols and Inceptisols. Different salinization (/alkalinization) modes can be distinguished on the basis of hydrogeopedological settings in different landscape units (landforms) as follows: **a.** The 'sierozemic' mode; **b.** Fan base-apex mode; **c.** Ingression of sea water through waves, wind and/or dew; **d.** Floodplain/terrace mode; **e.** Seep-water-based mode (Farshad 2008). About 4.3 million hectares of the 6.8 million hectares of the total arable land that is considered 'saline (to different degrees)', with EC > 4 dS/m (mmohs/cm), whereas the remaining 2.5 million hectares are not only saline but also having other limitations, such as soil texture, topography, erosion and high groundwater table (Pazira and Minami 1981; Momeni 2009; Pazira and Homaee 2010). Although quite a surface, mainly in the central plateau and Persian Gulf's coastal plains (including Khuzestan plains), is salt-affected, the remaining surface, including plains and valley bottoms within Alborz and Zagros, is free from salinity limitation. Other saline/alkaline surfaces, such as the Kavirs (Central and Lut deserts) , and the other scattered surfaces (like some dried-out old lake bottoms) are classified as miscellaneous land, with an approximate surface area of 20 million hectares (Momeni 2009).

Causes of soil salinity are often natural, depending on soil parent material, topography and groundwater depth, but, in many cases, soils have become saline because of mismanagement (human-induced). The important human activities, above those mentioned earlier, which accelerate the salinization/alkalinization processes are as follows:

- The way land is irrigated (mostly over-irrigation), while inadequate availability of water and a high evapotranspiration are very serious matters. The salty irrigation water is often another reason in salinization of soils (In a number of pilot projects such as the one in *Ahuchar*, Fars province (Momeni 1999) researches are carried out).
- Deep ploughing of the saline/alkali soils, at least those with a salic horizon lying below the topsoil (see Chap. 13). In this way, the saline/alkali layer is broken and brought up to the surface. Ploughing of lands where saline/alkali soils occur in scattered patches can lead to contamination of the neighbouring non-(less) saline/alkali soils.
- Removal of native vegetation causes salinization: bare soil surface increases direct soil evaporation, which causes upward movement of soil solution, followed by water evaporating from the soil and leaving the salts behind and, consequently, accumulation of salts in the surface layer of the soil.

12.4.4 Compaction

Improper use of agricultural machinery, ploughing when soil is too moist or wet, overgrazing and misusing chemicals, applying one and the same dosage of chemical fertilizers to all types of soils, no matter whether calcareous or saline/alkaline, have led to compaction and consequently to less yield (Farshad et al. 2005; See also Chap. 13).

12.4.5 Flooding and Ponding/Groundwater Table

Flash floods occur mostly locally in the mountainous areas and are very destructive. Floods, although in the first glance often seen as natural events, are accelerated by human activities such as deforestation, overgrazing, over-exploitation of vegetation for house consumption and fallowing. Niknami (2009) compared the flood occurrence in the period between 1960 and 2000 and concluded a threatening increase from 193 times in 1960 to 1341 times in the year 2000.

In 2012, Golestan province was affected most as compared to other provinces. Mazandaran and Gilan provinces were the next in terms of vegetation destruction and flooding.

Water ponding and waterlogging is mostly due to surface sealing caused by soil structural collapse and poor surface drainage. The ponding that is locally resulted from high groundwater table depth limits development projects in the provinces of Khuzestan and Gilan (Pazira 2012).

In arid regions, on the other hand, the problem of ponding is becoming a hot issue in the recent years. Some 1.4 million hectares of the arable land throughout the country, from Azerbaijan, Khorasan, through Hamadan, Kermanshah, Isfahan, Fars, Kerman to Khuzestan, Sistan and Baluchestan is reported to have groundwater table limitation. Water-logged soils, often human-induced, have become saline too, when compared with the maps of 24 years ago (Momeni 2009).

12.4.6 Aridification

In many places, in particular in the central plateau and in some of the rural areas located in the Alborz and Zagros mountain systems, deep water wells were (are) dug to compensate the water shortage. This has resulted to lowering of the local base groundwater table. Many ghanats were dried out and the soil moisture regime changed more and more towards an aridic regime (e.g. in the provinces of Isfahan, Hamadan and many other places). As an example, Water Organization of Hamadan reported that in the village of Shirinabad (Hamadan-Bahar), an observation on dug well showed a 13 m drop in groundwater level over a period of 6 years (between 1986 and 1991).

Anomalies and disruptions in socio-economic structure (landownership, etc.) do not only promote and/or accelerate the above-mentioned causes of degradation, but they also have led to migration of farmers to large cities, which means the end of agriculture.

12.5 Man-Induced Soil Degradation Map, Methodology

Although there exists quite a number of methods for assessing salinization, compaction and other types of degradation, the most attention has been paid to water erosion, which started with the Universal Soil Loss Equation (USLE); later on modified under the names RUSLE,

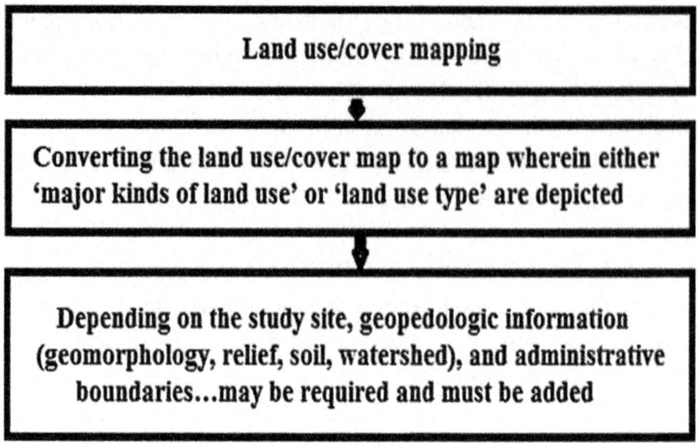

Fig. 12.1 Methodological flow-chart (after Farshad 2011)

MUSLE, etc. and/or being the basis for some other assessment methods, such as SLEMSA.

To map land degradation at the level of a country is certainly more complicated than mapping erosion or salinization, etc. Principally any method applied should include both bio-physical as well as social aspects of land (Figs. 12.1 and 12.2).

In other words, the method is backed by the geopedologic approach to soil survey (Zinck 1988; Zinck et al. 2016) and the landscape integral survey method (Zonneveld 1995). The former is a systematic and hierarchical classification system that integrates geologic(lithology), geomorphologic and pedologic components of the landscape. The different levels, at regional scale, include 'landscape' (e.g. mountain, piedmont, valley), 'relief-type' (e.g. hill, glacis, terrace), 'lithology' (e.g. rock type, alluvium, colluvium) and 'landform' (e.g. crest, footslope and tread).

Any of the above-mentioned four columns (as a tabulated legend) may be further specified by phases, for instance, 'very high rugged mountain', 'forested hill', 'severely eroded glacis'. The pedologic properties, depending on the scale of the study, will follow the precedent columns. Although a geopedologic map offers more attributes than the conventionally prepared soil maps do, it is still not an appropriate base-map for the degradation/conservation-oriented studies. Other attributes, mainly management-based, where some social aspects of the land come in, are required particularly when degradation is of human-induced type (accelerated erosion, compaction, salinization, etc.).

To satisfy the mapping unit requirements (Polous et al. 2011), modified version of the landscape integral survey method (Zonneveld 1995) can be employed. This approach suggests the term 'land unit', which results from incorporating other land attributes such as land use/cover and hydrology into the geopedologic units. Although 'land use' already to a certain extent covers some socio-cultural and economic aspects of the landscape, where degradation occurs, depending on the scale of the survey, additional data can still be added.

Following the same principles, but in larger lines, the soil degradation map of Iran was prepared (Fig. 12.2 after Farshad 2011) that shows a generalized picture of the distribution of different types of degradation, that is further specified with possible causes, rate and the extent.

In the framework of the GLASOD (Global Assessment of Soil Degradation) project, the status of human-induced soil degradation at world level (scale of 1:10 M) was prepared in three map sheets. Through the use of an expert system approach and following a general guideline (Oldeman 1988), the existing information was gathered and analysed. The map was meant to create awareness on the seriousness of soil degradation in a global perspective, in the same way that is done in the above Table 12.2, legend of Fig. 12.2. In each mapping unit in the map, the dominant types, causes, degree, extent of each type of degradation and the rate of degradation are shown by one or more symbols, depending on how pure the mapping unit is. As an example, the map unit symbol 'wt f/e 224' means moderately eroded (by water, abbreviated by Wt) terrain, caused by deforestation (f) and over-exploitation of vegetation for consumptive use (e); the terrain is still suitable for use in local farming systems, but with greatly reduced agricultural productivity (the first 2 after the cause); with a medium rate (the second 2 in the symbol); and an extent of 4, which means that the terrain is for 26–50% affected.

The Glasod approach, although used for such a small-scale survey (Fig. 12.2), can also be used for larger-scale surveys. In the present study, geological map, bioclimatic map and the contour map were used to prepare a physiographic map. This map was then used as a basis for the soil degradation map (Fig. 12.3). In addition, a database was made using the *DATAEASE* database management system, where different queries could easily be made. By means of the queries, additional maps such as distribution of dominant causes of degradation or dominant rates, etc. can be extracted. This database was also used to generalize the map units for the final edition of the GLASOD publication at scale 1:5,000,000 (Oldeman et al. 1990).

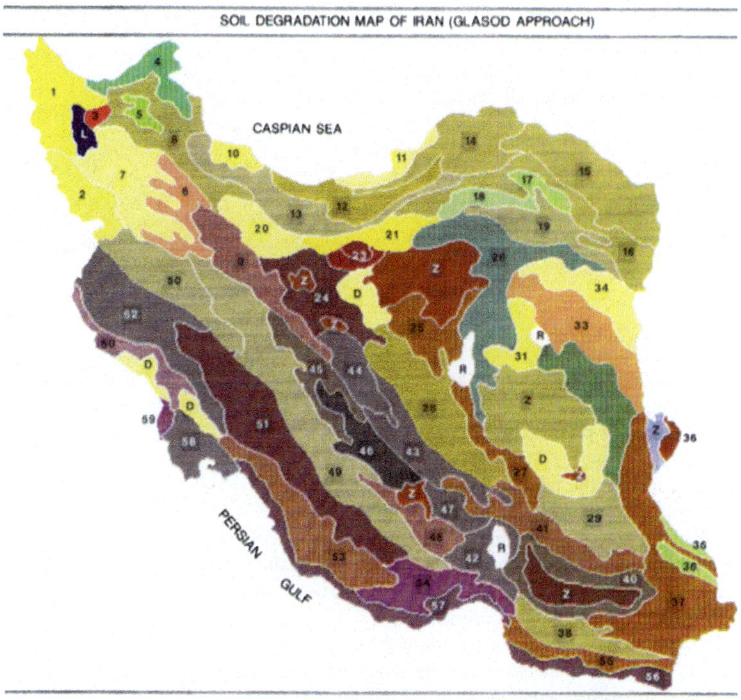


```
LEGEND :  SOIL DEGRADATION MAP OF IRAN (GLASOD APPROACH)
==================================================================
 1. Wt f/g 213        25. Ed  e  224             Pa f/i 123
 2. Wt f/g 222            Cs  e  215         48. Cs f/e 215
 3. Cs i/o 224            Wt  e  114             Ed f/e 224
 4. Wt g/i 213        26. Cs  e  313         49. Wt f/e 113
 5. Wt  f  112            Ed  e  215             Pa i/w 114
 6. Wt f/g 113        27. Ed  e  124         50. Wt f/g 114
 7. Wt g/i 223            Wt  e  114             Co  o  222
    Pa  i  114            Cs  e  113         51. Wt f/w 122
 8. Wt g/i 224            Pa i/w 113         52. Wt f/g 123
    Pa  i  124        28. Cs  e  214             Co  o  222
 9. Wt g/e 213            Ed  e  224         53. Wt f/e 114
    Cs g/e 111            Pa i/w 224         54. Wt f/e 113
    Pa  i  122        29. Ed  e  115             Et f/e 122
10. Wt f/i 123            Pa f/e 215             Cs f/e 212
11. Cs  i  224        31. Wt f/e 214         55. Wt f/e 113
12. Wt  f  112        34. Cs g/e 114             Ed f/e 112
13. Wt g/i 224            Wt g/e 113         56. Cs f/e 225
    Pa  o  122        36. Cs g/e 114             Ed f/e 224
14. Wt g/i 224            Wt g/e 113             Pw  e  112
    Pa  i  112        37. Et e/g 225         57. Cs f/e 225
15. Wt g/e 123            Wd e/g 225             Ed f/e 224
    Pa  i  112         .  Cs e/g 114             Co o/w 223
16. Wt f/g 113            Pa e/g 113             Pw  e  112
17. Wt g/i 224        38. Et e/g 225         58. Cs f/i 325
    Pa  i  114            Wd e/g 225             Ed f/g 224
18. Wt f/g 123        40. Ed e/g 224             Co  w  223
    Et f/g 113            Cs e/g 113             Pw  e  112
    Pa  i  113        41. Et e/g 224         59. Cs f/i 224
19. Wt f/e 214            Wt e/g 114             Co  w  233
    Cs  i  213            Cs e/g 113         60. Cs f/g 214
    Pa  i  113        42. Cs e/g 113             Co  w  233
20. Wt w/i 124            Et e/g 114             ================
    Pa  w  124        43. Cs  e  214         * Symbols indicate:
    Cp  w  223            Ed  e  224
21. Cs g/e 214        44. Wt f/e 114           -Degrad. type;
    Wt g/e 112            Cs f/e 113            (1st 2 letters)
    Ed g/e 112        45. Wt f/e 124           -Causat. factors;
    Pa  i  113            Cs f/e 223            (1 or 2 letters)
23. Cs g/e 222            Pa i/o 223           -Degree;
    Ed  e  224        46. Cs f/e 215            (1st figure)
    Pa e/i 225            Ed f/e 224           -Rate;
24. Et e/g 225            Pa f/i 123            (2nd figure)
    Cs e/g 224        47. Cs f/e 215           -Extent;
    Pa w/i 225            Ed f/e 224            (3rd figure)
```

Fig. 12.2 Glasod map (Farshad 2011)

To prepare the Watershed Atlas of Iran, Soil Conservation and Watershed Management Research Institute (SCWMRI) published a series of maps on, among others, soil erosion, landslide distribution and erosion features, where use was made of the topographic and some other relevant existing maps, remote sensing and field data that were worked out within GIS environment. Soil erosion classes were defined based on the EPM model (PSIAC 1968) (Figs. 12.4, 12.5 and 12.6). In this study, that is officially published in 2008, a soil loss of 1 billion tons, that is 5–6 ton/ha/year, is reported.

In a report, Siadat (no date), referring to some of the figures published by different researchers and organizations, warns policy makers on the threatening situations. Some of the figures are published and some are quoted. UNDP (1999) reports that soil erosion in Iran is doubled in the past few

Table 12.2 A summary of the human activities and the degree with which they accelerate degradation

Factors/causes												
Human-induced degradation	1[a]	2	3	4	5	6	7	8	9	10	11	12
A[b]	++	++	++	0	0	+	x	+	0	++	0	X
B	0	++	++	+	0	0	0	+	0	++	++	X
C	0	0	0	0	++	+	x	++	++	++	+	X
D	0	0	x	0	x	x	0	0	++	X	0	X
E	++	++	++	0	0	++	0	0	0	++	0	X
F	x	x	+	+	0	+	++	x	x	++	X	X

a

1: Cutting/uprooting trees for annual crop cultivation and charcoal making(in places in mountains)
2:Cutting natural vegetation for fuel and in some cases for medical uses
3: Overgrazing pastures
4:Long fallowing
5:Misirrigating the potentially saline/alkaline soils
6:Mismanagement in application of surface water
7:Misusing the aquifers through over-excavating deep wells
8:Deep ploughing of shallow/potentially saline/alkaline soils
9:Mismanagement in application of fertilizers/pesticides
10: Carelessness in land conservation
11: No erosion control on adjacent mobile sands (dune/sheets)
12: Industrialization regardless of the people's agricultural background (Note: farmers who cannot cope with the newly developed situations migrate to larger towns, attracted by the urban mirage)

b

A Water erosion; ***B*** Wind erosion; ***C*** Salinization/alkalinization; ***D*** Eutrophication/pollution; ***E*** Flooding; ***F*** Aridification
++: Direct strong influence; +: Direct moderate influence; 0: No-weak influence; x: Indirect influence

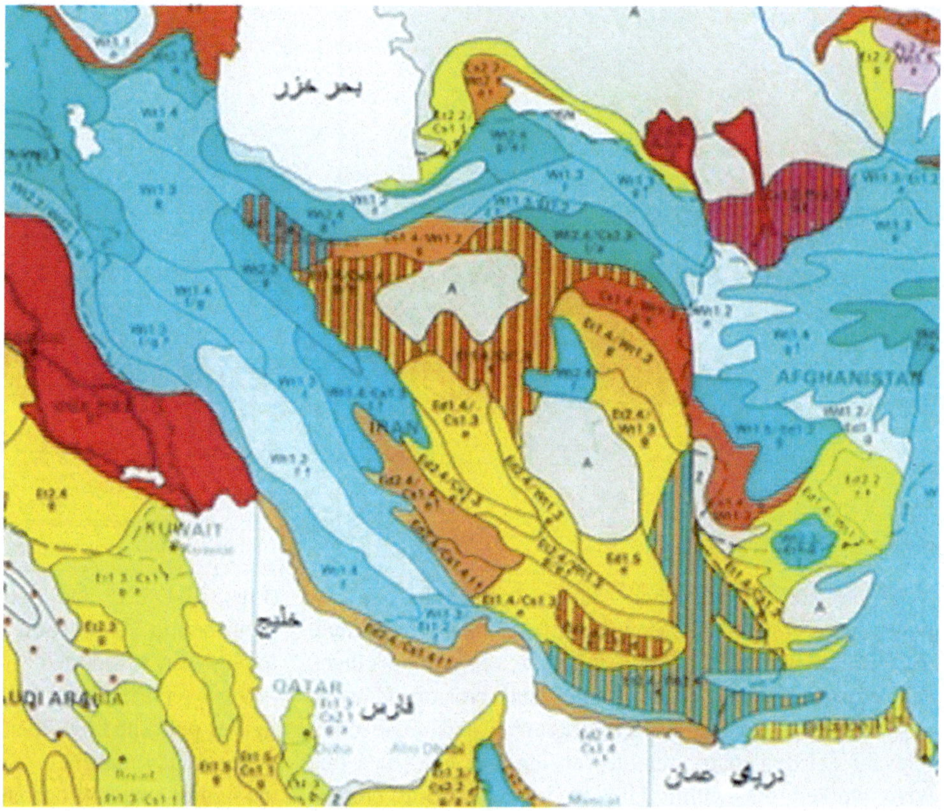

Fig. 12.3 Glasod map of Iran at scale 1: 5,000, 000 (cropped out of the original sheet) (Farshad 2011)

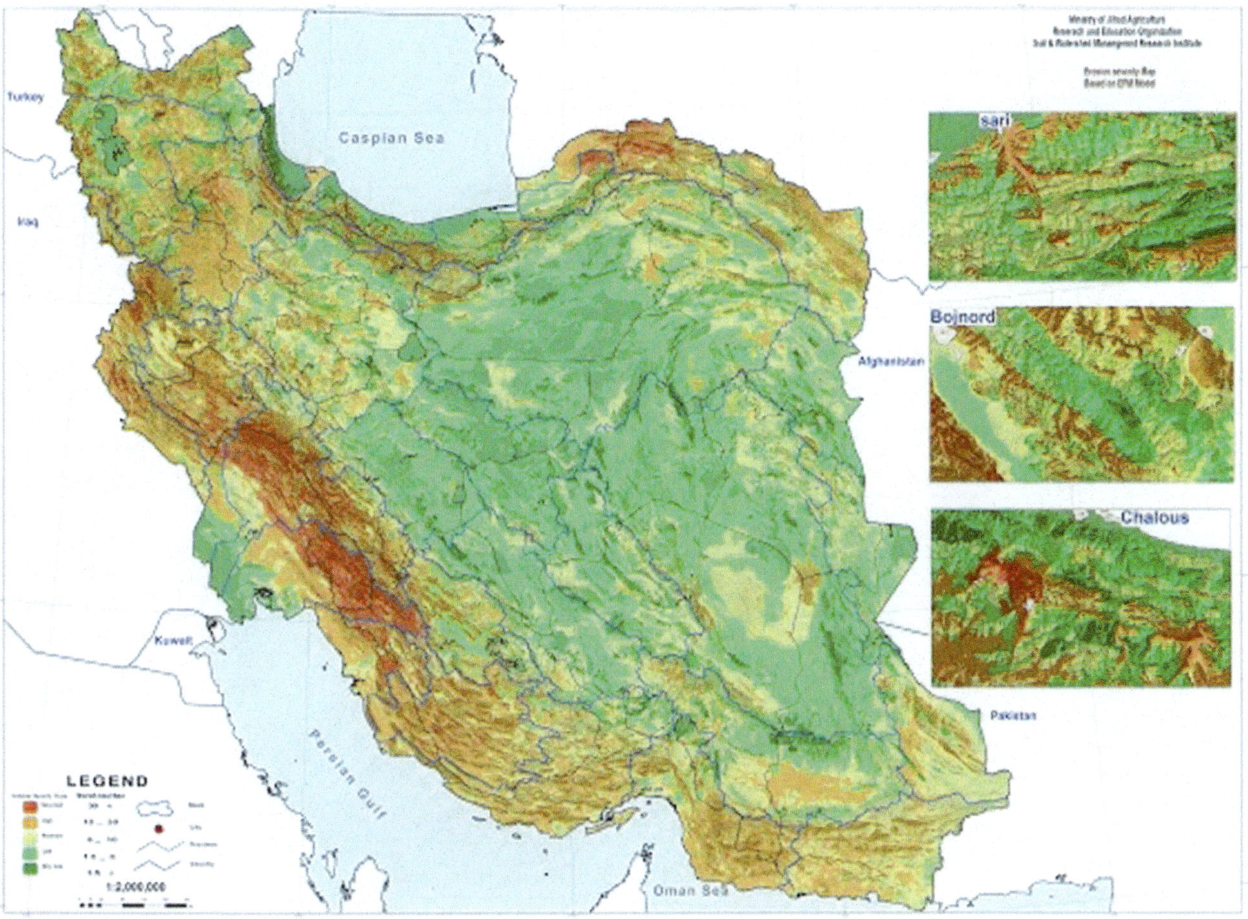

Fig. 12.4 Erosion classes map (SCWMRI 2008)

decades, reached the amount of 20 ton/ha /year. *ECONEWS* (2009) that quotes some statistics issued by the Ministry of Jehad-e-Keshavarzi has reported an amount of 2 billion tons soil loss. There are also some local studies such as that of Kalhor and Jalalian (2003) who believe that erosion in a small watershed in the Lorestan Province is 32–80 ton/ha. Whatsoever the true figure might be, the fact is that a considerable surface area is at risk, to different degrees. The people who worked in southern Iran (in *GharaAghach* project) report that 63% of the total surface area (1.2 million ha) is at risk. The estimation made in the various reports of UNDP, UNEP and FAO reaches 94% of the agricultural land that is at risk, to different degrees.

12.6 Strategies Followed to Improve the Situation

Considering the soil types/distribution and the limited arable land versus the population growth and the improvement in social welfare, amalgamated with political ambitions, agricultural systems were adapted to the newly developed conditions. The other measures that were taken are as following:

12.6.1 Land Reform

Land reform is known as a tool that can be employed to awake and restructure a stagnated agriculture, although 'in its generally accepted sense cannot be a patent solution for the agricultural policies of the developing countries' (Schiller 1972). Land reform started in 1962 with a very ambitious plan, in three phases. In the first phase, distribution of land was the principle. The permissible property was set on one village unit, while the rest of the great land owners' property was sold (with crediting via a specially established bank for the purpose) to the farmer tilling the land. Unfortunately, the first phase was not quite successful. The great land owners, who were disappointed, started to hesitate about their future and started to invest in other sectors rather than agriculture. The inter-farm managerial

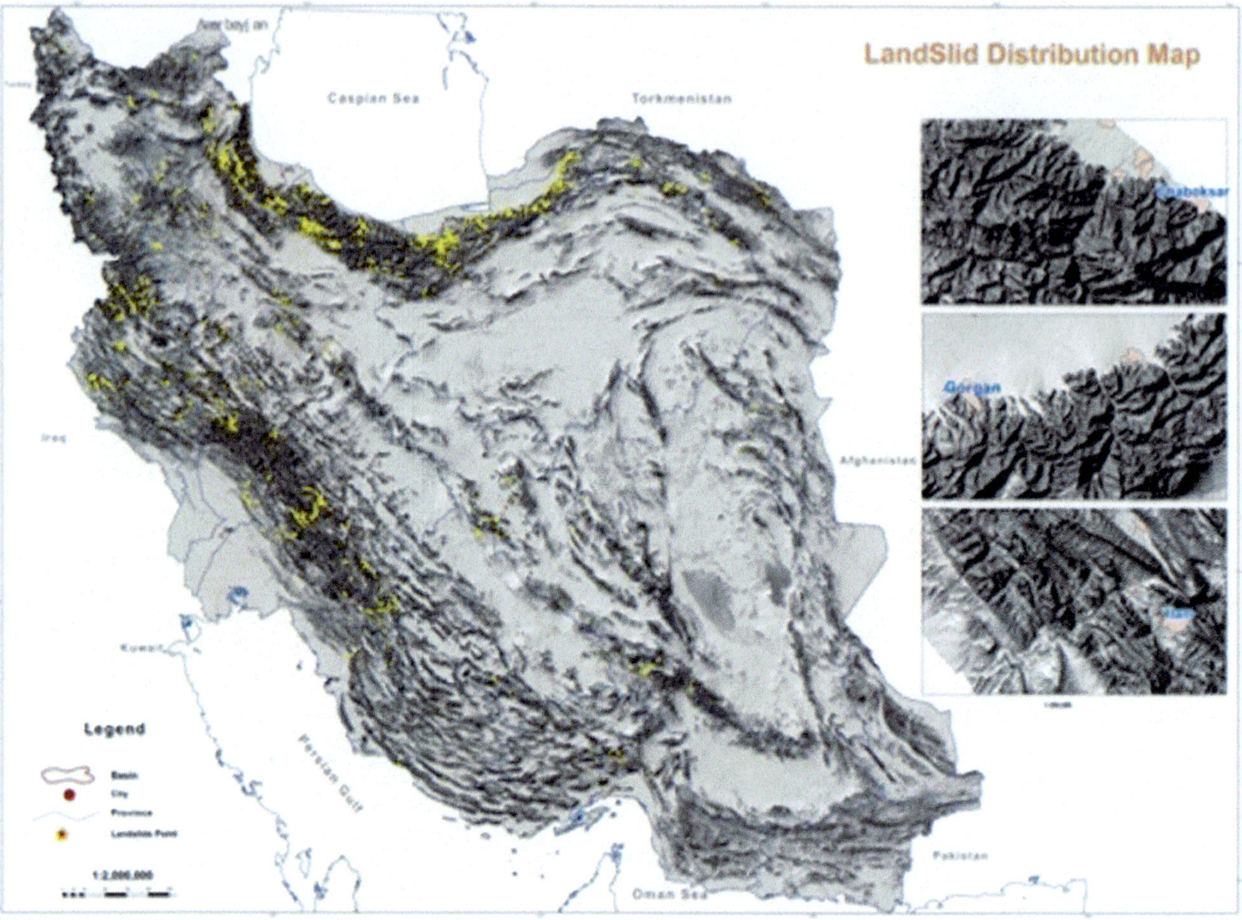

Fig. 12.5 Landslide distribution map (SCWMRI 2008)

role of the great land owner was not properly substituted, and the farmers often were not well prepared to cope with the new situation. As a result, maintenance was overlooked so that many ghanats were not cleaned in time and were soon collapsed. The second phase of the land reform was introduced, a modified Russian centralist approach. In this phase, the distributed lands were brought back together in form of corporation. By 1973, 15% of the total cultivable land was owned by agribusiness and farm corporations who employed 3% of the total rural labour force (Afshar 1985).

The third phase was a closing phase to a not successful programme. Lands still under dispute within the second phase options had either to be divided between the land owner and tenants or sold to the tenants (Denman 1978). This was taking place at a time when a considerable number of farmers were migrating to the larger towns, where full industrialization was in progress. However, whatsoever has taken place in this connection were some attempts towards such important measures as agricultural land consolidation and controlling land fragmentation.

12.6.2 Dam Construction

Studies on water resources development started in 1949 with the following objectives:

- Supply of water for agricultural uses, which should lead to food self-sufficiency,
- Generating the required electric power to run the national industries, and supplying water for a growing population, considering the fact that oil resources will not last forever.

Several irrigation projects such as *Karkheh*, *Karun* and *Doroodzan* were established. More than eighteen multi-purpose dams were in use by 1972 (e.g. *Karaj* dam 1962; *Golpayegan* dam 1958; Isfahan1970; Doroodzan Dam 1972). A total of 88 small and large dams were constructed in 2007, and as of 2010, Iran has constructed 588 (small and large) dams, with 137 more under construction, not forgetting a number of 546 planned ones (Wikipedia on list of dams). The capacity of water reservoirs under the

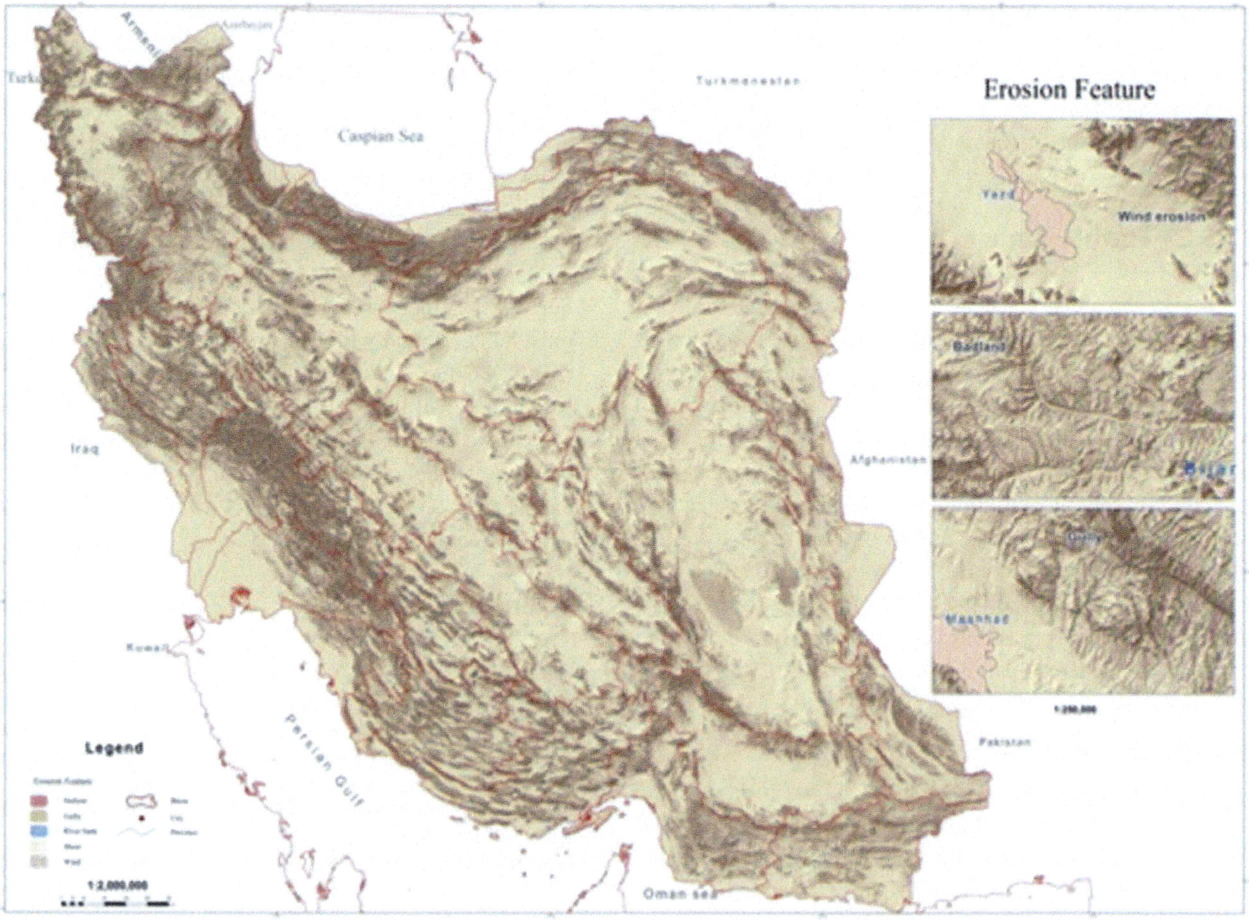

Fig. 12.6 Erosion feature map (SCWMRI 2008)

constructed dams was anticipated to be 20 billion m^3 by the year 1978. It is remarkable that the capacity has reached to 48 billion m^3 in 2013 (Persian. Comm. Siadat). Besides the enormous amount of energy (generating some 1800 million kilowatt power energy), their capacity to control about 30 billion cubic metres of floods was foreseen too (Pazira and Minami 1981).

12.6.3 Irrigation Methods/Drip Irrigation

Despite the shortage in water, basin irrigation was quite common. In traditionally irrigated crop lands (with estimated surface area of about 6.5 Mha), many farmers applied too much water to the field causing both drainage problems and loss of topsoil due to large flow rates. Hence, next to the studies on application of fertilizers and insecticides carried out in several pilot projects (e.g. Doroodzan in Fars Province and Dasht-e-Azadegan in Khuzestan Province), irrigation methods and land reclamation received attention too. It is interesting to know that Khuzestan was one of the most prosperous and populous parts of ancient Iran and is one of the oldest areas in the world where irrigation was first used. According to some documents, between 1500 and 2000 years ago Khuzestan during that time, very important irrigation schemes were set up. Some of the remains of irrigation canals and dams, which are more than 1500 years old, can still be seen in the area (Pazira and Homaee 2010).

The study results on drip irrigation were positive and well received by particularly for irrigating orchards, although studies on salinity-related problems were still going on.

12.6.4 Land Improving/Reclamation

Land preparation, increasing soil organic matter and proper fertilizer practices, in particular in slightly to moderately saline soil, have to receive attention, and are, to a certain extent, taken care of by various organizations, though not sufficiently. Draining saline soils takes place too, for example in Dasht-e-Ahoochar, near Shiraz (Momeni 2009) and in Khuzestan. The total area presently under the drainage system in Khuzestan province is about 120,000 hectares, which belong mainly to Sugar Cane Agro-Industry Companies, that is the most high water demanding crop, hence reclamation work; the subsurface drainage computation is done for this perennial and industrial crop.

12.6.5 Further Steps

Further to the steps taken in the direction of nationalizing forests and pastures, some areas were either afforested or conserved in terms of grazing and cutting. The Forest Department was responsible in carrying out the planned policy. At the same time, attention was also focussed towards stabilizing the sand dune lands mainly at the margins of the central and southern deserts (e.g. projects in Kashan, Yazd and Kerman), and to the areas that have lost their natural vegetation cover. Carbon sequestration was realized to be an appropriate way to increase soil fertility, and consequently preventing erosion and improving the hydrological system (Fahiminia and Afzali 2013).

12.7 Conclusions

Iran, being an oil producing country, chose for a dualistic development policy including agricultural and industrial sectors. This policy was not fully achieved partly due to management problems. Besides, there occurred a considerable migration of rural population to several large cities where industry was concentrated. On the other hand, agricultural inputs, such as machinery, improved seed, fertilizer, pesticides, were made available to the farmers without sufficient training on the proper use of these inputs.

Mismanagement of the land resources may be considered as one of the major causes of soil degradation, endangering the current and the future capacity of the soil. Thus, in order to reach equilibrium between the socio-economic and the physical environment, proper planning is required. Such a planning can never be copied from elsewhere, or planned based on dreams, but must be fully directed to the local social systems. Traditionally, relations between the peasants and the landowner were, for the most part, based on a muzara–eh (sharecropping) contract. This was the basic form of feudalistic exploitation of the peasant, which could not be accepted anymore.

No doubt that industrialization must not, and cannot, be stopped, but can be directed towards the agricultural needs, e.g. agro industries. Although agricultural planning based on the physical and socio-economic conditions is in progress, environmentalists, who expect faster actions, are worried about the future of the land, referring to the statistics reported on various land degradation types; the enormous soil loss that is resulted from erosion, salinization, etc.

Taking into consideration the causes and effects of the induced degradation, and having checked them against the listed strategies to overcome further environmental deterioration, it must be concluded that there still much to be done on people's awareness of the role of the soil in our life. In other words, extension service and the media must hand in hand with Soil and Water Research Institute (SWRI), Soil Conservation and Watershed Management Research Institute (SCWMRI), Forests, Ranges and Watershed Organization, universities, training institutions, Soil Science Society of Iran (ISSS) and others are expected to take care of people's awareness.

References

Afshar H (ed.) (1985) Iran, A revolution in turmoil. The MacMillan Press LTD, London

Banaei MH (1997) Soil Climate Regimes of Iran (in Persian). Soil and Water Research Institute of Iran, Tehran

Banaei MH, Moameni A, Bybordi M, Malakouti MJ (2005) The soil of Iran (Persian text), Soil and Water Research Institute (SWRI), Tehran. ISBN: 964-8871-00-0

Bergsma E, Charman P, Gibbons F, Hurni H, Moldenhauer WC, Panichapong S (1996) Terminology for soil erosion and conservation. ISSS-AISS-IBG, ITC, ISRIC. ISBN 90-71556-15-8. Recommended reference: ISSS

Buringh P (1970) Introduction to the study of soils in tropical and subtropical regions. Centre for Agricultural Publishing and Documentation Wageningen, The Netherlands, p 99

Denman DR (1978) Land Reforms of Shah and People. In: Iran Under the Pahlavis (see Lenczowski)

ECONEWS (2009) Soil degradation in Iran. http://www.econews.ir/fa/newscontent.aspx

Fahiminia M, Afzali SF(2013) Effects of wind erosion on soil organic carbon dynamics and other soil properties: Dejgah catchment, Farashband County, Shiraz Province. Department of Desert Management, College of Agriculture, Shiraz University, Shiraz, Iran

FAO (1976) A framework for land evaluation. FAO Soil Bulletin No. 32. FAO, Rome, Italy

Farshad A (1968–1978) Over 20 reports on soil survey and land evaluation of scattered regions in Iran. Soil Institute of Iran, Tehran

Farshad A (1990) A general overview on the effect of agricultural systems on the quality of soil in the semi- arid regions of Iran. In: 14th international congress of soil science: Transactions: Kyoto, Japan, August 1990: Vol. VI, Commission VI, pp. 219–220, 18 p

Farshad A (1997) Analysis of integrated soil and water management practices within different agricultural systems under semi—arid conditions of Iran and evaluation of their sustainability. ITC Publication 57, ITC Dissertation 45, 395 p. ISBN: 90-6164-142-X

Farshad A (2008) Lecture Notes on Salinization/Alkalinization. Monitoring and Evaluation, The Netherlands

Farshad A (2011) Mapping units in degradation/conservation oriented studies. Int J Nat Resour Mar Sci 1(2):125–144

Farshad A, Zinck JA, Stoops G (2005) Human impact on soil quality: Assessment of structure degradation using digital image analysis: A case study of Hamadan, Iran. In: Proceedings of the international conference on human impacts on soil quality attributes, Isfahan, Iran. 12–16 Sept 2005, pp. 624–627

Fisher WB (1968) The Cambridge History of Iran, Vol.: The Land of Iran. University Press, Cambridge

Hall AE, Cannell GH, Lawton HW (ed) (1979) Agriculture in Semi-Arid Environments. Springer Verslag Berlin, Heidelberg, NewYork

Kalhore M, Jalalian A (2003) Estimation of water erosion-induced soil loss in rainfedhilly land of Rimeleh watershed in Lorestan, Iran. Seventh International Dryland Development Conference. Tehran, Iran, 14–17 Sept 2003. Poster Abstracts http://www.icarda.cgiar.org/Abstracts/PDF/Theme P_1.pdf

Lahsaeizadeh A (1993) Contemporary rural Iran. Brookfield, USA, Aldershot

Momeni A (1999) Soil quality changes under long-term wheat cultivation in the marvdasht plain, South Central Iran. ITC Dissertation number 57, p 275. ISBN: 90-6164-236-1

Momeni A (2003) Impact of land utilization systems on agricultural productivity. Asian Productivity Organization

Momeni A (2009 = 1389 Iranian calander) Geographic distribution of saline surfaces and of soil resources in Iran. Soil Res J (Soil and Water Sciences)/ Alef/Vol.24/Number 3 (in Persian)

Niknami D (2009) Presentation (SCWMRI) at the "Work on Improving Agricultural Productivity, Water Use Efficiency and Strengthening Rural Livelihoods". BKK, Thailand

Oldeman LR (ed) (1988) Guidelines for general assessment of the status of human-induced soil degradation. ISRIC and UNEP in cooperation with Winand Staring Centre-ISSS-FAO-ITC, Global Assessment of Soil Degradation (Glasod)

Oldeman R, Hakkeling RTA, Sombroek WG (1990) World map of the status of human-induced soil degradation, an explanatory note. ISRIC and UNEP in cooperation with Winand Staring Centre-ISSS-FAO-ITC

Pazira A (2012) Conservation of physical land resources for agricultural production (Soil and Water). Iranian National Committee on Irrigation and Drainage (www.irncid.org). Publ. No. 50

Pazira A, Homaee M (2010) Salt leaching efficiency of subsurface drainage systems at presence of diffusing saline water table boundary: A case study in khuzestan plains, iran. XVIIth World Congress of the International Commission of Agricultural Engineering (CIGR). Hosted by the Canadian Society for Bioengineering (CSBE/SCGAB) Québec City, Canada, June 13–17

Pazira A, Minami I (1981) Irrigation and desalinization in Iran. Laboratory of water use engineering, Kyoto University, Japan

Polous K, Farshad A, Zarinkafsh M, Roozitalab MH(2011) Evaluation of the spatio-temporal of soil salinity variation using data mining approach. ICSDM2011 and BJ-IWGIS2011 Conference to be held at Fuzhou, Fujian, China from June 29 to July 1, 2011

PSIAC (1968) Report of the water management subcommittee on factors affecting sediment yield in the Pacific Southwest Area andselection and evaluation of measures for reduction of erosion and sediment yield. ASCE, 98, Report No. HY12

Roozitalab MH (1993) Soil survey in Iran. In: Zinck JA (ed) Soil surveys perspectives and strategies for the 21st century. ITC Publ. 21, Enschede, The Netherlands

Safinezhad J (1989) Boneh; traditional agricultural systems in Iran (in Persian). Sepehr Publisher, Iran

Schiller O (1972) Agrarian structures in a changing world. Publications of the research centre for international agrarian develpment, 1. Verlag der SSIP-Schriften—Saarbrucken

SCWMRI (Soil Conservation and Watershed Management Research Institute) (2008) National Atlas of Watersheds

Shrestha RP (2011) land degradation and biodiversity loss in Southeast Asia (Chapter 15). In: TrisuratY, ShresthaRP, Alkemade R (eds) Land use, climate change and biodiversity modelling: Perspective and applications. IGI Global

Sokouti OR, Farshad A (1999) Land-Use compatibility index for assessing agricultural sustainability in uromieh, Iran. Desertification Control Bulletin No. 35. United Nations Environment Program

UNDP (1999) Human development report of the islamic Republic of IRAN. Chapter 8, pp 113

Vafakhah M (2011) Mohseni Saravi (2011) Optimizing Management of Soil Erosion in Orazan Sub- basin. Iran. J. Agr. Sci. Tech. 13:717–726

Siadat H (no date) Induced soil degradation in Iran that is resulted from mismanagement, and the mitigation ways (In Persian)

Wikipedia http://en.wikipedia.org/wiki/List_of_dams_and_reservoirs_in_Iran

Zinck JA (1988) Physiography and soils lecture notes, ITC, Enschede, The Netherlands

Zinck JA, Metternicht G, Bocco G, del Valle HF (eds) (2016) Geopedology; AnIntegration of Geomorphology and Pedology for Soil and Landscape Studies. Springer, p 550

Zonneveld IS (1995) Land ecology. An introduction to landscape ecology as a base for land evaluation, land management and conservation. Amsterdam, SPB Academic Publisher, The Netherlands

Author Biographies

Dr. A. Farshad is a senior and retired scientific member of the ESA Department of ITC, Twente University, in The Netherlands. Backed with over 10 years of field experience in Iran, he joined ITC in 1980, where he taught soil-related topics, doing research and supervising students for 30 years. His job also made him cooperating with various institutions in different countries, namely Australia, Bangladesh, Belgium, Botswana, Cape Verde, Egypt, Ghana, Iran, Kenya, Mexico, Morocco, Nigeria, Portugal, Spain, Sri-Lanka, Thailand, Tunisia, Turkey, UAE, and USA.

Dr. Ebrahim Pazira Professor of land and water resources engineering and head of soil science department at Islamic Azad University in Tehran. His main research topics are reclamation of salt affected-soils, soil physics, drainage engineering, water quality evaluation, soil and water conservation engineering.

Dr. Ali Akbar Noroozi is senior researcher at the Soil Conservation and Watershed Management Research Institute (SCWMRI). He is Head of Watershed Management Department in the SCWMRI. His main research interests are application of remote sensing in soil, particularly erosion, salinity, moisture, organic matter and soil mapping and land evaluation.

Land Management in Space and Time

13

Abbas Farshad, Mohammad Hassan Roozitalab,
and Mohammad Reza Jalal Kamali

Abstract

This chapter begins with a brief description of the general concepts of soil/land management, followed by a historical overview of soil and water management in Iran from the beginning of agriculture to nowadays. Making use of geopedologic indicators, carbon dating, mineralogy, archeological studies, and history has led to propose interesting theories about the genesis of the various soilscapes, which have controlled the land use and soil and water management. The chapter also reports the land management practices devoted to soil and water conservation, many of them invented in Persia. Furthermore, the ongoing efforts toward the adoption of conservation agriculture technologies are shortly described.

Keywords

Soil/land management • A historical overview • Geopedologic/archaeologic indicators
Conventional agriculture • Soil and water conservation • Conservation agriculture

13.1 Generalities

Land management is defined as the process of managing the use and development of land resources to prevent degradation, with all its consequences. The long history of agriculture in Iran (see Chap. 2) clearly shows that people have always struggled to find ways to produce their needs (food, fiber, etc.) under a less to an unfavorable environment. The achievements, though not always sustainable, have led to considerable innovations in fabricating devices (e.g., *khish*, *Persian wheel*) and water harvesting (e.g., ghanat, various forms of *dam*). Considering the advances in science and in technology that have resulted in different socioeconomic and ecological environments, management approaches have continually been adapted.[1] This is indeed the subject of attention here, with an obvious emphasis on the soil quality. It is true that one speaks of 'soil management,' but in practice, a wide range of physico-socioeconomic attributes, i.e., the constituents of the 'land,' are involved in the management. Land being a wider concept than soil or terrain is defined as the physical environment, including climate, relief, soils, hydrology, and vegetation, to the extent that these attributes influence potential for land use, that is based on the functional dimension of land for different human purposes or economic activities. 'The fitness of soils for land use cannot be assessed in isolation from other aspects of the environment, and hence, it is land which is employed as the basis for suitability evaluation' (FAO 1976).

Managing soil does not merely refer to control degradation, or to maintain what there is, but it also implies enhancement of soil quality. This may simply be expressed in a pseudo-equation: '*Soil management = maintenance + quality enhancement.*' Management is materialized in a set of practices

A. Farshad (✉)
Earth System Analysis Department (ESA) of ITC,
Twente University, Enschede, The Netherlands
e-mail: farshad@itc.nl; abbasfarshad@gmail.com

M.H. Roozitalab
Soil and Water Research Institute (SWRI), Tehran, Iran

M.R.J. Kamali
Seed and Plant Improvement Institute (SPII), Karaj, Iran

[1]https://en.wikipedia.org/wiki/Agriculture_in_Iran#History.

M.H. Roozitalab et al. (eds.), *The Soils of Iran*, World Soils Book Series,
https://doi.org/10.1007/978-3-319-69048-3_13

that are adapted according to soil conditions, considering that not only soils are dissimilar, but also other contexts of the land, including climatic conditions, topography, accessibility, and the size of holding, vary from place to place. Variation in soils and landforms is the basis for forming land units that are used to evaluate land use suitability. It is indeed for this reason why soil information is very determinant (FAO 1976). Regarding soil variability and the fact that not all soils react similarly under a given use, various practices can be expected to comply with the possible tasks that might be prescribed in the above equation. Highly integrative generalizations are made for what are called '*management groups*' (USDA Soil Survey Staff 1993: page 284). Management groups identify soils that require similar kinds of practices to achieve acceptable performance for a soil use (see Chap. 9). To name a few examples, *Vertisols* (rich in swelling-shrinking type of clay) or *Gleysols* (wet soils) cannot be managed in the same way as *Xerosols* (dry soils of arid regions) or as *Lithosols* (shallow soils, with *lithic* contact). Obviously, variations within each of these soils will also urge the user to select and apply the appropriate practice.

Fig. 13.1 Salids (near Malayer) with the salic horizon at a depth of ±50 cm (Farshad 1997)

13.2 Why Soil Management?

The increasing demand for intensification in cultivation, especially in the areas with less favorable conditions, intensifies the role of the land management. An appropriate equilibrium must be achieved between human factors, socioeconomic conditions, and the factors of the physical environment (Sys et al. 1991). Various examples of soil mismanagement can be recalled answering the question (Lal and Stewart 2013; Farshad 2015). A very clear example is the way a Solonchak (Fig. 13.1), with the salic horizon occurring at ±50 cm depth, was managed in a rain-fed agricultural system to cultivate barley. Using the traditional animal-drawn wooden plow—*khish* (Fig. 13.5a, d), topsoil was scratched, deep enough to broadcast the seed. Obviously, the yield was not so high, but it was something, say better than nothing. The same soil, by the time that tractor came in farmers' life, was plowed using long metal blades, drawn by a tractor. The result was that broken fragments of the salic horizon were mixed with topsoil so that the area was turned into a bare-land, where nothing could grow.

Another example, which clearly shows not only the role of the soil management, but that of the land management, is that of the vast area in the southwestern province of Khuzestan (southwestern Iran), which was brought under the hydroelectric DEZ dam water supply, in 1960s (https://en.wikipedia.org/wiki/Dez_Dam). A vast area was brought under irrigation system, not because the dominant soils were either already salt-affected or being prone to. Therefore, the salinity of the soils in the area was increased (Fig. 13.2), and installation of a drainage system becomes imperative in

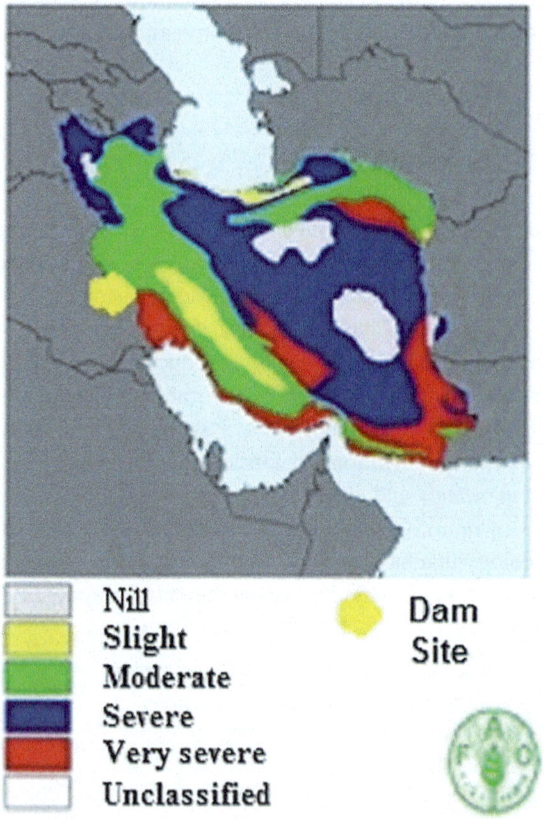

Fig. 13.2 Soil degradation because of mismanagement (see Chap. 12)

some places for an economic agriculture. Harsh climatic conditions, occurrence of salt-bearing lithology, insufficient or no vegetative cover, and lack of fresh water are the major causes in a great part of the country, not to forget the problem of groundwater fluctuations mainly in coastal areas (Pishkar 2003) and increasing depletion of the groundwater in many plains of the country. According to the information provided by the Ministry of Energy, in 2010 due to overdraft of the groundwater resources 499 out of 609 agricultural plains of the country are increasingly facing either water scarcity or crisis. Studies also show that 277 plains in the country are in a critical condition and declining groundwater table that has led to an annual land subsidence in some places (Madani 2014). The total overdraft of the groundwater resources in country in 2010, as reported by the Ministry of Energy, was more than 9 BCM compared to about 5 BCM a decade before. An example of a bad management that has led to the decline of groundwater table in Yazd Province, demonstrates how aridification (Farshad 1997) is accelerated.

Soils are generally characterized by an ochric epipedon (poor in organic matter content) which either directly overlies a C layer (Entisols) or, in a more developed form, is underlied by a subsurface horizon such as calcic, salic, gypsic or cambic (Aridisols). In general, Aridisols, if have cambic horizons, are usually more fertile, and if having any of the salic, gypsic, petrocalcic, and/or petrogypsic subsurface horizons, they then may be toxic or difficult to manage for food, feed, and fiber production (see Chap. 7).

Unfortunately, soil surveyors are often reluctant when coming across the salt-affected soils, as they are quickly classified as Salids, not sufficiently expressive for further interpretations. The fact is that besides such subgroups as Typic, Lithic, Aquic, Sodic, and Fluventic, which give important criteria for interpretation purposes, family and eventually series differentiae such as particle size class, mineralogy, horizon thickness, and arrangement can be vital when interpreting soil survey results for irrigation and reclamation purposes.

The provincial capital Yazd, about 600 km southeast of Tehran, is an ancient town within the fringe of the Central Desert, known for its traditional irrigation systems (*ghanats*[2]). The mean annual rainfall is below 100 mm, with high intensity. The mean annual temperature is about 20 °C, with a maximum of over 45 °C in July and a minimum of −16 °C in January and February, with a frost period of over 40 days. Summers are very long with very hot days. The annual evaporation is very high, varying from 2500 to 3500 mm (Ghobadian 1982). Thus, farming in Yazd cannot be done without irrigation. Considering the climatic conditions, surface water is very limited. In other words, all water needed in agricultural and industrial sectors must come from non-renewable aquifers. Although from about 13,000,000 ha the total surface area of the province, some 110,000 ha (about %1) is used for farming, many deep wells have been excavated to satisfy the needs of the above-mentioned sectors. The excavation of too many wells has caused not only the extinction of the traditional irrigation systems (ghanats), but it also has led to drying up of aquifers. This is evident in many plains in Khorasan, Fars, Isfahan, Kerman, Tehran, Qazvin, Mashhad, and other provinces by land subsidence, leading to the formation of deep cracks (Fig. 13.3), which have damaged infrastructure and led to obvious gully and tunnel erosion (Farshad 1997).

Our studies in the *Sharra* district, in Hamadan Province, showed that the salt-affected soils in this area are of anthropogenic origin, in line with its historic background, supported by archeological and carbon-dating studies (Farshad 1997, 2013). In contrast to the Hamadan–Bahar area, where salt-affected soils occur only in a small depression—called Nahalestan, the Sharra area looks like desert's periphery. Being endowed with the *Sharra* river water, the lithofunctional soils have been mismanaged by over-irrigation. The misuse of the Sharra river water for centuries has led to salinization.

Modernization of farming practices, associated with the use of machineries to prepare fields, harvest and transport crops (Fig. 13.4), and the use of chemicals demands appropriate management. Over the last 5 to 6 decades, use of heavy machineries in some places has led to soil compaction and consequently to yield reduction. In a study, the effect of traditional and advanced tillage, and the effect of converting the initially dry-farmed land into irrigated cultivation, on soil porosity and structure was investigated (Farshad 1994;

Fig. 13.3 Deep cracks as a result of land subsidence (Farshad 1997)

[2]قنات (in Persian) Also written as *Qanat*.

Fig. 13.4 Using heavy machines to level land (Aridisols) in Khuzestan (*photo* N. Ebrahimi)

Farshad et al. 2005). A comparison between the pore space (surface area) in a thin section prepared from a soil managed traditionally and that in a soil under mechanized agriculture with the soil under pasture clearly showed the adverse effect of agricultural machineries on soil compaction. The microscopic study of the thin sections also revealed that converting dry-farmed land into irrigated cultivation led to increased soil compaction. This plus the use of chemical fertilizers on calcareous and in places on saline soils has led to yield reduction (Farshad 1997).

13.3 A Historical Overview of Indigenous Knowledge and Technologies

The history of civilization and agriculture to produce food needed for urban population in some places of Iran such as Susa (Shush) goes back to more than 4000 BC (https://en.wikipedia.org/wiki/Susa). Till the Safavids dynasty (sixteenth century), indigenous agricultural knowledge was reasonably intact as it was not yet influenced by imported techniques. Information on how to make the best use of water, soil, seeds, and manure may be traced back to the fourteenth century in a limited number of documents (Chap. 2).

13.3.1 Soil-Based Measures

(a) **Testing soil moisture content**

To estimate the depth to groundwater and assess whether a soil could be used for dry-farming, a pit with approximate dimensions of 3 × 3 × 3 m was excavated wherein a large cook-pan was placed. Before placing the pan, its inner side was smeared with oil, and a piece of cotton-wool was fixed in the center of the pan's inner bottom. The pan was then placed upside down in the pit and covered with the original soil. After one night, early in the following morning, the pan was uncovered to check whether the piece of cotton had become moist, in which case farming without irrigation could be practiced.

(b) **Check on physical condition**

To assess the quality of soil structure, a 1 cubic meter pit was excavated and immediately refilled with the original disturbed soil material. The following cases were possible to occur:

(i) When the excavated soil materials were more than what was needed to refill the pit, it was concluded that the soil was well structured. A well-structured soil was classified as a fertile soil.
(ii) When the soil material was insufficient to refill the pit, the soil was considered poorly structured.
(iii) When the soil material was just enough to refill the pit, the structure was considered moderately developed.

(c) **Fertility maintenance and manuring**

Migrating tribes took care that the land parcels planned to be cultivated in the following season be manured by stabling animals on the parcels. This was a common practice when land extent was not restricted. In situations of land scarcity, productivity was increased through fertilization by crop rotation, including nitrogen-fixing crops such as clover, peas, and mung beans. Traditionally, four kinds of manure were distinguished, namely bird excrements, cattle manure, green manure, and the manuring through leaves. A mixture of a few kinds was recommended.

Animal raising was an important component of traditional agriculture. Cow dung was mostly used for fuel, but also applied to the land, particularly in orchards. Chicken, pigeon, horse, and human excrements were also used in summer crops (=*seifi*) such as vegetables and melons. Fertility in exhausted wheat and barley fields was restored mainly through fallow and rotation practices, but also by adding ash, house wastes, and materials from old house roofs and abandoned living places. Some decades ago, villagers practically ignored the use of chemical fertilizers (Wilber 1950). The application of agrochemicals grew from little to 32,000 tons per annum in the early 1960s, to 675,000 tons by the mid-1970s (Karshenas 1990). Since then, the application of chemical fertilizers has increased to more than

1.8 million tons in 2013 (Ministry of Jihad-e-Agriculture, unpublished data).

13.3.2 Farming Practices

(a) **Tillage (manual and/or animal-drawn wooden plow)**

Traditional land preparation for wheat and barley cultivation started with plowing at the beginning of spring, around the Persian New Year (*nowrooz*[3]), after the land had been under fallow for one year. The same tract of land was again plowed in autumn, followed by the broadcasting of barley seed. In the case of wheat, seed broadcasting took place about 20 days after the second plowing. Nowadays, traditional tools are only used in places where no machinery can reach. Animal-drawn plows, diverse types of harrow, and scoops for preparing furrows are abandoned or only applied in very remote sloping areas. The traditional draft animal was the ox, but mules, horses, and donkeys which were common too in the past.

(b) **Tools**

Khish, the traditional animal-drawn wooden plow, is pulled by two oxen while the farmer who walks behind controls the tracking. Depending on the soil type, an adapted type of *khish* was used (Fig. 13.5). Some other tools used in various phases of farming are illustrated too (a–e).

13.3.3 Irrigation Water; Supply and Management

In the past, water supply varied with the physiographic conditions of the area. In mountains, runoff water was harvested; in hilly areas, a neighboring river water was diverted; in piedmonts, underground tunnels (*ghanat*) were excavated; and in valleys, water was directly hauled from river channels (Farshad and Zinck 2001). Storing water was an important issue, where various ways were invented. Arabkhedir and Kamali (2006) give a long list of the ways (e.g., *Khoshab*, *bast*, *sooma*, *tirband*, *hootak*) of storing water in different places, adopted according to the physiographic condition.

[3]Also, written as Nowruz; usually occurs on March 21 or the previous/following day, depending on where it is observed (https://en.wikipedia.org/wiki/Nowruz).

(a) **Underground tunnels (*ghanat*)**

An underground tunnel dug into alluvial deposits on the mountain skirt taps the aquifer and brings the water to the surface by gravity flow (Figs. 13.6, 13.7, and 13.8). The construction of a ghanat system was a time-demanding engineering work, costly and risky for builders (Safinejad 1979; Ghobadian 1982; Kardavani 1990; Farshad 1997). An experienced senior ghanat builder (called *moghanni* in Persian) with his team is charged to do the work. According to Safinejad (1979), 14,500 man days' work is required to construct a ghanat chain of 6 km length. According to the information provided by the Ministry of Energy, the total numbers of ghanat were about 39,500 with a total annual delivery of about 6.3 BCM in 2012 (Daemi 2014)

(b) **Springs (*cheshmeh*)**

Spring, cheshmeh in Persian (Fig. 13.9), is still an important water source in many places throughout the country. As an example, about 20 years ago, the annual water delivery from 389 springs in the Hamadan–Bahar plain was 27 million cubic meters. Within roughly a decade, the number of active springs decreased to 21 and the water discharge to 0.64 million cubic meters (Djamab 1990; Farshad 1997). In 2012, the total number of springs in the country was about 160,000 with a total annual delivery of about 17.5 BCM (Daemi 2014). In recent years, the excavation of deep and semi-deep wells led to drying out of many ghanats and springs.

(c) **Harvesting runoff water (*seilaub*)**

In the sloping areas of the upper catchments, making use of runoff water is a common practice in the traditional agriculture, although in an inefficient way and without any effort to improve the technique. In some places, water is stored in roughly constructed pools (*estakhr*) with a hole at the bottom, which opens into a channel (*djoob*). The hole is kept closed with a piece of wooden beam (*dirak*), which is pulled off to start irrigation.

(d) **Hauling water from rivers**

The Persian wheel (*doolaub*) (Fig. 13.10) is made of wooden lathes and a number of buckets fixed all around, allowed hauling water from river channels. Usually, animals were used to turn the wheel (Farshad 1982).

Taking irrigated wheat farming as an example (Farshad 1997), activities included: (1) land preparation (plowing and leveling) and sowing. Land was plowed three times using *khish* (a traditionally made wooden device pulled by two oxen). The first time was in mid-April, the second plow was roughly one month later, and the third time was in the last

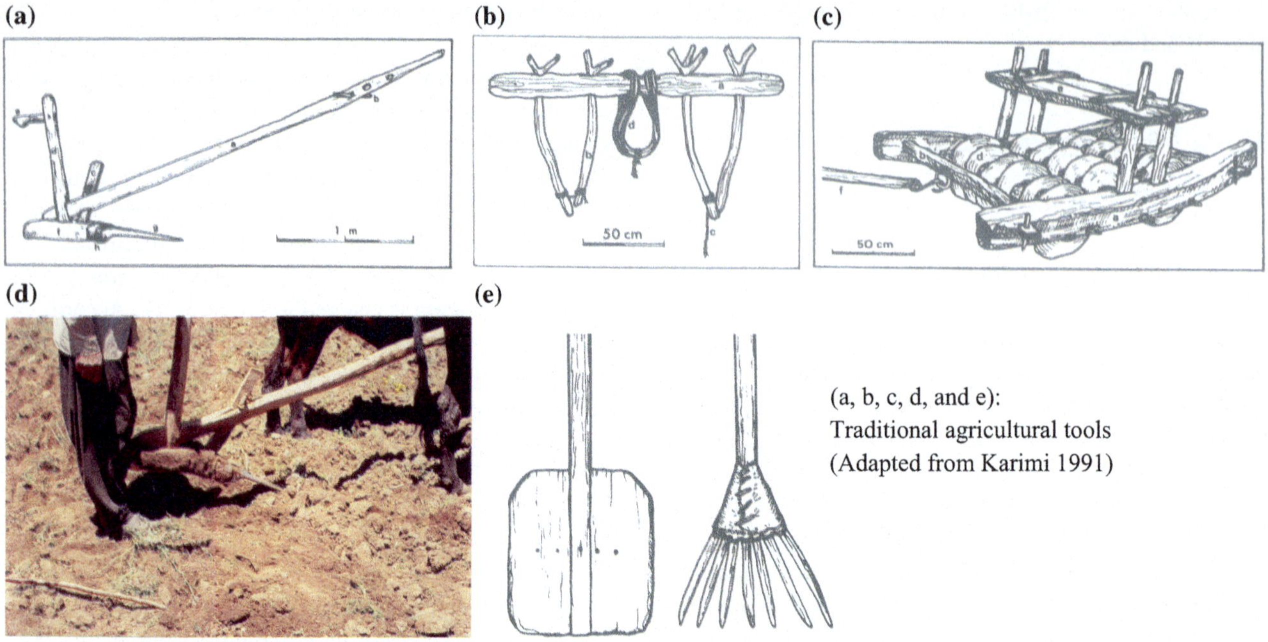

Fig. 13.5 **a** *Khish*, used in Hamadan; **b** *Yough* for fastening khish on to the two oxen; **c** *Kharman-koob or yowashen*, used to separate straw from grains; **d** original *khish*, fixed on the oxen; **e** types of fork used to harvest grains, separating straw from grains

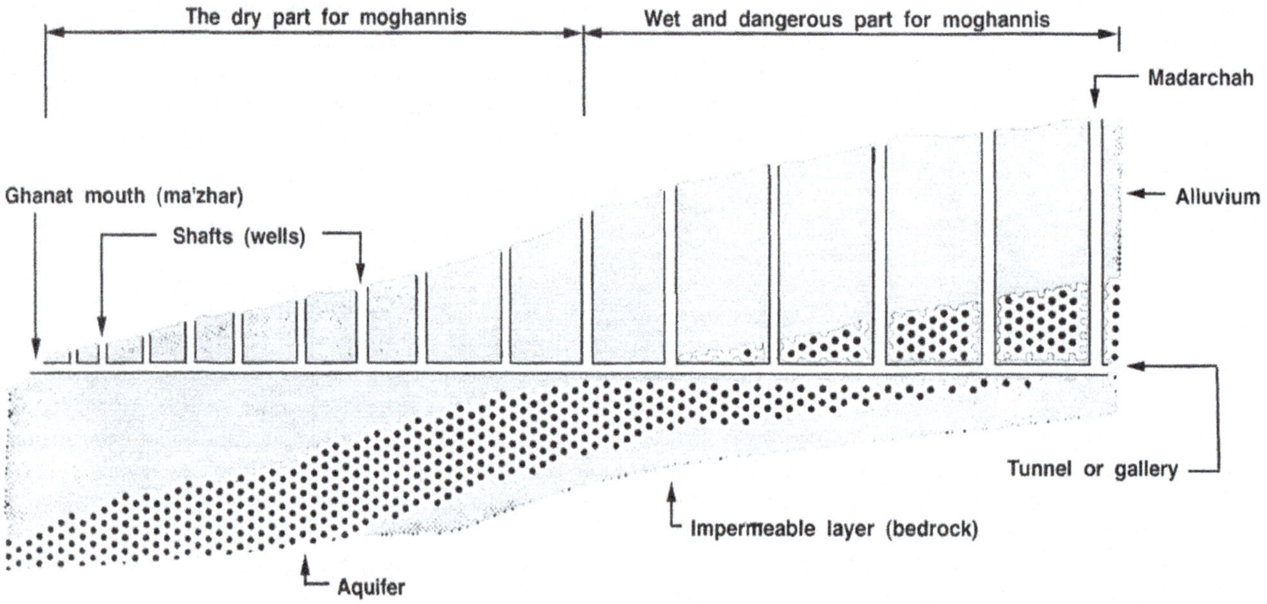

Fig. 13.6 Schematic cross section of ghanat system (adapted from Farshad and Zinck 1998)

week of September, after a lengthy period of fallow. Before the third plowing, after seeding, the land was irrigated and left to reach field capacity; (2) irrigation was done in three times with 12 days' interval; (3) fertilizing is done using manure, which was a farm system product. The manure produced by ten cows/year was enough for 3 hectares; (4) harvesting included cutting, pounding, out-blowing for grain-straw separation, and sieving. All these activities were performed either by hand or by animals; although once farmers got acquaintance with tractor, machine was used for pounding.

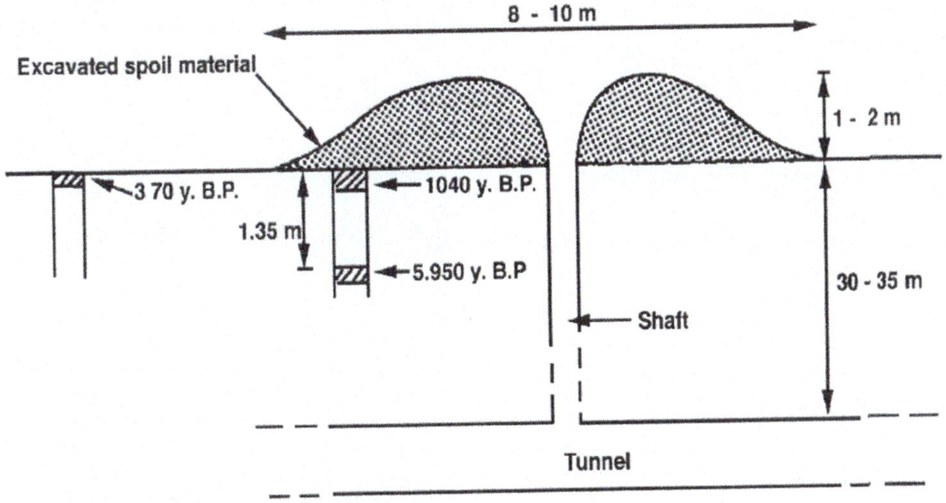

Fig. 13.7 Schematic shaft/mound, also demonstrating sampling for C14 dating (adapted from Farshad and Zinck 1998)

Fig. 13.8 Collapsed ghanat system depicting a few shafts and underground tunnel (Farshad 1997)

Fig. 13.9 Karst spring (Farshad 1997)

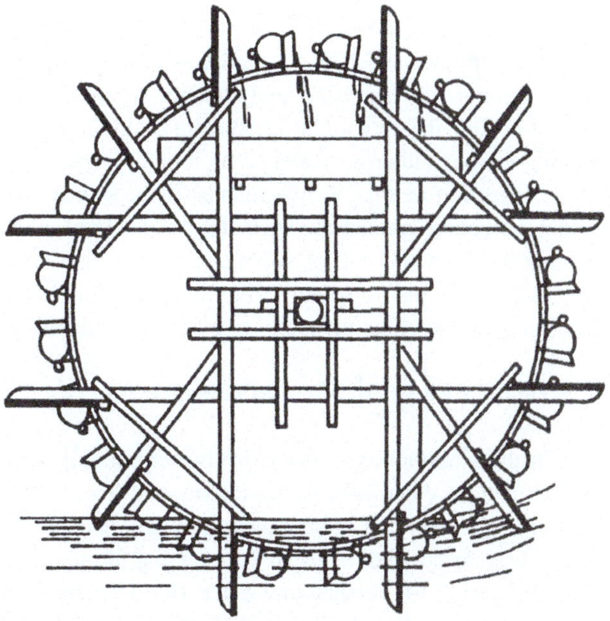

Fig. 13.10 Persian wheel (adapted from Farshad 1982)

13.4 Modified (Conventional) Agriculture; (Semi-)Mechanized

With the introduction of new sources of energy, technology and machinery and the changes in the socioeconomic norms, began roughly in the 50s, a new agricultural paradigm was born.

The well coevolved input–output relationship in the traditional agriculture, backed by a long history, could not secure the needs and hence was replaced by market-oriented, specialized, and independent production sectors (Farshad 1997; Farshad and Zinck 1993).

13.4.1 Contemporary Common Activities

As an example, activities in the case of irrigated wheat, that is widely grown in Iran, farming under a semiarid condition can be summarized as follows: (1) land preparation (plowing in spring and leveling); (2) sowing is done in October; (3) crop care (fertilizer application, spraying insecticide, irrigation); (4) Harvesting; and (5) marketing.

(a) **Farming; tillage and tools**

Plowing is done in two rounds by a tractor with four blades, each 30 cm wide. The plowed parcel is then leveled using a 4 -m-wide leveler (*maleh* in Persian) installed on a tractor. For sowing, a deep row crop cultivator (*amighkar* in Persian) is used. Fertilizing is done using some 200 kg/ha phosphate fertilizer and 100–150 kg/ha urea. No manure is used, but commonly harvested land is grazed by animals that are kept outside day and night. Depending on the farm size, spraying is done either manually or by a pump installed on tractor. Irrigation is done in five rounds: twice shortly after sowing, sometime around the second week in April, about two months from the day of sowing (in June), and about 75 days from the day of sowing. Harvest, depending on the farm size, is done either manually or by means of combine.

(b) **Irrigation water; supply and management**

With the introduction of the groundwater pumping technologies and the changes in the socioeconomic structure, the traditional groundwater withdrawal through ghanats and springs was abandoned, replaced by pumping through deep wells (Madani 2014; Farshad and Zinck 1998). In 2012, the total number of wells was about 690,000 with a total water withdraw of about 47 BCM (Daemi 2014). This resulted in drying up of many springs and ghanats (see Chap. 12). No doubt, this change is associated with a different management, controlled by Water Department of the Ministry of Energy (Figs. 13.11 and 13.12).

Fig. 13.11 Pumping water using gasoil-driven pumps (*photo* Farshad)

Fig. 13.12 Pumping water, using wind energy (*photo* Farshad)

13.4.2 Conservation Agriculture (CA)

Conservation agriculture is not a new science, but it is a set of defined principles that can be used by all agronomists/crop managers to better insure the development of sustainable crop management practices for diverse cropping systems. According to FAO (http://www.fao.org/ag/ca/1a.html), CA is defined by the following three principles: (1) continuous minimum mechanical soil disturbance; (2) permanent organic soil cover; and (3) diversification of crop species grown in sequences and/or associations, which indicate an obvious emphasis on soil care and health. On the basis of the same principles, Sayre (2014) gives a comprehensive presentation (http://www.slideshare.net/CIMMYT/), concluding that CA, where farmers contribute to solve their problems, does not only lead to sustaining crop production (Fig. 13.13), but it also improves soil quality through increasing soil organic matter and reducing soil and environmental degradation (Fig. 13.14).

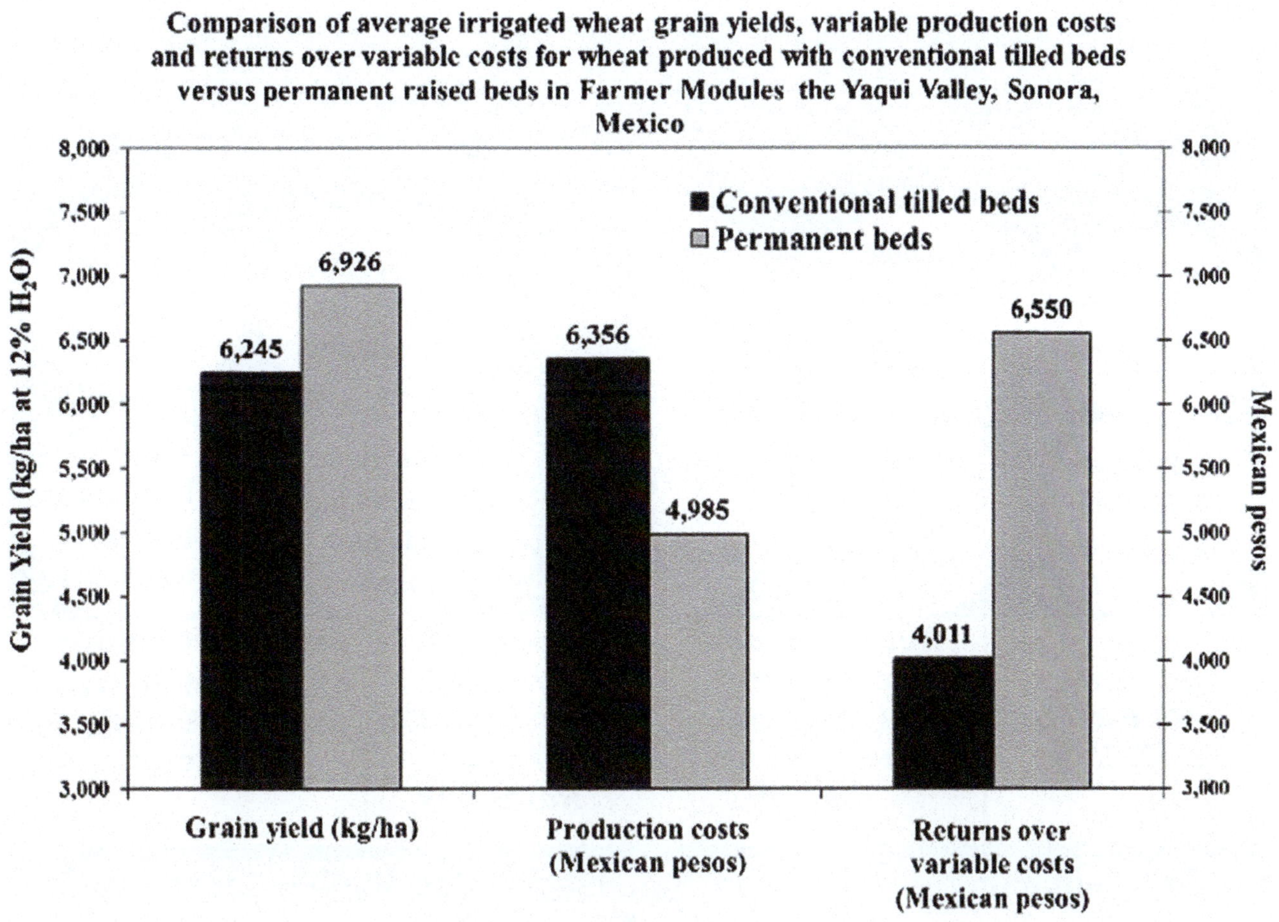

Fig. 13.13 One way to encourage farmers to go for CA (Sayre 2014)

Comparing the soil management-related activities in the traditional and the modified (conventional) farming, and considering the degree of degradation that is caused by bad management (see Chap. 12), concerns about the food security, not only in Iran but also worldwide, are increasing (FAO 2012; WASWC 2007). In the past when the world was less populated, land was utilized in a friendlier way, respecting fallow periods and other traditional soil and water management practices. With the population growth and changes in the social structure, land is used intensively, no time for resilience that should allow disturbed or depleted agroecosystems to recover.

Conservation agriculture involves changes in many conventional crop management practices as well as the mind-set of stakeholders including scientists, extension agents, farmers to overcome the conventional use of many tillage operations. Although adoption of CA is increasing worldwide and has reached globally about 157 M ha, corresponding to about 11% of field cropland, in some regions, it is progressing either slow or has not been practiced yet. 'Adoption has been intense mainly in North and South America, Australia, Asia, and recently in Europe and Africa' (Kassam et al. 2015; Table 13.1).

In his study, Sayre (2014) also concludes that there is a lack of information on the effects and interactions of key CA components, which affect crops' yields and technology's adoption.

For adoption of CA-based crop management technologies, farmers should believe that these practices would bring him/her economic benefits as well as sustainability in his cropping system (Sayre 2014).

Conservation agriculture was considered in Iran since 2007, and national efforts have been made to promote and develop CA-based crop management technologies for cropping systems in rain-fed and irrigated areas. In the Fifth National Development Plan (2010–2015), it was proposed to develop and apply CA-based management practices on 3 million hectares (2 million hectares in rain-fed and 1 M ha in irrigated areas) in the country. A national campaign was launched, and concluded, among other things, that many CA-based machineries have been imported from Western Europe and Brazil, which were heavy and expensive, and not suitable for many farmers in Iran (Fig. 13.15).

The latest statistics show that in 2015 there are more than 780,000 and 175,000 of so-called reduced and no-till tillage

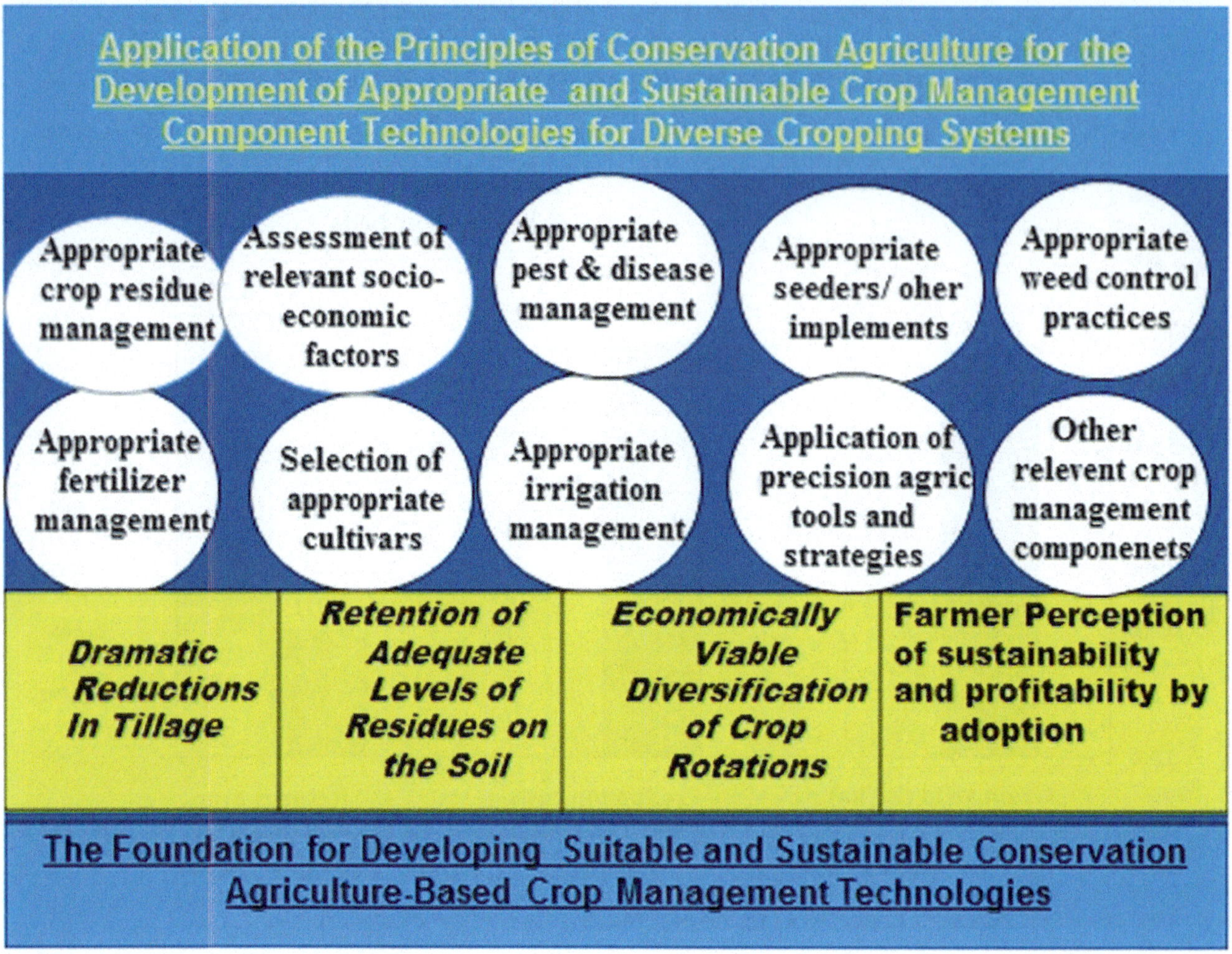

Fig. 13.14 Principles of conservation agriculture-based and some of its crop management components (*courtesy* Sayre 2014)

Table 13.1 Cropland area under CA (M ha), CA area as % of total cropland, and CA area as % of cropland by continent, in 2013

Continents	Cropland under CA (M ha)	Percent of global CA area	Percent of cropland
South America	66.4	42.3	60.0
North America	54.0	34.4	24.0
Australia and NZ	17.9	11.4	35.9
Asia	10.3	6.6	3.0
Russia and Ukraine	5.2	3.3	3.3
Europe	2.0	1.3	2.8
Africa	1.2	0.8	0.9
Global total	157.0	100	10.9

systems practiced in Iran, respectively (unpublished information from the Deputy for Crop Production, The Ministry of Agriculture). However, the actual use of the three principles of CA-based crop management practices is not fully respected, since almost no crop residues are being retained on the soil surface, and the system is interrupted using mold board plows or disk harrows, after a directly seeded crop is harvested. This is mainly due to baling off loose wheat straw after combine harvesting or full removal of residues from crops like chickpeas by hand harvesting followed by heavy grazing by sheep and goats, particularly in the rain-fed areas. Some of the grazing involves the farmer's own animals or grazing rights that the farmer sells to the sheep/goat owners. However, in many cases, it is by uncontrolled

Fig. 13.15 Examples of the imported direct seeders used in direct seeding practices in Iran (*courtesy* Sayre 2014)

grazing by nomads' animals. Suitable crop rotation is also not practiced, particularly in rain-fed areas.

Similarly, in irrigated areas, crop residue removal for most crops by baling off after harvesting is common as does residue burning to facilitate more timely seeding of the next crop, particularly if there are sizable amounts of loose residues, and there is no enough time to remove and apparently too much to directly seed into with available direct seeders and/or with the gravity irrigation systems used after seeding (Fig. 13.16).

Furthermore, in the rain-fed areas, one to several tillage operations, mainly using chisel-based tillage implements that combine some kind of spiked rollers in the back as well as some having additional disks that mainly 'smooth-out' the soil raised by the relatively large chisels operated fairly deep. However, on the other hand, this may be a potential positive step forward compared with the conventional tillage that farmers practice using mold board plowing followed by varying numbers of disk harrowing and other potential tillage operations to cope with 'hard soils' and weeds problems.

There are some Iranian manufacturers (e.g., in Tabriz and Arak) that are involved in developing CA-based machineries (http://www.agmachine.com/pmmd2.htm). Significant progress has been made in developing and manufacturing of high quality CA seeders for dry-farming, where farmers use 70–80 hp tractors.

Fig. 13.16 Baling off (left) and burning (right) wheat residues in Darab region in Fars Province in Iran (photographs from: Manoochehr Dastfal)

Some CA-based activities are practiced in various parts of the country, however, mostly led by agricultural engineers. Lack of multidisciplinary research for development of CA-based activities is an issue that has hindered the development and promotion of CA-based crop management technologies in Iran. The photographs below show some so-called CA-based activities in different provinces in Iran. However, weeds phobia is a mind-set that has been spread among farmers, particularly by weed scientists, without any side-by-side trails on farmers' fields or appropriate weed management practices in so-called CA-based systems. Integrated weed management is a key crop management component of CA-based crop management practices that has been neglected (Figs. 13.17, 13.18, 13.19, and 13.20).

CA-based crop management technologies are in developing stage in Iran, and cares must be taken as it is appropriately understood, developed and implemented before scaling out on large areas. Appropriate CA-based machines, multidisciplinary working, and changing mindsets are key elements for promotion and wide adoption of this CA-based crop management practices in Iran. CA-based crop management technologies are a potential to be used for saving water, time, improving soil quality, and preventing soil degradation (Fig. 13.21) that are determinant factors in crop production in Iran, should it is appropriately adapted and adopted by farmers.

While in the Fifth National Development Plan (2010–2015), 3 million ha of land (2 million rain-fed cropping systems and 1 million ha of irrigated land) was expected to be under minimum tillage or conservation agriculture (Kamali 2011), till 2016 only less than 1 million ha was actually brought under the above-mentioned farming system.

The government has recently prioritized resources for research, development, and capacity building in the areas of soil and water management, drought and watershed management, and conservation agriculture. The results and experiences obtained from conservation agriculture in pilot areas in Tehran, Khorasan, Kerman, Golestan, Fars, Isfahan, Hamadan, Gazvin, Dezful, and Ardabil are supposed to be used to prepare a national development project on conservation agriculture. In short, as previously indicated, the key farming activities (including land preparation, sowing, crop care, and harvesting) in various agro-climatic regions should be based on the three principles of "no soil tillage, maintenance of vegetative soil cover, and crop rotation" (Fig. 13.22).

13.4.3 Remarks

Studies to identify and evaluate tillage methods for sustainable crop production under different climatic and soil conditions, with contradictory results, are progressing (Hemmat 2004; Javadi et al. 2009). Wheat is cultivated at least two times a year, and the time for preparing soil is short. Comparing the conventional (using moldboard plow and disk harrows) with the surface-tillage, the former demands more time and energy besides the fact that it causes more implements depreciation. It is argued that as soil preparation with the surface tillage (no- and mulch-tillage) is faster, and there is no difference in yield; comparing the two methods, the surface tillage can be recommended in the irrigated wheat.

Fig. 13.17 Direct seeding of soybean in wheat residues (left) and established soybean crop in the residues of wheat (right) under irrigated conditions in Golestan Province, Iran (photographs from: Sohrab Sohrabi)

Fig. 13.18 Direct seeded maize in wheat residues (left) and established soybean crop in the residues of wheat (right) under irrigated conditions in Dezful in the north of Khuzestan Province, Iran (photographs from: Seyed Gholam Reza Ashrafizadeh)

Fig. 13.19 Direct seeded maize in wheat residues in Marvdasht (left) and established sesame crop in the residues of wheat (right) under irrigated conditions in Darab in Fars Province, Iran (photographs from: Manoochehr Dastfal)

Some other experiments conclude that surface tillage is a promising measure, with the remark that the effects of tillage methods differ over the regions. This technique, though reducing erosion, offers habitat for pests, meaning that heavier amounts of pesticides are required if crop rotation is not respected. The question about what would be the constraints in long term when applied under farmers' conditions has still to be answered. Regarding the yield, experiments have shown that some degree of soil disturbance is necessary, and hence, chisel plowing is recommended in most provinces of Iran (Javadi et al. 2009).

In short, we share the concluding points reported in WOCAT publication (Wocat 2007) that there are no simple 'silver bullet' solutions to the complex problems of land degradation. Understanding the ecological, social, and economic status in any given area, under study, is vital to set up the right land management, which should mitigate degradation. Obviously, comprehensive researches, where land

Fig. 13.20 Direct seeding of chickpea in wheat residues (left) and blooming chickpea crop in the residues of wheat (right) under rain-fed conditions in Maragheh field station in East Azerbaijan Province, Iran (photographs from: Iraj Eskandari)

Fig. 13.21 An abandoned village (Farshad 1997)

Fig. 13.22 Residue crasher device (e.g., after harvesting maize) (*photo* Jalal kamali)

users, scientists from different disciplines, and decision makers are involved, must be carried out before widely implementing any land management system.

Acknowledgements The authors are grateful to Dr. Ali Akbar Noroozi and Dr. Hamid Siadat for providing some relevant research papers/reports (in Persian) and for contributing useful comments on earlier drafts of this chapter.

References

Arabkhedir M, Kamali K (2006) Traditional methods of soil and water conservation in Iran (in Persian). Soil Consevation and Watershed Management Research Institute, Tehran, Iran

Daemi AR (2014) Managing water crisis ahead (in Persian). 2nd national conference on rainwater catchment systems, Mashhad, Iran, 19–20 Feb 2014

Djamab Consulting Engineering Co (1990) A general study of water resources at national level (several reports in Persian). Ministry of Power of Iran. Tehran, Iran

FAO (1976) A framework for land evaluation. FAO Soils Bulletin 32. Rome, Italy, p 87

FAO (2012) http://www.fao.org/docrep/017/i3028e/i3028e.pdf: The state of food and agriculture, and http://www.fao.org/docrep/016/i3027e/i3027e.pdf: The state of food insecurity in the world

Farshad M (1982) The history of engineering in Iran (in Persian). Naghsh-e-Djahan, Nashr-e-Balkh. Tehran, Iran, p 492

Farshad A (1994) Spectral image analysis in soil micromorphology with ILWIS. ITC Journal 2:168–171

Farshad A (1997) Analysis of integrated soil and water management practices within different agricultural systems under semi—arid conditions of Iran and evaluation of their sustainability. ITC Publication 57, Dissertation 45, p 395

Farshad A (2013) Geopedology reports historical changes in climate and agroecology: a case study from Northwestern Iran. ECOPERSIA 1(2):145–159

Farshad A (2015) Soilscape speaks out; an account of the past, present and future. ECOPERSIA 3(1):881–899. See also http://www.plr.ugent.be/50-anniversary/boekje.pdf

Farshad A, Zinck JA (1993) Seeking agricultural sustainability. Agr Ecosyst Environ 47:1–12

Farshad A, Zinck JA (1998) Traditional irrigation water harvesting and management in semiarid Western Iran: a case study of the Hamadan region. Water Int 23(3):146–154

Farshad A, Zinck JA (2001) Assessing agricultural sustainability using the six—pillar model: Iran as a case study. In: Gliessman SR (ed) Agroecosystem sustainability, developing practical strategies, CRC, Boca Raton ISBN 0-8493-0894-1. pp 137–151

Farshad A, Zinck JA, Stoops G (2005) Human impact on soil quality: assessment of structure degradation using digital image analysis: a case study of Hamadan, Iran. In: Proceedings of the international conference on human impacts on soil quality attributes, Esfahan, Iran, pp 624–627, 12–16 Sep 2005

Ghobadian A (1982) Central Iranian plateu; natural resources of Yazd province in relation to desert problems. Jundi Shapur University, Ahwaz, Iran, p 480

Hemmat A (2004) grain yields of irrigated and dryland continuous winter wheat under conservation and conventional tillage systems in Iran. American Society of Agricultural and Biological Engineers, St. Joseph, Michigan www.asabe.org

Jalal Kamali MR (Workshop Organizer) (2011) Iran's ambitious plans for conservation agriculture. http://blog.cimmyt.org/irans-ambitious-plans-for-conservation-agriculture/

Javadi A, Rahmani MH, Tabatabaeefar A (2009) Sustainable tillage methods for irrigated wheat production in different regions of Iran. Elsevier, Soil Tillage Res 104(1):143–149

Kardavani P (1990) The problems of water in Iran, vol. 1, 2nd edn. (in Persian), Tehran University Publ. 2029. Tehran, Iran, p 520

Karshenas M (1990) Oil, state and industrialization in Iran. Cambridge University Press, Cambridge, UK, p 308p

Kassam A, Friedrich T, Derpsch R, Klnzle J (2015) Overview of the worldwide spread of conservation agriculture (http://factsreports.rcvucs.org/3966)

Lal R, Stewart BA (eds) (2013) Principles of sustainable soil management in agroecosystems. CRC Press, ISBN: 978-1-46-651346-4. Hardcover, p 568

Madani K (2014) Water management in Iran: what is the looming crisis? J Environ Stud Sci, Springer, doi:10.1007/s13412-014-0182-z (see http://www.kysq.org/docs/Madani_Iran.pdf)

Ministry of Jihad-e-Agriculture (No date) A comprehensive report of conservation agriculture in Iran, Part 3

Pishkar A (2003) Analysis of the relationship between soil salinity dynamics and geopedologic properties; A case study of the Goorband area, Bandar Abbas, Iran. Unpublished MSc thesis, ITC, Enschede, The Netherlands, p 143

Safinejad J (1979) Traditional irrigation systems in Iran (in Persian). Publi. 89, University of Tehran, Tehran, Iran, p 277

Sayre K (2014) http://www.slideshare.net/CIMMYT/strategies-to-develop-and-extend-conservation-agriculturebased-crop-management-practices-to-farmers-cabased-applied-research-and-delivery-hub

Sys C, Van Ranst E, Debaveye J (1991) Land evaluation part I: principles in land evaluation and crop production calculations. International Training Centre for Post-Graduate Soil Scientists, University of Ghent, Belgium, p 273

USDA Soil Survey Staff (1951) and (1993) Soil survey manual. Agriculture handbook No. 18. USDA, Washington D.C., p 503

WASWC (World Association of Soil and Water Consevation) (2007) No-till farming systems. Special publication No. 3, p 540

Wilber DN (1950) Past and present. Princeton University Press, Princeton, New Jersey. second edn. Printed in the United States of America, p 243

WOCAT (2007) Where the land is greener; case studies and analysis of soil and water conservation initiatives worldwide. A Wocat (World Overview of Conservation Approaches and Technologies) publication, co-authored by CTA, FAO, UNEP and CDE. ISBN 978-92-9081-339-2, p 364

Author Biographies

Dr. A. Farshad has his B.Sc. degree in geology from the State University of Tehran (1965), M.Sc. in Soil Science (1975) and Ph.D. in Earth Sciences, with emphasis on agricultural sustainability (1997) from the State University of Ghent, Belgium. He is a senior and retired scientific member of the ESA Department of ITC, Twente University, in The Netherlands. Backed with over 10 years of field experience in Iran, he joined ITC in 1980, where he taught soil-related topics, using remote sensing (RS) and geographic information system (GIS), doing research and supervising students for 30 years. His job also made him cooperating with various institutions in different countries, namely Australia, Bangladesh, Belgium, Botswana, Cape Verde, Egypt, Ghana, Iran, Kenya, Mexico, Morocco, Nigeria, Portugal, Spain, Sri Lanka, Thailand, Tunisia, Turkey, UAE, and USA.

Dr. M. H. Roozitalab is a distinguished soil scientist with wide range of scientific and working experiences in Iran and abroad. He received his M.

Sc. (1974) and Ph.D. (1978) from Oklahoma State University, Stillwater, Oklahoma, USA. His major research and teaching activities at Tehran University have focused mainly on soil genesis and classification and clay mineralogy. He has contributed to the advancement of soil science in Iran as President of Soil Science Society for 17 years from 1992 to 2009 and Director General of Soil and Water Research Institute from 1983 to 1995. Dr. Roozitalab also served as Deputy Head of Agricultural Research, Education and Extension Organization (AREEO), in Iran from 1995 to 2006. He has contributed to strengthening the regional and global agricultural research collaboration as President of Association of Agricultural Research Institutions in the Near East and North Africa, AARINENA (1998–2000), Coordinator of Regional Highland Research Network of ICARDA (2008–2011), and Chairman of the Global Forum on Agricultural Research (GFAR) from 2002 to 2006.

Dr. Mohammad Reza Jalal Kamali is principal scientist and wheat breeder at Global Wheat Program of International Maize and Wheat Improvement Center (CIMMYT) in Iran. He has wide experiences on wheat breeding research activities in the national wheat program and has worked as wheat breeder at Seed and Plant Improvement Institute (SPII) in Iran before his retirement.

Soil Resources of Iran: Challenges and Recommendations

14

Hamid Siadat, Mohammad Hassan Roozitalab, and Abbas Farshad

Abstract

Agriculture has been practiced in Iran for millennia. The agricultural land of the country is presently fragile and rather limited, as only about 8% of the total land area is under annual crops and orchards. Soil resources of Iran are now facing big challenges which need to be seriously considered in the country's sustainable development plan. These challenges include lack of an effective national soil policy, land degradation and desertification (erosion, salinity, declining soil fertility and organic matter, deforestation, soil pollution, deterioration of soil's physical and biological condition), climate change, water shortage, and land-use changes. To combat these challenges, some recommendations are presented including formulation and implementation of an appropriate national soil policy, improving public awareness on the importance of soil resources, enhancing research quality, providing different thematic soil maps, improving agricultural extension and advisory services, and enabling the soil scientists to contribute and collaborate with other scientists in resolving the challenges facing the country's soil resources. These recommendations need to be urgently taken by the government to ensure sustainability of soil resources, environmental protection, and, consequently, well-being of the next generations.

Keywords

Climate change • Land degradation • Land use change • National soil policy Sustainability

14.1 Introduction

Agriculture has been practiced in Iran for millennia. Therefore, deterioration of soil resources of the country is now posing serious threats and challenges which need to be dealt with in the country's sustainable development plan. Sustainable development plan for any country requires good soil maps, which provide comprehensive information on the kinds and properties of the soils developed on various landscapes needed for their proper use and management by different land users (farmers, herders, foresters, water engineers, ecologists, urban developers, and other stakeholders). The soil map is also required for carrying out good research activities and extrapolating the research findings or technologies developed in a research station or a field to other areas with similar soils.

With all the importance that a soil map holds, there are certain points regarding soil resources that cannot be "mapped" and need to be dealt with. These are mainly the threats and challenges as well as the proper practices that can affect the future sustainability of these resources. In the case of Iran, these threats and challenges stem from the fact that soil resources have been utilized since the dawn of the Persian civilization to provide food, feed, and fibers needed

H. Siadat (✉) · M.H. Roozitalab
Soil and Water Research Institute, Tehran, Iran
e-mail: hamidsiadat@yahoo.com

M.H. Roozitalab
e-mail: mroozitalab@gmail.com

A. Farshad
Department of ITC, University of Twente, Enschede, The Netherlands
e-mail: farshad@itc.nl

M.H. Roozitalab et al. (eds.), *The Soils of Iran*, World Soils Book Series,
https://doi.org/10.1007/978-3-319-69048-3_14

for the growing population. Therefore, it is natural to expect that soil resources have been degraded to different degrees; particularly in recent decades, when intensified utilization of soils not only in agriculture, but also in industry and urban development was started.

The agricultural land of the country is fragile and rather limited, as only about 8% of the total land area is under annual crops and orchards. There is little potential for expanding the arable land in the future, due to shortage of water resources and constraint of the climatic conditions (see Chaps. 3 and 9). The soils of the country are now facing major challenges that threaten this ancient civilization, mainly due to overexploitation, mismanagement, and unwise intervention of the users in recent decades. Therefore, protection and proper management of the present soil resources, particularly agricultural land, are highly essential to assure the food security, environmental sustainability, and wellbeing of the next generations (see Chap. 13).

Later in the chapter, we suggest the proper measures needed to be urgently taken by the government and the people to combat the following serious challenges facing the soil resources.

14.2 Lack of an Effective National Soil Policy

The country still lacks an effective national soil policy to be implemented by all stakeholders at the national and provincial levels. Therefore, formulation and implementation of an appropriate soil policy and its elements for protecting soil resources and promoting soil management and monitoring soil degradation and land-use changes are a high priority. Presently, several soil research institutions, many soil science departments at the universities, and various executive agencies or organizations at national and provincial level are responsible for various activities related to soil without any close partnership and coordination. The scattered activities of these institutions and organizations have not yet led to a tangible impact on the proper conservation, use, and management of the soils in the country.

14.3 Land Degradation and Desertification

Human-induced land degradations are the main challenges facing the soil resources, environment, and the well-being of the society (see Chap. 12). These processes are accelerating soil erosion and salinization of soil and water resources as well as depletion of soil fertility and organic matter, deforestation, pollution, and deterioration of soil physical and biological conditions.

14.3.1 Soil Erosion

Large areas of the rangelands are degraded due to overgrazing, conversion to cultivated rainfed production systems, and mismanagement. Although many projects have been recently implemented to rehabilitate the rangelands and stabilize the sand dunes (Fig. 14.1), large areas of rangelands in the arid and semiarid regions of the country are still facing severe water and/or wind erosion (see Chap. 12 and Fig. 14.3).

There is limited policy and effective socioeconomic approach to provide the needed incentives to farmers and herders for successful implementation of projects to control soil erosion. The traditional low-productive pastoral and nomadic practices to graze the rangeland are still very common in many places (Fig. 14.2). High unemployment rate and low annual income of the rural households due to inadequate infrastructures and small-to-medium size enterprises in the small towns and villages for creating more jobs

Fig. 14.1 Sand dune stabilization in Kashan, Yazd, and Kerman: good projects, but not properly coordinated or continuously supervised (Farshad 1997)

Fig. 14.2 Overgrazing of rangelands is a major challenge for soil conservation (Farshad 1997)

Fig. 14.3 Salinity and loss of the vegetation cover are the main causes of land degradation and desertification (photo by M H Roozitalab)

are the main reasons for the overgrazing and the prevailing excessive utilization of the rangeland (Aw-Hassan and Noori 2007; Roozitalab et al. 2013).

Soil erosion is a common phenomenon under both rainfed and irrigated farming system due to lack of proper tillage practices, unsuitable crop rotation, and absence of conservation measures.

14.3.2 Soil Salinity

A serious threat to Iranian agriculture in many parts of the country is salinity of soil and water resources. The area of total land affected by salinity has been estimated between 16 and 23 Mha (Dewan and Famouri 1964; Kovda 1970; Roozitalab 1987; Siadat et al. 1997). These figures include both cultivated and barren lands. No exact data is available on the extent of this problem in the farmlands, although it is well known that by far the greatest extent of the salinity problem prevails in the irrigated fields (Siadat 1998). The main difficulty in determining the extent of salt-affected areas is the temporal variations of salinity due to the effects of irrigation water, which add or leach the salts, as does the rain (Noroozi 2011; Polous et al. 2011).

There are many factors that cause the development of salinity problems in the soil resources. Siadat et al. (1997) classified these factors in two groups: natural factors and human-induced factors. The natural causes of soil salinity in Iran are geological conditions, climatic factors (evaporation, rainfall, and wind), salt transport by water, and intrusion of saline bodies of water into the coastal aquifers. However, what seems to be of greater concern and importance is the human-induced salinity. Salinity of soils due to improper water management or misuse of saline water to irrigate land has led to increased salinity of a large area of the cropland in the arid and semiarid regions of the country (Fig. 14.3). The high amount of evapotranspiration rate in many regions has led to the accumulation of more salts on the soil surface (see Chaps. 6 and 7).

Therefore, appropriate irrigation systems and soil management are needed to prevent excessive salt accumulation.

Special attention should be paid to soil salinity in expanding the pressurized irrigation system, which has been promoted and subsidized by the government to increase irrigation efficiency at the farm level. In any irrigation method, the farmers should learn to prevent salt accumulation by applying proper amount of water and practicing salt leaching out of the root zone. Reclamation of salt-affected soils by leaching and installation of drainage systems should be regarded as a priority in the irrigated lands of the country.

14.3.3 Declining Soil Fertility and Organic Matter

Depletion of soil fertility due to improper soil management and lack of a suitable crop rotation or incorporating leguminous crops in the cropping system are also among the challenges facing agricultural land in irrigated and rainfed areas. Depletion or loss of soil organic matter is also prevalent in rainfed and irrigated farming systems due to lack of soil conservation measures by farmers (see Chap. 12). There is some evidence that more than 50% of cultivated soils suffer from the deficiency of one or more nutrients. Furthermore, the decreasing trend in soil organic matter content is a serious threat to sustainable soil fertility and crop production. Recent surveys have estimated that organic matter content in about 60% of the country's soils is less than 1%. As to plant nutrients in soils, nitrogen is perhaps the most limiting element for crop production. Improving agronomic management practices such as use of proper crop rotation, incorporation of crop residues in soil, adoption of conservation tillage, and use of organic fertilizers will certainly help improve soil fertility in the future.

Undoubtedly, part of the present soil fertility problems has emerged as a result of misuse, overuse, and unbalanced use of fertilizers. Thus, the future calls for proper fertility management of soils, which depends on innovative research and better education and training for farmers in this respect.

14.3.4 Deforestation

Deforestation and degradation of the unique forests in the Caspian Sea region and the scattered forests of the Zagros Mountain have been among the main concerns of the government. In 1963, the total area of the forests was estimated to be around 18 Mha, i.e., 11% of total area of the country. Since then, the total area of the forests has been reduced to about 12 Mha (see Chap. 5). The reduction of the forestlands in the Caspian Sea region is quite noticeable and has changed from 3.4 Mha in 1962 to around 1.8 Mha in 1977. Deforestation has been among the main sources of soil erosion and land sliding in these areas (Fig. 14.4).

Thus, conducting innovative research and promotion of local residents' participation in the stewardship of forest lands and increasing public understanding and awareness of the importance of sustainable use of forests and how to minimize the risks of flooding and soil erosion due to deforestation are the key to preservation of these lands. To this end, provision of higher and experiential education in forestry and conservation is of utmost importance.

14.3.5 Soil Pollution

Generally speaking, soil pollution has been caused by urban and industrial activity, agricultural chemicals, or improper disposal of waste materials. In Iran, the more extensive and important source of soil pollution is the use of wastewater and chemical fertilizers, while geological formations have also a role in contaminating the soils in certain limited regions.

Using wastewater for irrigation has become a common practice in many areas, particularly in farms adjacent to the big cities. From around 1000 cities in Iran, only 200 cities have operating or under-design wastewater collection network. In many cities, wastewaters from sewage wells infiltrate through deep layers of the ground and ultimately reach the aquifers (Tajrishy, no date). Direct use of untreated

Fig. 14.4 Deforestation in the Caspian Sea region and Zagros Mountain is a major source of land degradation (Photos by FKhormali andMHRoozitalab)

wastewater from sewage outlet is not widespread, but direct use of treated or partially treated wastewater for irrigation without being mixed or diluted is more common.

Results of a study on the negative impacts of untreated wastewater irrigation on a soil in suburb of Shiraz and the building up of the heavy metals such as Cd, Pb, and Ni above the maximum permissible limits and contamination of vegetables and wheat are alarming (Qishlaqi et al. 2008).

While agrochemicals have been commonly used for fertilizing soils and controlling pests and crop diseases, they are a major factor affecting the environment and human health. Farmers often see application of fertilizer as a key factor in increasing productivity (see Chap. 10). In several cases, application of excessive chemical fertilizer on crops such as potatoes, onions, and other vegetables has led to severe damage to soil nutrient balance. Overuse of some fertilizers, particularly P, has led to relatively high contents of elements such as Cd. Also, ingestion of Pb deposited on the forage by fallout is thought to be the source for high Pb in manure, thereby contaminating some fields that had used animal manures (see Chap. 10). However, sporadic contaminations such as As, Co, and Cr in some western parts of the country have been attributed to geological formations.

These contaminants cannot be easily removed from soils, but actions are needed to reclaim polluted lands in the future. This issue deserves due attention to prevent expansion of the problem and ensure safety of the food produced in such lands.

14.3.6 Deterioration of Soils' Physical and Biological Conditions

In many areas of the country, soil's physical and biological conditions are deteriorating due to improper use of agricultural machineries, tillage practices, lack of crop rotation, and depletion of soil organic matter (see Chaps. 10 and 12). These have led to soil compaction, poor aeration, and decrease in soil porosity, thereby limiting the root zones expansion and proliferation.

Besides, soils biological conditions have been under threat of pollution, salinity, and soil physical degradation.

Future actions need to address rehabilitation of such conditions and improve the soil environment to allow for healthy microbiological activities of the indigenous soil fauna.

14.4 Climate Change and Water Shortage

In the last two decades, the country has been suffering from shortage of water due to below-average precipitation and above-average temperatures (Siadat and Shiati 2002). Cyclic drought has hit the Middle East, including Iran, and water shortages have led to catastrophic consequences to agricultural resources, particularly soils and water resources

Fig. 14.5 Rainfed fig plantation in Estahbanat, Iran, wilted and withered due to the drought in early 2000 (Siadat and Shiati 2002)

(Fig. 14.5). The impacts of drought, especially if it persists over long periods, are probably the worst in the case of natural vegetation, which removes the cover crop and leaves the soil surface exposed to erosion agents.

Due to lack of adequate water resources, about 5.7 Mha or 40% of the present cultivated land is under low-productive rainfed (see Chap. 7). Excavation of about 750,000 authorized and unauthorized wells and building more than 450 dams have provided just enough water for irrigating areas of about 6.7 Mha. However, water resources have not been properly managed, and consequently, many lakes and marshlands are dried up or have been severely damaged. On the other hand, the total renewable water resources are not sufficient to expand irrigated land much further. Underground water resources in most of the 609 agricultural plains have been depleted in such a way that more than half of the plains are now in crucial state. Therefore, water crisis remains a big challenge for agricultural and socioeconomic development of the country.

As reported by IPCC (2015), the impact of incipient climate change and greenhouse effect on water, soil, and agricultural productivity of Iran and other countries in West Asia and North Africa is considerably more than other regions (IPCC 2014). A higher earth temperature will adversely affect soil organic matter and soil–water–plant relationship and, thus, reduce agricultural productivity and diversity of fauna and flora. The rate of soil erosion and salinity and frequency of drought will increase. Therefore, new strategy and approaches on natural resource management and agricultural development are needed to mitigate the impact of the climate change.

14.5 Land Use Change

Although up-to-date statistics are very limited, it is easy to notice that the urban and industrial areas have exponentially expanded in recent years, while the areas of the fertile soils

Fig. 14.6 Prime agricultural soils are used in brick-making factories, an example from Hamedan Province (Farshad 1997)

and prime agricultural lands, particularly around the cities, are declining. A case study indicated that during 1955–2001, an annual average of about 3000 ha of suitable agricultural land around the six major cities of the country (Fig. 14.6) has been permanently converted to other uses (Momeni 2005). The high prices of the land around the cities for urban development provided incentives to many farmers to sell their lands and abandon their long-standing profession to migrate to the cities.

During 1963–2014, the total agricultural land of the country decreased from about 18.8–16.5 Mha (Statistical Center of Iran 2014), mainly due to urban development and industrial use. The continuation of this process is a big threat to the growth of agricultural production and food security of the next generation.

14.6 Conclusion and Recommendations

To a great extent, sustainable development of Iran depends on its soil resources. However, the future status of these resources is threatened by the aforementioned challenges and factors. The future development plans of the country must properly deal with these threats by adopting suitable policies and paying due attention to the following recommendations:

- Formulation and implementation of an appropriate national soil policy approved by the highest government authority for studying and protecting soil resources, promoting soil management, monitoring soil quality and land-use changes, and other responsibilities are high priority.

- It is essential to urgently improve public awareness on the importance and unique role of soil in sustaining life and destiny of a nation. Better educations and training of farmers, technicians, and students at the universities as well as increased awareness of policy makers and managers are highly needed.

- Enhancing research quality and implementing site-specific and customized research and development projects for improving soil quality with emphasis on increasing soil fertility and organic matter, promoting conservation agriculture and carbon sequestration, practical and economic measures to improve soil and water productivity, soil health, efficient use of agricultural inputs, monitoring soil erosion, pollution and salinity, as well as prevention of deforestation are highly important.
- In soil survey activities, serious attention must be paid to providing different thematic soil maps that can be usable and comprehended by agronomists, engineers, landscape planers, farmers, and others. Advanced technologies such as remote sensing, geographic information system, and soil database management must be used properly to make the soil data usable to digital applications, modeling of soil erosion, salinization, etc. The presently available maps do not provide the required information to various stakeholders and users of the soil resources. Moreover, conducting soil survey in a conventional way is too expensive and has been gradually left out in development projects with consequences of being all neglected when planning a development project.
- Strengthening the national research and technology development system and creating more incentive for the various stakeholders to participate in the national dialogue to promote sustainable use of soil resources.
- Strong collaboration and coordination of the national and provincial research institutes and centers working on soil with many soil science departments of the universities to promote joint studies and research projects and to generate suitable technologies for sustainable use and management of soil resources are highly imperative.
- Improved and more effective agricultural extension and advisory services to farmers on promoting soil management and conservation measures, implementation of conservation agriculture, and alternative environmentally friendly practices are highly essential for sustainable use of soil resources.

- The future Iranian soil scientists are expected to give major contributions in the five following areas:
  - Spatial planning of the landscape,
  - Enhancing agricultural productivity, food safety, and protection of the environment,
  - Ecological risk and natural resources management
  - Integrated soil and water management
  - Mitigating the impact of the climate change on agriculture and soil productivity as well as generating economic and effective technologies needed to face the incipient environmental changes.

Iranian soil scientists need to adopt new approaches combining in-depth specialized soil knowledge and basic information of soil as a natural and continuous body, which is closely related to other natural resources (water, geomorphology, ecology, forests, ranges, biological diversity, etc.) and the prevailing environment. This will enable the soil scientists to contribute and collaborate with other scientists to resolve the future challenges facing the country and succeed in what Richard W. Arnold referred to as "helping our mother earth" (Zinck et al. 2016).

References

Aw-Hassan A, Noori K (2007) Characterization of production and livelihood systems of rural communities in upper KRB. In: Proceedings of international workshop on improving water productivity and livelihood resilience in Karkheh River basin. Karaj—Iran, ICARDA—AREEO, Report 2, 10–11 Sep 2007

Dewan ML, Famouri J (1964) The soils of Iran. FAO, Rome

Farshad A (1997) Analysis of integrated soil and water management practices within different agricultural systems under semi-arid conditions of Iran and evaluation of their sustainability. Ghent-Belgium/ Enschede, the Netherlands. Dissertation 45, ITC Publication 57, 395 p. ISBN: 90-6164-142-X

International Panel on Climate Change, IPCC (2014) Climate change synthesis report, summary for policymakers

Kovda V (1970) Prevention of salinity and reclamation of saline soils of Iran. Soil Institute of Iran. Publication No. 227

Momeni A (2005) Investigating spatial and economic dimensions of land use change around big cities in Iran. Final research report 1228. Soil and Water Research Institute

Noroozi AA (2011) A new model for predicting soil salinization based on integrated multi-temporal-spatial satellite imagery. Ph.D. thesis, Tarbiat Modares University (TMU), Faculty of Agriculture, Soil Department, Tehran, Iran

Polous KH, Farshad A, Zarinkafsh M, Roozitalab MH (2011) Evaluation of the spatio-temporal of soil salinity variation using data mining approach. In: ICSDM 2011 & BJ-IWGIS2011 conference at Fuzhou, Fujian, China from 29 June to 1 July 2011

Qishlaqi A, Moore F, Forghani G (2008) Impact of untreated wastewater irrigation on soils and crops in Shiraz suburban area, SW Iran. Springer Science Business Media B.V., Environ Monit Assess 141:257–273

Roozitalab MH, Serghini H, Keshavarz A, Eser A, De Pauw E (2013) Elements of research strategy and priorities for sustainable agricultural development of highlands in Central, West Asia and North Africa. International Center for Agricultural Research in the Dry Areas (ICARDA), Review Report

Roozitalab MH (1987) National soil policy and its technical and administrative organization in Iran. Soil and Water Research Ins. Publication No. 725 (in Farsi)

Siadat H, Bybordi M, Malakouti MJ (1997) Salt-affected soils of Iran: a country report. In: International symposium on "sustainable management of salt affected soils in the arid ecosystem", Cairo, Egypt

Siadat H (1998) Iranian agriculture and salinity. In: Proceedings of the international symposium held in Tehran, Iran 12–15 October. http://archive.unu.edu/env/Land/iran-1/index.htm#papers

Siadat H, Shiati K (2002) Overview of drought in Iran: impacts, lessons, and recommendations. In: Regional seminar on "drought mitigation", Tehran, IR, Iran, 28–29 Aug 2001. UNDP-Iran Technical Series 4

Statistical Center of Iran (2014) Results of detailed census on agriculture in 2014

Tajrishy, M (no date). Wastewater treatment and reuse in Iran: situation analysis. Environment and Water Research Center (EWRC), Department of Civil Engineering, Sharif University of Technology, Tehran, Iran (Tajrishy@sharif.edu)

Zinck JA, Metternicht G, Bocco G, Del Valle HF (eds) (2016) Geopedology; an integration of geomorphology and pedology for soil and landscape studies. Springer. ISBN: 978-3-319-19158-4 or 978-3-319-19159-1 (eBook)

Author Biographies

Dr. H. Siadat is a distinguished soil scientist and editor-in-chief of two scientific periodicals in Iran: Journal of Soil Research and Journal of Water Research in Agriculture. He is a retired research professor and has been an invited professor in The Academy of Sciences, Islamic Republic of Iran, for more than 20 years, and a member of the recently formed Council of Luminaries in Agricultural Research, Education and Extension Organization. He served as Deputy Director General of Soil and Water Research Institute from 1983 to 1995. His main field of expertise is soil–water–plant relationship.

Dr. M.H. Roozitalab is a distinguished soil scientist with wide range of scientific and working experiences in Iran and abroad. He received his M.Sc (1974) and Ph.D. (1978) from Oklahoma State University, Stillwater, Oklahoma, USA. His major research and teaching activities at Tehran University has focused mainly on soil genesis and classification and clay mineralogy. He has contributed to the advancement of soil science in Iran as President of Soil Science Society for 17 years from 1992 to 2009 and Director General of Soil and Water Research Institute from 1983 to 1995. Dr. Roozitalab also served as Deputy Head of Agricultural Research, Education and Extension Organization (AREEO) in Iran from 1995 to 2006. He has contributed to strengthening the regional and global agricultural research collaboration as President of Association of Agricultural Research Institutions in the Near East and North Africa, AARINENA (1998–2000), Coordinator of Regional Highland Research Network of ICARDA (2008–2011) and Chairman of the Global Forum on Agricultural Research (GFAR) from 2002 to 2006.

Dr. A. Farshad is a senior and retired scientific member of the ESA Department of ITC, Twente University, in The Netherlands. Backed with over 10 years of field experience in Iran, he joined ITC in 1980, where he taught soil-related topics, doing research and supervising students for 30 years. His job also made him cooperating with various institutions in different countries, namely Australia, Bangladesh, Belgium, Botswana, Cape Verde, Egypt, Ghana, Iran, Kenya, Mexico, Morocco, Nigeria, Portugal, Spain, Sri Lanka, Thailand, Tunisia, Turkey, UAE, and USA.

Index

M.H. Roozitalab et al. (eds.), *The Soils of Iran*, World Soils Book Series,
https://doi.org/10.1007/978-3-319-69048-3

The manufacturer's authorised representative in the EU is Springer Nature Customer Service Centre GmbH, Europaplatz 3, 69115 Heidelberg, Germany. If you have any concerns regarding our products, please contact ProductSafety@springernature.com

Printed and bound by CPI Group (UK) Ltd, Croydon, CR0 4YY
01/10/2026
02225991-0001